PAYMENT FOR ENVIRONMENTAL SERVICES IN AGRICULTURAL LANDSCAPES:

ECONOMIC POLICIES AND POVERTY REDUCTION IN DEVELOPING COUNTRIES

NATURAL RESOURCE MANAGEMENT AND POLICY

Editor:
David Zilberman
Dept. of Agricultural and Resource Economics
University of California, Berkeley
Berkeley, CA 94720

EDITORIAL STATEMENT

There is a growing awareness to the role that natural resources such as water, land, forests, and environmental amenities play in our lives. There are many competing uses for natural resources, and society is challenged to manage them for improving social well-being. Furthermore, there may be dire consequences to natural resources mismanagement. Renewable resources such as water, land, and the environment are linked, and decisions made with regard to one may affect the others. Policy and management of natural resources now require interdisciplinary approach including natural and social sciences to correctly address our society preferences.

This series provides a collection of works containing most recent findings on economics, management, and policy of renewable biological resources such as water, land, crop protection, sustainable agriculture, technology, and environmental health. It incorporates modern thinking and techniques of economics and management. Books in this series will incorporate knowledge and models of natural phenomena with economics and managerial decision frameworks to assess alternative options for managing natural resources and environment.

The Series Editor

Recently Published Books in the Series
Haddadin, Munther J.
Diplomacy on the Jordan: International Conflict and Negotiated Resolution
Renzetti, Steven
The Economics of Water Demands
Just, Richard E. and Pope, Rulon D.
A Comprehensive Assessment of the Role of Risk in U.S. Agriculture
Dinar, Ariel and Zilberman, David
Economics of Water Resources: The Contributions of Dan Yaron
Ünver, I.H. Olcay, Gupta, Rajiv K. and Kibaroğlu, Ayşegül
Water Development and Poverty Reduction
d'Estrée, Tamra Pearson and Colby, Bonnie G.
Braving the Currents: Evaluating Environmental Conflict Resolution in the River Basins of the American West
Cooper, Joseph, Lipper, Leslie and Zilberman, David
Agricultural Biodiversity and Biotechnology in Economic Development
Babu, Suresh Chandra and Djalalov, Sandjar
Policy Reforms and Agriculture Development in Central Asia
Goetz, Renan-Ulrich and Berga, Dolors
Frontiers in Water Resource Economics
Just, Richard E., Alston, Julian M. and Zilberman, David
Regulating Agricultural Biotechnology: Economics and Policy

PAYMENT FOR ENVIRONMENTAL SERVICES IN AGRICULTURAL LANDSCAPES:
ECONOMIC POLICIES AND POVERTY REDUCTION IN DEVELOPING COUNTRIES

Edited by

LESLIE LIPPER
FAO, Rome

TAKUMI SAKUYAMA
FAO, Rome

RANDY STRINGER
University of Adelaide, Glen Osmond

DAVID ZILBERMAN
University of California, Berkeley

Published by the

Food and Agriculture Organization of the United Nations
and
Springer Science+Business Media, LLC

Springer

Disclaimer: The designations employed and the presentation of material in this publication do not imply the expression of any opinion whatsoever on the part of the Food and Agriculture Organization of the United Nations concerning the legal status of any country, territory, city or area of its authorities, or concerning the delimitation of its frontiers or boundaries. The mention of specific companies or products of manufacturers, whether or not these have been patented, does not imply that these have been endorsed or recommended by the Food and Agriculture Organization of the United Nations in preference to others of a similar nature that are not mentioned. The views expressed herein are those of the authors and do not necessarily represent those of the Food and Agriculture Organization of the United Nations.

ISBN 978-0-387-72969-5 e-ISBN 978-0-387-72971-8

FAO ISBN 978-92-5-105830-5

Library of Congress Control Number: 2008939917

This book is dedicated to the memory of Karma Kesang, who died in a tragic river accident, aged 35, while completing the survey for the Bhutan chapter. May his last work be dedicated to the benefit of all sentient beings.

Preface

Agricultural development is widely recognized as crucial for poverty reduction. At the same time, agricultural expansion and ever more intensive practices are widely recognized for their contribution to ecosystem degradation. Less well recognized is that, in many cases, agriculture offers the potential to generate both poverty reduction and better environmental outcomes. The studies presented in this volume look at one policy tool that may address this gap: payments for environmental services (PES). PES programs offer the potential to reduce poverty by focusing on low-income producers who inhabit, manage, consume, and produce important agro-ecosystem services.

The relatively long-term contracts in many PES programs can stabilize household income flows, providing greater opportunities for investment. Such programs can also enable producers to diversify their income sources, thus reducing the risks to economic shocks. In many cases, incomes could grow further over time as improvements in soil, water, and nutrient quality slow land degradation and enhance agricultural productivity.

Other potential benefits range from better access to finance (based on better land quality and secure cash flows) to lower demand for children's labor, resulting in higher school attendance and eventually increases in human capital. Local, national, and global welfare could improve, too, as those producers sequester carbon and reduce soil erosion, protect wildlife buffer zones, and provide refuge for endangered species, control watershed runoff, and safeguard the quality of streams, rivers, and wetlands.

Much of the work presented in this volume results from findings of a research program on the Socio-Economic Analysis and Policy Implications of the Roles of Agriculture in Developing Countries (ROA Project). The research project, funded by the Ministry of Agriculture, Forestry and Fisheries of the Government of Japan, and managed by the United Nations Food and Agriculture Organization from 2000 to 2007, aims at extending current thinking about the environmental, social, and economic roles of agriculture.

This volume complements existing outputs from the ROA project (available on the project website at http://www.fao.org/es/esa/roa/) and other deliverables from ROA and the ongoing research work on payments for environmental services and poverty reduction at the Agricultural Development Economics Division (ESA) of the FAO. These include (1) a Special Edition in *Environment and Development Economics* (Vol. 13, Issue 3, June 2008); (2) a full-day Learning Workshop as part of the 2006 International Association of Agricultural Economists Conference in Australia's Gold Coast, August 2006; (3) FAO's flagship publication, *The State of Food and Agriculture 2007: Paying Farmers for Environmental Services*; and (4) a website on PES in agricultural landscapes at http://www.fao.org/es/esa/PESAL/index.html. An important aim of the ROA project and the ESA research program is to help make contributions to sustainable development goals, concepts, and results, acting as a catalyst for change and promoting conditions in which the rural poor are able to enhance environmental outcomes, raise their incomes, and live longer, healthier, and more productive lives.

A key motivation for the ROA project was to provide policy guidance for improved development strategies, especially for more sustainable rural development. The ROA research findings call attention to a diverse set of indirect environmental, social, and economic contributions of agriculture. The evidence suggests that these indirect contributions are not well understood, seldom analyzed in the context of development, and rarely reflected in national and rural development policy strategies. This innovative research initiative reflects the FAO priorities of poverty reduction, food security, and sustainable rural development – key goals shared by the international community.

In addition to the financial and administrative support provided for the ROA project by the Government of Japan and FAO, many people contributed to making this volume possible. The editors express their thanks and appreciation to Prabhu Pingali, Kostas Stamoulis, Keith Wiebe, and Monika Zurek from FAO's Agricultural Development Economics Division (ESA). We thank Amor Nolan for her excellent editing and relentless dedication to this project, and Jenny Aker, Jennifer Alix-Garcia, John Forgách, Brian Gross, Thomas Koellner, Robin Marsh, and David Roland-Holst for ideas and support throughout the process.

Randy Stringer
Leslie Lipper
Takumi Sakuyama
David Zilberman

Contents

Contributors

Khalil Allali
Department of Rural Economics, Ecole Nationale d'Agriculture de Meknès Morocco, kallali@enameknes.ac.ma

Jennifer Alix-Garcia
Department of Economics, University of San Francisco, San Francisco, CA, USA, jmalixgarcia@usfca.edu

John M. Antle
Department of Agricultural Economics and Economics, Montana State University, Bozeman, MT, USA, jantle@montana.edu

Erwin Bulte
Department of Economics, Tilburg University, Tilburg, the Netherlands, e.h.bulte@uvt.nl

Oscar Cacho
School of Economics, University of New England, Armidale, Australia, ocacho@une.edu.au

Alain de Janvry
Department of Agricultural and Resource Economics, University of California at Berkeley, Berkeley, CA, USA, alain@are.berkeley.edu

Stefanie Engel
Institute of Human-Environment Systems, Swiss Federal Institute of Technology (ETH) Zürich, Zürich, Switzerland

Nobuhiko Fuwa
Department of Agricultural Economics, Chiba University, Matsudo City, Chiba, Japan, nfuwa@faculty.chiba-u.jp

Benjamin M. Gramig
Department of Agricultural Economics, Purdue University, West Lafayette, IN, USA, bgramig@purdue.edu

Amir Heiman
Department of Agricultural Economics and Management, Hebrew University, Rehovot, Israel, heiman@agri.huji.ac.il

Richard D. Horan
Department of Agricultural Economics, Michigan State University, East Lansing, MI, USA, horan@msu.edu

Yanhong Jin
Department of Agricultural Economics, Texas A&M University, College Station, TX, USA, YJin@ag.tamu.edu

Leslie Lipper
Agricultural and Development Economics Division, Food and Agriculture Organization, United Nations, Rome, Italy, Leslie.Lipper@fao.org

Nancy McCarthy
International Food Policy Research Institute, Washington, DC, USA, N.McCarthy@CGIAR.org

Imelda Nalukenge
Faculty of Agriculture, Makerere University Kampala Uganda

Charles Palmer
Environmental Policy and Economics, Swiss Federal Institute of Technology (ETH) Zürich, Zürich, Switzerland, charles.palmer@env.ethz.ch

Elisabeth Sadoulet
Department of Agricultural and Resource Economics, University of California at Berkeley, Berkeley, CA, USA, sadoulet@are.berkeley.edu

Asa Jose U. Sajise
University of the Philippines Los Baños, College, Los Baños, Laguna, Philippines, asajise@yahoo.com

Takumi Sakuyama
Agricultural and Development Economics Division, Food and Agriculture Organization, United Nations, Rome, Italy, Takumi.sakuyama@fao.org

Jason F. Shogren
Department of Economics and Finance, University of Wyoming, Laramie, WY, USA, jramses@uwyo.edu

Jetse J. Stoorvogel
Department of Environmental Sciences, Wageningen University, 6700 AA Wageningen, the Netherlands, jetse.stoorvogel@wur.nl

Randy Stringer
School of Agriculture, Food and Wine, University of Adelaide, Adelaide, Australia, randy.stringer@adelaide.edu.au

Karma Ura
Centre for Bhutan Studies, Langjophakha, Thimphu, Bhutan, drphilos@yahoo.co.uk

David Zilberman
Department of Agricultural and Resource Economics, University of California at Berkeley, Berkeley, CA, USA, zilber@are.berkeley.edu

List of Abbreviations

AHEAD	Animal health/emerging animal diseases
AR	Afforestation and reforestation
CCF	Carbon content factors
CCT	Conditional cash transfer
CDCF	Community Development Carbon Fund
CDM	Clean development mechanism
CERs	Certified emission reductions
CIDE	Centro de investigación y docencia económica
CITES	Convention on International Trade in Endangered Species of Wild Fauna and Flora
CO_2e	CO_2 equivalents
COLPOS	Colegio de posgraduados
CONAFOR	Comisión Nacional Forestal/National Forest Commission
CONAPO	National Population Council
CRP	Conservation Reserve Program
CRSP	Collaborative Research Support Program
CV	Contingent valuation
DH	Dirham
DNA	Designated National Authority
DOE	Designated operational entity
ECB	Economic benefit
ENB	Environmental benefit
EQIP	Environmental Quality Incentives Program
EQIP	Environmental Quality Improvement Program
ES	Environmental services or ecosystem services
FFM	Mexican Forest Fund
FWI/GFW	Forest Watch Indonesia/Global Forest Watch
GEF	Global environmental facility
HYVs	High-yielding rice varieties
ICMS	Imposto sobre Circulação de Mercadorias e Serviços
ILIAD	International Lookout for Infectious Animal Disease
IMF	International Monetary Fund

INE	Instituto Nacional De Ecología
INEGI	National Institute of Geography, Statistics and Information Systems
IPM	Integrated pest management
IRRI	International Rice Research Institute
JNF	Jewish National Fund
Kg	Kilogram
MBC	Mesoamerica biological corridor
MD	Minimum data
MEA	Millennium ecosystem assessment
MgC	Mega-gram or metric ton of carbon
NBS	Nash bargaining solution
NCV	Net calorific values
NEMA	National Environment Management Authority
NEMP	National environment management policy
NGOs	Non-governmental organizations
NPV	Net present values
Nu	The Bhutan currency ngultrum
O.I.E.	Office International des Epizooties
PCF	Prototype carbon fund
PDD	Project design document
PGR	Plant genetic resources
PES	Payments for environmental services or ecosystem services
RES	Rewards for environmental services
ReNED	Research Network for Environment and Development
RHS	Right-hand side
RUPES	Rewarding the upland poor for the environmental services
SHCP	Secretariat of Hacienda and public credit
SINAP	National system of protected areas
UIA	Universidad Iberoamericana
UNFCCC	United Nations Framework Convention on Climate Change
VSF	Vétérinaires sans frontiers
WCL	Wildlife conservation lease
WRI	World Resources Institute
WTP	Willingness to pay

Chapter 1
Introduction and Overview

Randy Stringer, Leslie Lipper, Takumi Sakuyama, and David Zilberman

There is growing concern about natural resource degradation, sustainable development, and poverty reduction. In recent years, payments for environmental services (PES) programs have emerged as major components of sustainable development policies. These programs are in place in several developing countries. However, the implications of PES programs for the rural poor, the optimal design of programs to contribute to economic development, and how these initiatives integrate into international treaties to address global warming and biodiversity loss, are still not clear. This book attempts to fill this gap.

To date, the vast majority of theoretical, methodological, and empirical PES research in developing countries has focused on forests or water resource systems (FAO, 2007; Landell-Mills & Porras, 2002). This book turns attention toward the role of environmental services in agricultural landscapes because the poor in developing countries are concentrated in rural areas and earn their livelihoods from agriculture.

Small-scale farmers and pastoralists are the most likely groups of poor to be heavily dependent on the environmental assets embodied in agroecosystems. The rural poor tend to live and work in ecologically fragile, economically marginal, and environmentally degraded areas. More than one billion people in developing countries live in fragile ecosystems (World Bank, 2003). Many researchers and conservationists question whether ecosystem conservation and anti-poverty goals can or should be combined in the same PES program. Some warn against over-burdening PES projects with anti-poverty objectives and raise concerns about pushing the environmental problem elsewhere. Others question whether PES projects can actually reach the poorest populations, whether the costs of delivering payments to large numbers of smallholders are prohibitively high so as to outweigh the benefits, and even whether introducing PES could damage traditional systems of community-based conservation stewardship. Still others point out the dangers involved in ignoring the inherent trade-offs between economic growth, poverty reduction, and the environment.

This book attempts to answer these and related questions by providing policy insights, scrutinizing analytical tools, and stimulating debate on poverty-environment linkages in agricultural landscapes. An important objective is to encourage

L. Lipper et al. (eds.), *Payment for Environmental Services in Agricultural Landscapes*,
Natural Resource Management and Policy.
doi: 10.1007/978-0-387-72971-8_1;

further debate about how to coordinate policies in ways that simultaneously address poverty and improve environmental outcomes. The book combines conceptual overviews that summarize and expand the economic analysis of the impacts and design of PES programs with case studies that highlight their diversity and identify key factors for consideration in assessing their value as tools for poverty reduction. While much of the literature emphasizes design and impacts of PES programs, this book also includes a discussion of the demand for these programs and how to obtain resources to fund them.

The book intends to serve as a reference for academics including ecologists, economists, resource management specialists, as well as policy analysts. The issues, ideas, concepts and lessons will also be of value to a wide range of policy advisors, non-governmental organizations (NGOs), and the general development community. The academic audience includes upper level undergraduates and graduate students in economics, development, resource management, and environmental studies. The writers of most of the chapters are economists and the analysis applies economic tools and methods, but we attempt to minimize technical, theoretical discussion, so readers with only basic knowledge of economics should be able to follow the arguments presented.

The book includes 13 chapters. The first four are conceptual, with a methodological emphasis, addressing general problems. The remaining chapters address specific aspects of PES programs in the context of agricultural development, and have a strong empirical emphasis.

Chapter 2 presents an overview of the economic rationale behind PES programs, principles for effective design, and the conditions necessary for PES to both improve environmental quality and reduce poverty. One rationale for PES programs arises from the recognition that humanity is a beneficiary of the *natural capital* embodied by nature, which provides environmental services or ecosystem services (ES). PES are part of an arsenal of environmental economic policy tools that includes command and control, cap and trade, and pollution taxation. PES programs may arise in situations where polluters are either too powerful or too impoverished for compliance under alternative policy regimes. In other words, PES programs are a “carrot” that may work in situations where “sticks” or direct control are ineffective. In addition, PES can function as mechanisms for transfers of income that enhance distributional objectives and reduce poverty. We present a framework for categorizing PES programs: some aim to address negative externalities (reduce pollution), while others are designed to support the provision of public goods (e.g., biodiversity), and still others to resolve dynamic inefficiencies and maintain conservation. Since agricultural producers are the largest group of ecosystem managers in the world, their activities may produce (or reduce) ES. Farmers may provide ES by either diverting land from farming activities (land diversion programs) of modifying their resource use and technologies (working-lands programs). Different types of programs vary in their environmental and distributional consequences. We distinguish between programs where the source of payments should be the private sector as opposed to those which require government support, those that should be managed at the local level, compared with those that require involvement of national governments or global arrangements. We then assess the issue of targeting payments to achieve their objectives efficiently, with attention to the equity implications of PES programs. The final section addresses issues of monitoring and enforcement of PES contracts, with a summary of the key findings in the conclusion.

Frequently, ES in developing countries are provided in regions where local stewards of the land and resources have weak property rights, while other agents, like multinational corporations, are interested in commercial exploitation of the resources. In this case, design of PES programs has to account for the interaction between local and external interests. Chapter 3 provides a game-theoretic conceptual framework to address this challenge and apply the tools to a case study in Indonesia. In particular, the chapter develops a conceptual framework for PES design where forest-dependent communities, which have only weak property rights over the forest, interact with logging companies. A strategy that emphasizes distributional objectives may focus on setting up PES programs that benefit the poorest communities, or those that face the lowest expected payments from logging deals. However, these communities may lack the capacity to enforce a PES agreement and preserve the forest. Therefore, it is important to consider the overall capacity of communities to enforce contracts, enhance conservation, and work together with logging companies to reach environmentally sustainable outcomes efficiently. The analysis and case study suggest that to attain environmental objectives efficiently, under some circumstances it will be optimal to allow individual communities to negotiate individually the best deal with logging companies they can. However in this case, some of the poorer communities will be excluded from the benefits – which suggest that resolving differences between environmental and distributional objectives will require multiple tools.

Someone has to fund PES programs, and lack of funding is a major barrier to their development. While there is much research on the design and management of PES programs and their impacts, there has not been much study on alternative sources of demand for ES and their implication for the design of PES programs. Chapter 4 aims to fill this void. Since the benefits of ES are realized by diverse groups, they can attract a diverse set of buyers. Thus far, much of the demand for ES has originated from the public sector, under government programs, that pursue either environmental or distributional objectives, or both. International agreements to address global problems like climate change are another source of demand for ES. There are some industries that may use ES as a least-cost approach to solving environmental problems or a source of revenue appealing to consumers desire to contribute to environmental objectives. Another source of demand may be individuals who view PES as a source of direct consumption benefits, altruism, pro-social behavior, or even "warm glow." Marketing strategies should be targeted for the specific characteristic of various market segments. This suggests the need for a mixture of institutional arrangements and strategies to create and institutionalize the demand for ES. In some cases, there is an advantage to commoditizing ES and marketing them through large exchanges. In other cases, the ES are unique, and special efforts are needed to find the appropriate buyers that are needed. In all cases, increasing consumer awareness of availability, value, and benefit of ES is important and leads to increased demand. Furthermore, in all cases, buyers have to be assured of product reliability, which requires explicit mechanisms for monitoring, enforcement of contracts, and insurance.

An important source of demand for ES comes from cap and trade environmental regulations. Concern about climate change has led to establishment of a cap and trade carbon emission regulations such as the clean development mechanism of the Kyoto Protocol. Under these regulations, countries and industries can meet at least some part of their emission reduction compliance through the purchase of offsets. Carbon sequestration in forest and agricultural ecosystems is one form of offset the agricultural sector can provide. However, there are several barriers to participation in carbon-sink projects, especially those involving smallholders in developing countries, which is the focus of Chap. 5.

These are mostly transaction costs associated with designing and developing projects, measurement and monitoring, certification, selling carbon emission offsets, and distribution of payments. Based on interviews, and on reviews of existing programs conducted in Mexico and Indonesia, the chapter examines five different types of transaction costs. It suggests that viability of these projects requires aggregation and a large scale of operation, and identifies threshold values for three project-design parameters for various transaction costs. This chapter clearly illustrates the need to go beyond simply identifying supply and demand for ES in designing a successful PES program, which may require formation of new regional institutions to allow better capture of economies of scale and reduce overall transaction costs.

A major objective of PES is to provide incentives for wildlife conservation. The most serious threats to wildlife are habitat fragmentation and infectious diseases. Many PES programs involve the purchase of development rights and expand the habitat of species at risk. However, frequently these programs are implemented in locations where neighboring livestock production may be a source of infectious disease that can also threaten the species. Chapter 6 presents a modeling framework for design of management systems to address issues of fragmentation and disease risk and enhance wildlife conservation. This framework consists of a dynamic, bioeconomic model, and its application requires interdisciplinary effort to obtain key biological and economic parameters. In particular, the chapter focuses on designing PES for private landowners and ranchers, for the provision of various risk-reducing ecological investments. In particular, it examines payments for habitat connectedness, livestock vaccination, and reduced movement of infected livestock. It identifies cost-effective dynamic strategies, where in the early stages, most of the funds are allocated to purchase land rights to connect habitat. Over time, once habitat is sufficiently connected, disease risks increase and the optimal investment focuses on payments for disease prevention. The conservation payments result in significantly increased wildlife abundance, increased livestock health and abundance, and increased development opportunities. This chapter emphasizes the political challenge of implementing PES programs that vary in their emphasis because the groups that benefit from them may change during the course of the program life.

While Chap. 5 investigates PES in a forest context, Chap. 7 analyzes the potential impacts of PES programs compensating farmers for soil carbon sequestration. Implementation of such programs is challenging because of issues of monitoring, enforcement, and contract design, especially because some of the sequestration gain can be easily reversed. For example, if farmers are paid for sequestering carbon for several years by engaging in minimal tillage, the gains can be reversed by deep plowing. The chapter develops analytically several hypotheses regarding the technical and economic factors affecting adoption of practices that increase soil carbon and their impacts on poverty. Using trade-off simulation software with data from Senegal, Peru, Kenya, and other countries, the evidence suggests that carbon payments could have a positive impact on the sustainability of production systems while also raising incomes and reducing poverty. However, carbon contracts are found to have only modest impacts on poverty, even at relatively high carbon prices. Moreover, the participation of poor farmers in carbon contracts is likely to be constrained by the same economic and institutional factors that have inhibited their use of more productive, more sustainable practices in the first place. Thus, PES are most likely to have a positive impact on poverty and sustainability when they are implemented in an enabling economic and institutional environment.

Many of the original PES programs in developed countries have been payments for hydrological services. For example, the water supply for New York City is partially guaranteed by subsidized conservation efforts of farmers in the watershed that feeds the metropolis, an effort that began in the 1980s. In recent years, such programs have increasingly been introduced in developing countries. Despite their increasing number, there is scarcity of rigorous studies analyzing their effectiveness in providing ES and their impact on communities receiving the payments. Chapter 8 intends to partially address this gap by presenting an analysis of the first 2 years of the Mexican PES program for hydrological services where payments were made to individuals and communities as incentives to preserve existing forest. The chapter provides background information on forests, deforestation, and the ES potential, as well as political economy of the processes that led to establishment of the program. The analysis shows that the payments in the early years did not necessarily achieve the goals of the program. For example, they were largely allocated to lands that were not within critical watersheds, they were not targeted at forests at-risk, and there were few behavioral changes induced by the payments, yet they increased participation in conservation activities. However, this program served distributional objectives as the majority of the payments went to poor and very poor forest holders.

Rural landscape aesthetics are another form of ES that farmers can receive payment for via ecotourism. Chapter 9 presents an empirical analysis of agricultural landscape externalities and indications of their potential impacts for poverty reduction in Morocco's Western High Atlas Mountains. The chapter first identifies the various types of benefits provided by rural landscapes, and relates them to agro-ecological activities of farmers. For example, farmers can provide landscape externalities by stone clearing, water rehabilitation, terracing, and planting fruit trees. The chapter includes an empirical study of 134 farms, quantifying the benefits provided by agro-tourism farms, as compared with returns of the traditional farms. It also suggests that payments for landscape aesthetics can serve as a valuable approach to reduce rural poverty. It identifies various forms of agricultural policy interventions and compensation schema to induce farmer provision of landscape ES.

The design of PES to improve farmer well-being must account for the specific features of the crop production system, and the local ecosystem and political-economic situation. One of the major agricultural crop systems is rice farming in Southeast Asia. Chapter 10 designs a hypothetical scheme of green payments to induce agrobiodiversity in rice farming in the Philippines. The chapter develops a methodological approach that consists of discrete choice modeling and simulation that can be applied to a wide range of circumstances. It first estimates the probability of adoption of various practices, including preservation of old varieties, as a function of socioeconomic and ecological variables, and then simulates the consequences of alternative (hypothetical) PES schemes under a fixed budget constraint. The chapter finds a clear trade-off between enhancing biodiversity and poverty reduction. Even the totally untargeted lump-sum subsidy would have a larger poverty reduction impact than would the first-best conservation subsidy payment scheme. The chapter also identifies a trade-off between the efficiency of targeted conservation payment and the information requirement for implementing subsidy schemes.

One of the major objectives of PES is to provide incentives for protection of environmentally sensitive areas, such as wetlands, that may be endangered by encroachment of farming or hunting, resulting from population growth and enforcement failures. PES can serve as a mechanism to enhance profitability and sustainability of agricultural practices on existing lands to reduce the incentives for encroachment on pristine lands. Chapter 11 reports on a study of the potential for PES to encourage the adoption of more sustainable agricultural practices in the Pallisa district in southeastern Uganda. Due to low productivity and population pressure, the subsistence agriculture that dominates the upland areas is increasingly encroaching on wetland areas critical to a many ecosystem services. While encroachment is illegal, enforcement has not been effective, raising the possibility that a positive incentive mechanism might be a more effective approach to wetlands protection. This study began with a workshop designed to learn about the potential importance of wetlands and their services from local and national stakeholders, and to assess the legal and institutional setting in which environmental policy is being implemented. The next step was to implement a quantitative analysis of ecosystem service supply, to estimate the possible rates of participation by farmers in contracts for wetlands conservation and the impact on farmers' incomes. The analysis suggests that PES could be a viable alternative to conventional environmental regulation if local institutions can manage contracts with farmers at a reasonable cost, and if national and international beneficiaries are willing to pay for wetlands protection.

Much of the literature on wildlife valuation is associated with benefit-cost analysis of conservation policies for specific sites and more commonly focused on consumptive use values such as hunting or fishing or general ecotourism. Surveys extract information from individuals on the willingness to pay (WTP) to provide for the management of species conservation. In contrast to the WTP assessments, the Bhutan case presented in Chap. 12 examines the costs of conservation imposed on small producers by the country's overall strategy to protect forests and wildlife. While this information provides only part of the picture, it allows for a more adequate assessment of the mix policy options and how to target those policies and programs to compensate effectively small producers.

In Bhutan, policy debates focus increasingly on whether most of the conservation costs are borne directly by the small producers and rural poor through crop losses and labor time diverted to guarding crops and livestock. Many producers argue that too much farm output and too much of their incomes are sacrificed due to the country's commitment to conservation. This chapter attempts to quantify the extent of wildlife damage to crops and to livestock in Bhutan. The study provides an assessment of the extent of the problem around the country, presents results of a survey of 526 households, and outlines the extent of wildlife damage to their crops during a 12-month period.

Wildlife cause damage by eating crops and killing livestock, resulting in: (i) lost income and destroyed and damaged assets; (ii) large cost in time and money attempting to protect crops and livestock; (iii) a disincentive to plant and invest in rural production; and (iv) greater levels of rural-urban migration. The growing concern is whether and how much conservation benefits are taking place at the expense of basic food security and poverty reduction. The study concludes that wildlife damage to crops and livestock limits severely the agricultural livelihoods of many small producers; in addition, the increased presence of forest cover near agricultural land due to conservation policies has not necessarily provided significant increased benefits to local residents. While other benefits, such as improved watershed quality, could still prove to be a significant benefit, the increase in services due to growth in forest coverage is unlikely to outweigh the increased costs observed for Bhutan's small-scale producers.

The analysis proposes seeking a balance between benefits and costs borne by producers, communities, and the nation. At present, the majority of the conservation costs are borne by producers. A different type of cost, for example, requiring communities to pay in part for fencing or corralling solutions, might benefit directly both rural livelihoods and conservation efforts while operating within the framework of PES programs.

References

FAO (2007) The state of food and agriculture: Paying farmers for environmental services. Food and Agriculture Organization of the UN, Rome, Italy.

Landell-Mills N, Porras IT (2002) Markets for forest environmental services: Silver bullet or fools' gold? Markets for forest environmental services and the poor, emerging issues. International Institute for Environmental Development, London, UK.

World Bank (2003) World development report 2003: Sustainable development in a dynamic world. World Bank, Washington, DC.

Chapter 2
Putting Payments for Environmental Services in the Context of Economic Development

Leslie Lipper, Nancy McCarthy, and David Zilberman

Abstract Paying for the provision of environmental services (ES) is a recent policy innovation attracting much attention in both developed and developing countries. Linking payments for environmental services (PES) to economic development and poverty reduction is important since they may represent a new source of finance to developing countries that are potentially important suppliers of global ES. In this chapter we apply concepts from natural resource and environmental economics to a wide range of issues associated with the introduction of PES programs in the context of economic development. We introduce an economic typology of PES, showing how they can provide a solution to externalities and public good problems within the bounds of political economic constraints. Secondly, we focus on the problem of who should, and will, pay for ES. Third, we will turn to issues of program design. We assess the issue of targeting payments to achieve their objectives efficiently, with attention to the equity implications of PES programs. The final section addresses issues of monitoring and enforcement of PES contracts, and we summarize the key findings in the conclusion.

2.1 Introduction

Paying for the provision of environmental services (ES) is a recent policy innovation that is attracting much attention in both developed and developing countries. The innovation involves a move away from command-and-control environmental policies, harnessing market forces to obtain more efficient environmental outcomes. It also involves the notion of rewarding the providers of ES, which until now have supplied services for free.

Linking payments for environmental services (PES) to economic development and poverty reduction is an issue of importance in developing countries for several reasons. PES may represent a new source of finance to support both the environmental and economic development policy objectives of the country through the

L. Lipper et al. (eds.), *Payment for Environmental Services in Agricultural Landscapes*, Natural Resource Management and Policy.
doi: 10.1007/978-0-387-72971-8_2;

mobilization of funds from the global community to suppliers of ES in developing countries. Developing countries are potentially important suppliers of global ES, as they may be low-cost producers of the service or a unique source of the services which are location specific. Biodiversity conservation is an example of the latter, where developing countries are uniquely and richly endowed with species and ecosystems not found in the developing world. Carbon sequestration is an example of the former type of service: Its production is not location specific, but developing countries may be competitive suppliers due to low opportunity costs of labor and land. PES programs for the provision of global environmental goods can contribute to economic development by increasing employment and income generation possibilities, as well as diversification of livelihoods among the suppliers.

The development of payment schemes for the provision of local-level ES could also be an important contributor to economic development. The impacts of the payments on employment and incomes are likely to be important here too, but in addition there could be significant economic development benefits associated with the ES itself. In many cases environmental problems create major barriers to economic development. For example, degraded soils result in reduced agricultural productivity, and poor water quality causes disease and health problems in many parts of the developing world. PES could be an effective means of dealing with these problems.

However, PES programs may also pose problems for developing countries, most of which are highly dependent on using, and often depleting, their natural resource base in the process of economic development. Diverting natural and environmental resources to the provision of ES rather than short-term economic growth could generate high opportunity costs from foregone development in the short run. The poorest people within developing countries are the most highly dependent on direct use of natural resources and the environment for their subsistence, although they often are not the owners of these resources (World Resources Institute, 2005). If PES programs result in the reduction or loss of access to the resource by the poor, they could exacerbate poverty. Despite their low production costs for ES, developing countries may not be very competitive in the market for their supply due to poor institutional and physical infrastructure that results in high transaction costs.

The objective of this chapter is to apply economic concepts, particularly those from natural resource and environmental economics, to a wide range of issues associated with the introduction of ES programs in the context of economic development. First, we introduce a typology of ES based upon economic reasoning, showing that PES provide a solution to externalities and public good problems within the bounds of political economic constraints. Secondly, we focus on the problem of who should pay for ES: to what extent are payments likely to be covered within a global framework, rather than within a national or regional framework? Third, we will turn to issues of program design. We present some answers to the questions of how to target payments to efficiently achieve their objectives, and what the implications of alternative design schemes are. In particular, we focus upon the equity implications of ES programs and how they can affect poverty alleviation.

The final section addresses issues of monitoring and enforcement of ES contracts, and we summarize the key findings in the conclusions.

2.2 What are ES?

ES can be thought of as a set of biophysical outcomes generated by the management of natural resources and the environment. These outcomes have impacts on human well-being, as well as on wider natural processes. For example, standing forests may regulate water quality and storage capacity. Deforestation thus results in the elimination or reduction of these services, generating soil erosion and contamination of water quality. The conservation of wetland areas may provide significant flood protection, enhance water quality, and protect wildlife habitat, all of which would be lost if wetlands were to be converted to other land uses.

When moving to the concept of PES, the role of humans as both providers and beneficiaries is key, as is the concept of externalities. Humans, as economic agents, manage natural resources and the environment in ways that frequently generate either positive or negative externalities[1] that impact others in society or future generations. The basic premise of PES is rewarding economic agents for managing the environment and natural resources to generate environmental goods and services that benefit others. These goods and services may also have private benefits, but the primary impetus of PES programs is the provision of private incentives to generate a positive externality. Thus, when someone plants a tree that provides carbon sequestration, eliminates a source of pollution affecting a wider community, or generates a habitat for wildlife, she or he is a provider of ES.

We identify three categories of ES, based on their justification and potential source of funding. These categories are (1) pollution control, (2) conservation, and (3) amenities creation. We will discuss each of these categories, provide an economic rationale for their emergence, and discuss some implications with respect to the schemes that will be used to pay for them.

2.2.1 Pollution Control

Traditionally, pollution has been viewed as a negative externality caused by a missing market for a side effect of production activities, with the logical response being the introduction of remedies such as pollution taxes (Baumol & Oates, 1974; Pigou, 1932). However, in many cases, the "polluter pays" principle is difficult to introduce and implement effectively. Then it was recognized that pollution is

[1] Externalities occur when the supplier or consumer of a good does not bear all the costs of its production or consumption. Marginal private costs are less than marginal social costs of production or consumption.

actually the outcome of a property rights problem (Coase, 1960). Coase's analysis suggests that in a world with well-defined and enforced property rights and relatively low transaction costs, side payments will be established; and optimality will prevail whether polluters own the right to pollute or the polluter owns the right for a clean environment. Quite frequently, society explicitly or implicitly gives the polluter the right to pollute. To solve the pollution problem in this case, the beneficiaries from pollution reduction need to pay polluters to not exercise their rights to pollute. If pollution harms a large number of individuals and the cost to enforce pollution regulations is very high, it may be more expedient and efficient to introduce incentive measures – including payments – for pollution reduction.

While the assumptions behind the Coase theory are restrictive, the spirit of analysis applies to a wide arrange of circumstances. Historically, farmers have been given the implicit right to produce food and fiber in ways that are known to cause environmental problems. It is quite clear, for example, that intensive production systems of milk and other livestock products result in problems of animal waste management. Residues from such production systems have ended up in nearby bodies of water, causing damage to wildlife and reducing water quality. The problem was not very significant when the human population and livestock production density was low, but population growth coupled with the increase in intensive production systems has resulted in increased problems. Increasing recognition of the environmental problems generated from such systems has resulted in calls for changes in policies, including increased regulation as well as the use of economic instruments such as PES (Food and Agriculture Organization of the UN, FAO, 2007).

Externalities represent a transfer of welfare between groups and individuals in society; thus, political economic considerations are important in determining their solution (Buchanan & Tullock, 1975). Polluters may not have the ability to pay or may have sufficient political muscle to force legislative outcomes in which they are paid for pollution reduction rather than being subjected to the polluter-pay principle. This is particularly important in developing countries where environmental regulations restrict the use of natural resources that poor people rely upon for subsistence. For example, penalizing low-income farmers who cause soil erosion and downstream siltation by cultivating on steep slopes is not likely to be an effective means of controlling the problem, unless some alternative income-generating strategy is possible.

In the developed world, particularly the United States and European Union, PES is often used to subsidize pollution reduction in agriculture. For example, the Environmental Quality Incentives Program (EQIP) in the United States is being used to subsidize the introduction of technologies that reduce the damages associated with animal waste.

2.2.2 Conservation

ES may be used to pay for the conservation of natural resources, ecosystems including forest resources and wetlands as well as wild flora and fauna species, and agricultural crop and livestock species. Payments for conservation are generally introduced to correct a situation of dynamic social inefficiency where the owners of the resource, or the people who control its utilization, are likely to modify use patterns in a way that would reduce or eliminate the potential of supplying public goods to future generations. For the purposes of this chapter, we consider the goal of conservation to be the protection of a sufficient quantity and distribution of natural resources themselves, so they may provide amenities to future generations. The service is the preservation itself, as compared with other categories of ES, where the goal is to generate positive externalities or mitigate negative externalities in the present.

One example of how this service is generated is through payments to reduce the incentives for deforestation, with the objective of conserving habitats and biological diversity as well as maintaining carbon sinks found in forest ecosystems. Payment to the "de facto" owners of the resource, be it a single company or a forest community, may be used to assure that the forest resources are left in their natural state or managed in a way that provides conservation. Where the existing bundle of ownership rights allows holders the option to modify the use of owned resources, PES are means of motivating owners to not exercise these rights, while maintaining the existing property rights structure. The obvious examples are payments to individuals, forest communities, and even governments (e.g., the debt for nature program) to control deforestation. The establishment of wildlife conservation corridors through various payments to rights holders is another example of growing importance in both developed and developing countries (Reto-o-Reto Policy Brief 1, n.d.). Payments to farmers to maintain traditional crop varieties in production are yet another form of conservation payment that has been implemented (Global Environmental Facility, 2007).

Conservation programs are justified on several grounds. The primary justification is the presence of uncertainty about future needs for environmental goods and services, and the constellation of natural resources needed to provide them. For example, the genetic resources contained in the current global set of plants and animals, even if they are not utilized at present, could provide an important future use as further information becomes available or as conditions change (Zilberman & Lipper, 1998). Maintaining a bank of such resources results in a wider set of future options for the utilization of genetic resources, and this is an important ES. Conserving genetic resources *in situ* generates not only the preservation of the genetic resources, but also a dynamic evolutionary process, combining the effects of human and natural selection.

Besides options value, the conservation of natural resources generates existence values, where humans derive benefits and are willing to pay for the preservation of natural resources, even if there is no current or potential future use. For example, conservation efforts that protect endangered wildlife (e.g., the black rhino) generate existence values. The benefits of conserving a species and the values they generate are dependent on the individual preferences of humans that are highly variable. This implies significant variability in the willingness to pay for ES among potential consumers.

2.2.3 Amenities Creation

In this category of ES, the services are a means of generating current public benefits at local, regional, and global scales. Examples include planting trees to sequester carbon to reduce greenhouse gases in the atmosphere (a global public good), and/or to regulate water flows and soil erosion to improve watershed function (a local or regional public good). Payments are made for the adoption of activities that generate amenities – such as contour tillage and bunds that reduce soil erosion and improve watershed functions. In another example, overpumping of groundwater from a shared aquifer is a common-pool resource problem that may well have negative spillover effects not only on those with recognized rights to use the water for irrigation, but also to urban consumers of drinking water. PES programs that stimulate better management of water resources and restore the aquifer functions can be very effective means of generating benefits to urban water consumers.

2.3 The Demand for ES

Most of the research regarding PES addresses the challenging questions of design, assessment, and utilization to attain the desired environmental objectives. A fundamental question that remains to be answered is how to finance PES programs, particularly those that are intended to also contribute to poverty reduction. In particular, who should and would pay to support these programs, and how can we increase the resources available for the PES if they are underfinanced?

Welfare theory provides us some suggestions of where funding should be sought. One important question in determining the source of funding to pay for the provision of ES is the type and distribution of benefits they generate, be they at local, national, or global scales. A second key factor is identifying the sources of effective demand. In the next section, we identify three broad categories of ES purchasers – private firms, governments, and non-governmental organizations (NGOs) – and consider the scale at which demand might operate.

2.3.1 Public Sector

National and regional government agencies are paying for a variety of programs for environmental quality improvement and restoration. The Conservation Reserve Program (CRP) in the United States and various payment schemes for multifunctionality in Europe are models of government-financed agricultural ES programs. While the official objectives of these programs are environmental, they also serve as mechanisms to transfer payments to support the farm sector. This is not surprising, since legislative packages require construction of coalitions that carry majorities (Fischhendler & Zilberman, 2005), and these programs can thus cater both to environmental and agricultural interests. This is an increasingly important issue as pressures mount to cut agricultural subsidies under trade liberalization. ES payments to the agricultural sector that are not linked to production or price of a commodity are not constrained by the World Trade Organization rules (FAO 2007; Economic Research Service, ERS, 2001). Government-funded PES programs in developing countries are also likely designed to meet multiple objectives, as can be seen in the case of China's sloping land conversion program, which has been cited as a means of subsidizing the ailing State Grain Bureau (Bennett & Xu, 2005).

In addition to being part of large national agricultural programs in developed countries, PES programs are being used as tools for implementing local and regional environmental policies. For example, in the United States state funds, sometimes in the form of tax incentives, are used to facilitate purchases of resources to protect endangered species via wetland banking and conservation easements (FAO, 2007). In Brazil, the *Imposto sobre Circulação de Mercadorias e Serviços* (ICMS) is a state level tax on the sales of goods and services which some states are using to fund watershed protection and land retirement to conserve biodiversity (Grieg-Gran, 2000).

The capacity of governments in the developed countries to tax a relatively rich population allows them to establish large funds for PES, but this is not the case in many developing countries. Zilberman and Parker (1998) argue that environmental policy tools used by richer governments with large tax revenue bases are necessarily different than the environmental policies of poorer countries, so caution is required when drawing lessons from those well-documented experiences.

One potential source of public sector funding for PES programs in developing nations is overseas development aid, although this is controversial. Developing countries have raised concerns about the potential problem of PES diverting overseas development assistance from their priorities for economic development to the priorities of the donors. One example of how this concern has been manifested is in the design of the Clean Development Mechanism (CDM) under the Kyoto Protocol which requires certification that the public funds used for CDM projects are not diverted from official development assistance and that they contribute to the sustainable development of the country supplying emissions offsets (Clean Development Mechanism, CDM, 2001). Donor support for PES programs in developing nations is likely to involve yet additional pressures to meet multiple objectives. Programs that combine the pursuit of development and environmental objectives may have higher likelihood of support as they may appeal to several constituencies.

The Global Environmental Facility (GEF) is another major source of international public sector funds for projects that generate global environmental goods such as climate change mitigation, biodiversity conservation, and the management of international water bodies (FAO, 2007; Global Environmental Facility, GEF, 2007). GEF funds are being used for the purchase of ES in PES programs, as well as for capacity building for the establishment of PES programs with a wide range of purchasers (FAO, 2007).

2.3.2 Demand from the Private Sector

Private firms are already purchasing ES that result in higher profits by reducing production or environmental regulation compliance costs or increasing the sales value of their products on the market. Water utilities and bottlers pay nearby landowners to plant and maintain trees or reduce livestock grazing in the upper reaches of a watershed, generating improved water quality and flow (Heal, 2003). Some firms may contribute to environmental programs to generate *goodwill or improved reputation*, as the literature on voluntary compliance with environmental standards suggests (Anton et al., 2004). Goodwill reduces the costs firms face in obtaining a license to operate, if it results in reduced community opposition to the activities of the firm (ten Kate, 2005). Improved reputation can generate increased market share for the product, or a higher sales price per unit sold if consumers are willing to pay for the added value of the companies' support of ES activities. Additionally, firms may generate product value by meeting consumer demands for attributes associated with a production process that generates ES, such as organic or shade grown coffee, by developing niche markets through labeling, certification, and advertising. Other important potential private sector purchasers of ES are the developers of recreation facilities and tourism, who rely upon environmental amenities in the areas where they operate to attract clientele. Ecotourism obviously falls in this category, but even mainstream tourist operations rely upon environmental amenities and can be a potential source of payments, as are hunting operators.

Private firms may also be willing to pay for biodiversity conservation as a means of obtaining the rights to explore and exploit potential benefits from the environment, as in the case of bioprospecting, where pharmaceutical companies pay for the right to explore and develop the genetic resources contained in a reserve (ten Kate, 2005).

Private firms may purchase ES as a least-cost means of complying with an environmental regulation. Tradable permit schemes such as the CDM of the Kyoto Protocol facilitate this type of exchange by allowing firms in developed countries to purchase credits for reducing carbon emissions through activities in developing countries that reduce emissions or sequester carbon. In most cases, the purchase price of these credits is lower than the costs many firms face if they were to reduce the emissions from their own production processes, and this cost differential is the basis of their demand for such services from developing countries (Lipper & Cavatassi, 2003).

2.3.3 NGOs as Purchasers

Individuals with strong preferences for various kinds of environmental amenities have realized that governments will not provide the amount of the specific ES they desire, and in response have established NGOs that pursue their interest. Some of the most effective ES funds are managed NGOs that represent groups with specific environmental interests. The Nature Conservancy has invested millions of dollars in various programs that buy or lease land and purchase development rights and other assets in order to provide ES. American Farmland Trust and the Trust for Public Lands are investing in purchases of land and development rights to slow urban sprawl. Ducks Unlimited is another group that is interested in the development of wetlands or other reserves that provide ES for its members. World Wildlife Fund has an active program developing PES in both developed and developing countries as a means of attaining sustainable agricultural development and poverty reduction objectives.

2.3.4 Demand for ES from Wetlands Conservation: An Illustrated Example

In this section we illustrate the concepts described above on the scale and source or demand for ES by identifying the ES and their potential purchasers from the conservation of wetlands. Table 2.1 summarizes the various ES that wetlands can provide and their attributes. Where wetlands support birds that can be hunted, individuals or members of a local private club can be expected to pay for the conservation of the wetland. Wetland habitats supporting birds and other species that provide a utilitarian benefit to all the population generate local public goods that should be supported by local governments or a NGO; and wetlands that provide habitat and shelter to migrating birds – generating a global public good – should be supported by an international agreement or NGO. Recreational amenities may generate local, regional, or even global benefits to individuals who access wetlands for these services, and thus the private sector should be willing to pay. Wetlands also provide existence values by supporting endangered species and wildlife which existence provides utility to various people. In this case either government agencies or NGOs can pay for the ES.

Table 2.1 The dimensions of wetland services

	Local	National	International
Wildlife habitat	Public/Private	Public	Public
Flood control	Public/Private	Public/Private	Public/Private
Water purification	Private/Public	Public/Private	
Aesthetic value	Public/Private	Public	Public
Recreation	Private	Private	Private
Existence	Public	Public	Public

Source: Leslie Lipper, Nancy McCarthy, and David Zilberman

2.3.5 Developing the Market for ES

ES is an emerging market, thus engaging in marketing programs to generate awareness, and a willingness to pay for these programs is necessary. One of the challenges of the marketing effort is to develop new payment mechanisms and identify new sources of money that will be available for PES programs. Even more challenging is to develop PES programs that also aim to reduce poverty. Once resources are available for PES programs, it is crucial to develop mechanisms targeting these resources so they will attain the objective of these programs.

An important source of demand for ES is the establishment of tradable permit programs to address environmental problems. Tradable permits have been successfully used to meet air quality targets in the United States, and they are a crucial element of the Kyoto Protocol to reduce greenhouse gas emissions. Tradable permits programs establish an aggregate level of the activity to be controlled (pollution, emission of a gas) and distribute pollution permits, allowing trade in these permits. These programs provide the financial incentives to reduce pollution or provide environmental amenities, but unlike taxes or subsidies, they do not transfer financial resources from the affected industry to the government or vice versa. Thus, they may be preferable to taxes or subsidies on distributional grounds.

The demand for ES may also be derived from non-environmental objectives. ES provision may be a least-cost means of achieving a development goal such as maintaining irrigation efficiency, provision of clean drinking water, and disaster preparedness (FAO, 2007). For example, reducing siltation in major waterways in China under the sloping lands conversion program provides significant economic benefits to the country in terms of hydroelectric power, and improved navigability and obtaining these benefits through an alternative means, such as dredging, is likely to be more expensive and less effective. Thus, the demand for ES in the public sector may arise from either environmental or broader development objectives. Likewise in the private sector, the demand may arise from the existence of environmental regulations that allow market-based compliance, but demand may also arise for other motivations such as improving public image and relations, or reducing production costs. Quantitative understanding of the sources of demand for ES programs that can benefit the poor is an important area for future research. It is especially important for assessing the extent and potential of ES programs in financing development efforts.

There is not sufficient knowledge of the factors that affect the willingness of public and private sector firms (both in developed and developing countries) to support ES programs in the developing world, and this is an important area for future research. For example, reducing risk is an increasingly important source of demand for ES, but relatively little work has been done on the topic. The following section outlines some of the key issues.

2.3.6 Restoration and Risk Management as Sources of Demand

As discussed above, much of the rationale for ES programs is presented within a dynamic framework of natural resource management, particularly those concerned with conservation and reducing exposure to risks. In this section, we consider more fully the implications of dynamic externalities and mitigating exposure to risk, and the implications for the demand for ES, particularly from poorer producers in developing countries.

2.3.6.1 Restoration

Many development projects have tended to divert natural resources productive agricultural activities, at the expense of reducing the capacity of the agroecosystem to provide other forms of ecosystem services required for sustainability. For example, large-scale drainage projects that led to the conversion of wetlands to farmlands have increased societal vulnerability to floods (U.S. Department of Interior, 1994), and deforestation can reduce protection against floods. In addition to increasing the exposure of people to greater risk, such programs have often completely ignored the potential public goods benefits from conservation, and thus underestimated the opportunity costs of changing land and resource use in terms of existence and option values. Often, decisions about new investments in development projects have been made in a static framework, reflecting the current state of preferences, knowledge, and technology. However, as technology improves and the marginal benefits of ES increase, a society's demand for ES increases.

The use of incorrect decision criteria has led to the overinvestment in development projects (Arrow & Fisher, 1974), which may justify restoration efforts (Zhao & Zilberman, 1999). PES programs may be the least-cost option to achieving objectives of these restoration efforts. Zhao and Zilberman (1999) argue that in situations where uncertain but promising technological prospects are present, there is a case for restoration of environmental amenities. Critical evaluation of past choices, combined with increased environmental awareness, leads to demand for policies that will reverse past choices and restore natural capital.

Restoration projects, which obtain ES by removal of structures or capital goods that serve development objectives, may result in a significant loss of economic opportunities to the poor that have to be recognized. Compensation payments have to take into account all of the losses that affect both current and future income streams. In some cases, it may well then be cost effective for the compensation scheme to include the provision of new sources of income and new opportunities to the affected population. The latter is especially important if, without new sources of income, the population increases its reliance on natural resources not affected by the restoration program (e.g., leakage). Additionally, it is important to realize who benefits from the restoration efforts. Are they the poor or indigenous people who gain access to improved ecosystem functions or natural resources, or are they citizens of developed countries who accrue "existence value?" The value of the project and the computation and distribution of the ES payment should consider the distribution of benefits and costs.

An alternative interpretation of restoration, the restoration of functions provided by an ecosystem targeting those with the highest potential benefit in terms of economic development and poverty reduction, might be more consistent with the pursuit of poverty alleviation by ES programs. When the emphasis is on functional restoration (restoration of flood control, role of wetlands, or water purification role of forests), economic considerations and constraints should affect the design of the restoration projects. Functional restoration may mean redesign of ecosystems or environment, both to provide environmental amenities and to provide economic opportunities to the affected population. In designing projects aimed at restoring the function of an ecosystem, it could be of benefit to fund activities aimed at reducing the impact of the restoration effort on the livelihood of local populations.

2.3.6.2 Risk Considerations

Many of the policy tools (projects, regulations, incentives, and institutions) that societies utilize are aimed at reducing risks and adverse situations. Some of these tools are aimed at physical risks (dams against floods) and others against monetary risks (various insurance schemes against loss of incomes or assets). Some ES can be very valuable as tools for risk reduction. The assessment of their value and determination of PES should be derived from the value they provide in controlling risks. One obvious case cited above is wetlands that serve as buffer zones for flood control. The expectation of the losses they prevent is one measure of the value they provide. However, if flood protection can be achieved by other means, for example a dam, then the saved costs of the dam (construction, operation, and environmental side effects) provide a measure of the value of the ES of the wetland that replaces the dam. Knowledge of the expected damage prevented by ES and costs of alternative means for damage control are essential for assessment of the value of ES.

The value of ES may thus be affected by changes in the risks or damage they control or the costs of alternative means of risk and damage reduction. The value of water quality protection provided by wetlands or forests is enhanced as the value of water quality increases (incomes are rising and people are ready to pay more for it) or when the cost of alternative filtering systems increases. Crop genetic diversity conservation is another ES with an element of risk reduction. The genetic material conserved provides options for future, but currently unknown, crop-breeding needs. The value of their conservation should thus reflect the expected values of beneficial traits that they may contain. At least part of the computation of the conservation value is dependant on the likelihood that the benefits of the conserved resources will be utilized. Genetic resources that are less likely to be utilized because of constraints on access or management are likely to be less valuable than those that are accessible and well managed.

Recognition of the risk reduction value of ES is important in identifying entities willing to pay for ES. Insurance companies will be willing to pay for ES that will reduce the financial risks they are exposed to; so in principle, they have positive willingness to pay for the ES of a wetland that reduces their exposure to a flood. Governments and international agencies responsible for responding to natural disasters should also be willing to pay for ES to reduce exposure to these disasters.

Those living in locations most prone to such risks will also gain. In this case, if ES programs to reduce risk are to work, they must understand if and why local people "appear" to be underproviding ES that reduce their exposure to risks, both idiosyncratic and generalized. Coordination amongst a large number of smallholders is likely to be a key transaction cost that will need to be surmounted. Various property rights issues, as discussed above, may also lead to the underprovision of ES that generate local benefits in terms of reduced risk. Additionally, certain activities, such as wide-scale reforestation or construction of stone terraces or bunds, may simply require cash outlays that are far too great, given the imperfect – or entirely – missing credit markets in many developing country contexts.

2.3.7 Demand and Market Power

One issue that needs careful attention in the developing country context is the "market" power of the demander. If the hydroelectric company is the main – or indeed the only – purchaser of upstream ES, then the company can exploit its power as the sole purchaser, and will subsequently pay less per unit land area and enroll fewer hectares than would be the case if there was a competitive market for these services. This is a bit of an odd story, to be sure; we are discussing the creation of markets for ES for which, heretofore, have been entirely absent. The point is that creating an imperfect market on the part of purchasers may well increase inequality, even if all participants gain absolutely. There is also a question of "fairness" in an absolute sense, irrespective of whether those who voluntary enroll in an ES scheme receive enough benefits to leave them "at least as well off" as before, in addition to the fact that fewer units will be purchased than would be the case under a perfect market.

On a more global scale, the issue of whether to have several large funds versus many smaller ones is related to another issue raised by Wu et al. (2001). If the manager of the fund has a good understanding of industry behavior, and can use their market power in establishing a payment for resources that are diverted from production, it will lead to underpayment for the diverted resources and suboptimality of resource allocation that in this case will lead to overdiversion of land to ES provision.

2.4 Issues in the Design of PES Programs

2.4.1 *Targeting ES Funds*

Targeting is a critical element in designing programs that effectively meet the desired objectives of PES programs. This is true whether demand is from the private, government, or NGO sector, as in each case the purchaser will be interested in obtaining the highest returns to their investment, although how the returns are defined varies according to the objectives. As already noted above, PES programs frequently have multiple objectives, with some more explicit than others. In some cases PES programs have multiple environmental objectives; seeking to generate more than one service. Combining environmental with economic objectives, especially poverty reduction, is commonly found in the design of PES programs, particularly those in developing countries.

In this section we start with a discussion on targeting criteria for a program with a single environmental objective, then expand the discussion to consider programs that have multiple environmental objectives, finishing with those that have both environmental and economic objectives, focusing on poverty reduction for the latter. A framework for targeting ES payment in situations where only one environmental objective is being pursued has been developed by Wu et al. (2000b) and applied to the CRP by Wu et al. (2001). In their analysis the distribution of property rights over land and the heterogeneity of environmental conditions are important elements in designing the optimal targeting strategy. They consider the case where land or other resources are owned by many small producers, and analyze the spatial correlation of the economic benefit (ECB) of farming per unit of land and environmental benefit (ENB) per unit of land, arriving at four possible combinations of high and low benefits for each type of benefit. The authors give the conditions under which the "win-win" situations of high ECB and ENB are likely to be found, as well as conditions that generate the trade-off situations.

As illustrated in Fig. 2.1 below, Wu et al. (2000b) show that the gains from using any one of the possible targeting strategies depend on several elements, in particular, the degree of heterogeneity of ENB and/or ECB over land and the correlation between ENB and ECB. For example, consider the case with (1) significant heterogeneity of ENB, but ECB does not vary significantly across lands, and (2) a positive correlation between ENB and ECB. If the program has a land-maximization objective, the result is a low amount of aggregate ES generation, as the cheaper lands have low ENB and the transition to environmental benefit maximization or cost-benefit strategies may result in significant gains in ES. The cost of selecting the wrong strategy depends on the correlation between ENB and ECB. In cases where there is a strong negative correlation, the land-maximization strategy and the benefit targeting may be maximizing the ES obtained with the budget and be identical to the benefit-cost strategy. When the correlation is small or positive, the cost of the wrong targeting strategy may be substantial.

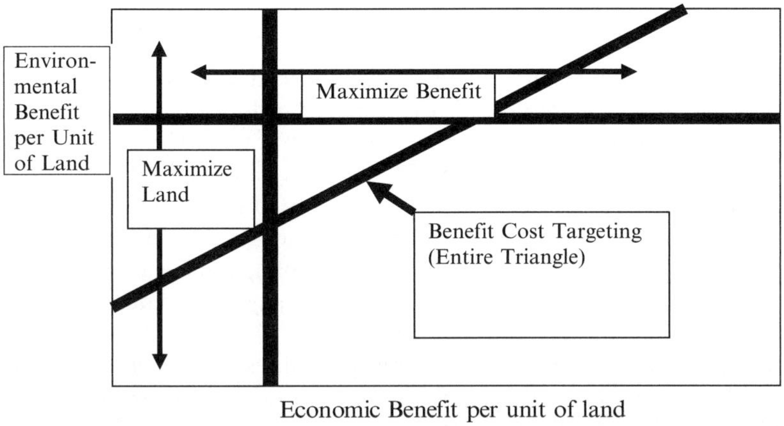

Fig. 2.1 Alternative targeting strategies for ES fund. Leslie Lipper, Nancy McCarthy, and David Zilberman

Wu et al. (2001) argue that the benefit-cost strategy provides the maximum environmental benefit for a given budget, but other strategies have merits as well. The land-maximization strategy will be most preferred by the landowners, as they will maximize the overall revenue (since the ES fund will be spent on the least productive land). Indeed, when farm interests dominate control of the ES fund, they may well target land of lowest productivity. Targeting land with the higher ENB will result in the smallest land area and may be pursued by policymakers who are looking only for the environmental crown jewels, without quantification of existing services, the highest efficiency results from paying according to the highest *expected* benefit-cost ratio. That is to say, payments should only be given for benefits that are at risk of being lost.

2.4.2 Targeting for Poverty Reduction and Environmental Objectives

A controversial issue is the extent to which poverty reduction and agricultural development objectives could, and should, be incorporated into PES programs. The controversy goes back to the issue of efficiency – programs with dual or multiple objectives are likely to be less efficient in achieving any one of the objectives than if they were pursued individually (Tinbergen, 1956). Pagiola et al. (2005) argue that if the policy objective is to reduce poverty, then a different set of tools may well be far more effective than ES payments – which are designed to address a specific type of market failure that results in the underprovision of ES. Nonetheless, there are possibilities to meet both goals and also a need to target such opportunities, as well as mitigate negative effects, where trade-offs do occur.

There are three groups of the poor that need to be considered when considering a program with joint poverty reduction and ES provision objectives: those who are owners or users of the resources that are needed to supply the ES, the landless or those without access to resources who cannot be direct suppliers but could be impacted by changes in labor markets, and the urban poor who could be impacted through general equilibrium effects of a PES program on food supply or prices (Zilberman et al., 2008). The poor are more likely to gain from ES programs that generate environmental amenities that are not luxury goods, provide more employment opportunities (through ecotourism, for example), or income for landless poor. Programs that take resources out of production by paying resource owners are less likely to be beneficial (Mayrand & Paquin, 2005).

Our primary focus here is on the poor as potential suppliers of ES. We consider two possible scenarios for the supply of ES from a poor landowner: The first is where land is sold or taken out of production (e.g., retiring lands from agricultural production), and the second is where production is continued but modified so as to provide ES (e.g., adoption of no-tillage systems of production). We assume that the landowner cultivates L hectares of land and receives a rent of R per hectare. Each hectare can generate B quantity of ES, which will be purchased for the price of V. The income of the farmer before participating in the ES program is L^*R. Poor farmers are assumed to have small quantities of land with low rents. If the land is retired or sold, then farm labor will be released which has a value of W. The landowner will sell or retire the land if:

$$VB + W > R. \tag{2.1}$$

Their gain will be:

$$VB - R + W. \tag{2.2}$$

The spatial correlation between wealth and land quality is a critical determinant of whether poor landowners can benefit from ES programs (Zilberman et al., 2008). If there is a positive correlation between the owner's wealth and the rent from agricultural production, e.g., poor farmers are located on poor quality land for agricultural production, but these lands have a high capacity to generate ES (e.g., high B), the poor could gain substantially from participating in an ES program. Gains will be higher, of course, if V and W are higher as well.

The poor are less likely to benefit from land diverting ES programs where there is no correlation between the rents from agricultural production and ES provision or where they are positively correlated. The poor are not likely to gain much from ES programs if they are operating on lands that generate a high rent from agriculture.

Now take the case where ES provision is made through a change of agricultural production, as opposed to land-use change. The payment per hectare is still VB, but here the landowner also experiences some change in returns to agricultural production R due to changes in production costs as well as output. The landowner will participate in the ES program if:

$$VB \geq P\Delta Y + \Delta C. \tag{2.3}$$

The poor will benefit if the ratio of changes in net revenues from agriculture to net revenues from ES provision is negative, e.g., if $P\Delta Y + \Delta C/VB$ is negatively correlated to income.

This analysis suggests that PES programs with dual objectives of poverty reduction and ES provision need to take into account the spatial distribution of poverty and land quality as it relates to the production of agriculture and ES. Situations where the returns to ES and agriculture are negatively correlated over space and that between poverty and the returns to ES production are positively correlated are the best candidates for targeting to meet the dual program objectives (Zilberman et al., 2008). This was verified by an empirical study of potential supply response to carbon payments in Costa Rica; however, land quality was not the only determinant of the supply response. Pfaff et al. (2007) found that the poor were more likely to be located on low-quality agricultural lands, but which had a high potential for providing sequestration in the form of avoided deforestation or reforestation. The analysis indicated, however, that the poor were likely to be less responsive to carbon payments due to a range of barriers in switching land uses that wealthier suppliers would not face. Pfaff et al. conclude that targeting the poor is not the optimal strategy if the efficient provision of sequestration is the only objective of the program. Given a program with dual poverty reduction and sequestration objectives however, the targeting strategy should concentrate on areas where the returns to agriculture are low, but ES high.

2.4.3 Factors Conditioning ES Supply Response

The targeting section above emphasizes the need to consider the conditions under which the dual goals of generating environmental and economic efficiency benefits, or environment and poverty objects, are complementary or face trade-offs. To highlight these potential regions, we ignored several other elements that may be very important in determining the benefits realized from ES programs, and/or the supply response to these programs. Below we consider a number of factors to be particularly important for designing ES programs in a developing country context; however, as above, we use developed country examples where empirically relevant.

2.4.3.1 Output and Price Effect and Slippage

When programs lead to significant reduction in the production of certain goods as land goes out of production, this reduction in supply may lead to increase in output price with unexpected negative consequences for consumers of those products (Wu et al., 2000b). In the larger developing countries, such as China and Brazil, large ES programs could well have these types of general equilibrium effects. In addition, certain ES schemes have led to changing land-use practices on lands not enrolled in the project, often with unintended negative consequences. Enrolling some land in the program can lead to opening up marginal land to cultivation – lands that were previously never cropped in the past because of low profitability. Such indirect impacts on land-use change may well reduce environmental quality, thereby mitigating overall gains from the program. Clearly, if the environmental benefits provided by these resources when they are idle are substantial, the net environmental benefit from the ES program may be negative. For example, PES programs, such as the China Sloping Lands Program that aims to reduce soil erosion through a variety of changes in land-use practices on specific hilly lands, may well initially reduce output supply enough to lead to price increases, which in turn may lead to the reutilization of previously idle erosive land not within the program area.

2.4.3.2 Scale or Agglomeration Effect and Location Specificity

Both economic and especially environmental benefits may be dependent on scale of resources allocated to these activities. For example, a certain amount of land is needed to provide the critical mass to support wildlife populations, and the spatial configuration of that land may also matter in addition to an absolute size. In other cases, marginal environmental benefits may be increasing over some size range before then decreasing. Both the discrete and marginal scale effects need to be taken into account in order to determine the benefits of the program, and also identify whether and which specific land resources need to be enrolled in order to generate these benefits. As Wu et al. (2000a) argue, in some cases it will be worthwhile for ES purchasers to concentrate purchases in a certain location in order to take advantage of scale effects. In terms of targeting, accounting for scale effects can clearly change priority areas; and furthermore, the rankings themselves can change with modifications in budget constraints. For instance, certain locations that generate the greatest ES for a given budget may well lose priority as larger budgets enable the purchasers to "switch" to locations that generate much greater environmental benefits but on a much larger scale.

Particularly in the case of biodiversity conservation, the biological and geographical requirements for the resources may be very explicit, implying a good deal less flexibility in allowing voluntary, self-selection of individuals into programs aimed at generating these ES. Designing a reserve is a challenging interdisciplinary exercise. As Parkhurst et al. (2002) argue, the spatial pattern of land needed to sustain various populations differs by ecosystem and species. In some cases it is useful to have a large contiguous critical mass of lands, while in other cases, having several separate locations is preferable. Sometimes, it is useful to establish corridors to allow movement among populations. Some species require being located near bodies of water and others on hills farther away, and then there are species that need to move between topographies.

In experimental studies Smith and Shogren (2002) have demonstrated that it is feasible to establish the incentive scheme that will induce landowners to sell lands with desired specifications to meet various requirements for natural reserves. Their studies showed the power of incentives but also demonstrated that obtaining the right land-use patterns may be a time-consuming exercise that will require continuous effort and significant adjustments. This suggests the need to establish an administrative structure that will enable negotiating effectively, will integrate bidding with negotiation, and enable obtaining land resources in a timely and reasonable manner.

2.4.3.3 Property Rights to Land and Water Resources

Another issue of particular relevance to developing countries hoping to provide ES is the issue of property rights to land and water resources. Many analyses assume that rights to the resources required to generate ES are characterized by well-defined property rights and by well-functioning markets in those rights. In practice, "ownership" of resources is often a prerequisite for entering into ES contracts (Grieg-Gran et al., 2005; Landell-Mills & Porras, 2002).[2] However, the bundle of property rights to various land- and water-based resources in many developing countries is often very complex, incorporating multiple layers of claims for access, use, exclusion, and management rights amongst both well-defined and very "fuzzily" defined groups; rights to alienate are often highly restricted. For instance, forest or pastures may be held in common (a well-defined, exclusive group of users over a well-defined resource, e.g., Mexican *ejido* land), or rights may be held by a tribal group where access, use, and exclusion rights differ among different tribal members, and indeed such rights might be both incomplete and conditional at the level of the individual or household (Niamir-Fuller, 1999; Livingstone, 1991; Sandford, 1983). Alternatively, the state itself is often the *de jure* owner of many forest, pasture, and water resources, though certain rights are also often devolved to different users, for instance, either informally through tacit recognition of long-standing claims or formally through long-term leases. State and tribal land can also degenerate into open-access situations when institutions responsible for managing and enforcing property rights break down.

The first difficulty posed by the lack of individual property rights is simply that negotiating with a group may be more difficult than negotiating with individuals; though there would be trade-offs with a number of negotiations that would need to be undertaken if a group could act on behalf of a large number of people. For instance, land degradation or deforestation upstream may lead to reduced water flow and quality for downstream users. If the land upstream is held by private individuals, then payments could be made to these individuals in order to change

[2] Even so, there are already PES programs targeting community groups and not individuals; c.f. FAO (2007); Munoz-Pina et al. (2005); Smith and Scherr (2003); Swallow et al. (2007).

practices; we assume that it is not in their own interest to provide the services without compensation. However, if the lands upstream are common pastures or forests, it may be the case that individuals themselves would benefit by reducing overgrazing or excessive cutting or, for instance, by making more investments in soil erosion control. Thus, there is an additional layer of complexity; given the property rights to resources, users may be under providing ES even against their own joint best interests because of externalities generated among users due to the non-private nature of the property rights to these resources. At this point, however, it is worth noting that groups of users often can and do surmount the problems of coordination and cooperation in managing these resources jointly and do provide the "local social optimum" of these services. However, ES payment programs aimed at communities that currently have difficulties providing the local optimum are going to have to address the incentive issues in joint management – incentives that do not arise when individuals hold a private title to the resource.

One possible solution – often proposed for other reasons – is to privatize the commons. This does happen, of course, often as a result of increasing population pressure and concomitant increase in relative scarcity of land, and ES service programs might be used as yet another rallying cry for privatization. There are many good reasons that such lands remain under some form of common tenure, including prohibitively high costs of issuing private title deeds and enforcing private claims,[3] the fact that such resources often provide a buffer or safety-net for many community members in times of either idiosyncratic or generalized shocks, the strength of socio-cultural norms embodied in non-private property rights systems, and the fact that flexible access to a wide range of resources can increase average production and reduce variability in production, as is often the case of livestock in semi-arid and arid environments. Many of these lands offer the potential for increased ES, and they are often used by some of the poorest people. For those concerned with finding opportunities to combine the twin goals of increasing ES and reducing poverty, understanding the added complexity involved when both the resources and the externalities generated by their use and management are not private "commodities," becomes crucial.

Under this more complex situation, it is necessary to evaluate how the efficacy of different mechanisms for promoting each type of ES (reducing pollution, conservation, amenities, etc.) is affected by the distribution of property rights to resources. Returning to the example of upstream-downstream users of water resources, consider a payment scheme to reforest or invest in agro-forestry break lines to reduce soil erosion and improve water quality and flow downstream. In the case of common land upstream – and where the demander is concerned that all those with claims to a resource are adequately compensated for a change in those claims – the demander would need to identify the primary, secondary, and tertiary

---

[3] For example, enclosure or defining individual rights can be prohibitively expensive; also, privatization can lead to a loss of access to land resources on the part of the poor as elites move to capture the increased benefits.

claims to various resources, which often poses significant problems. The group would then need to negotiate internally; incentives to break agreements and/or free ride on provision will simply be greater in groups – especially the heterogeneous group – than is the case with individual ownership of resources; and this increases transaction costs of providing the ES. Devising enforcement schemes and penalty clauses also poses additional difficulties – should the group be punished for any individual infraction, following the group-credit rationale? Unlike credit groups, where members choose to work together, communities have members with existing rights to resources, and self-selection into groups simply cannot follow a similar pattern; membership is likely to be more heterogeneous and power-relations are likely to be far more important. Thus, it remains an open question as to the circumstances in which ES mechanisms would increase self-monitoring and enforcement rather than engender conflicts and hasten a breakdown in collective management.

There are some interesting empirical studies evaluating different mechanisms for increasing ES in areas where many resources are not privately held. For instance, Alix-Garcia et al. (2005) developed a framework to analyze several targeting schemes to prevent deforestation in the context of *ejidos* in Mexico. They emphasize the importance of designing payment schemes that are based on variables that cannot be manipulated by the recipient. Their analyses also emphasize comparing both environmental outcomes with distributional outcomes. It's often not sufficient to consider targeting of ES funds for resources that are managed collectively in order to best achieve both objectives. In some cases, land may not have formally defined ownership, but nonetheless has been cultivated for a very long time. Understanding the traditional rules and institutions that have been used to control land can be used to structure ES programs to generate incentives compatible with the institutional structure and thus promote the provision of ES more effectively.

2.4.3.4 Risk and Supply Response

Taking land out of food production to set aside for conservation ES may well increase the variability of agricultural production on remaining cultivated lands. If PES are based only on the "average" returns to such lands, fewer people will enroll than expected; how many fewer will depend on such things as the proportion of new income generated by the "safe" ES payment, access to other risk-coping and management mechanisms, and risk preferences as well as the extent of the increased variability in agricultural production. The poor are likely to be more risk averse and have less options in terms of managing risk; thus, their supply response to risk-increasing activities can be expected to be lower than average. Additionally, dynamic considerations may well come into play. To the extent that PES programs promote a change in management and perhaps input use (e.g., switching from pesticides to an integrated pest management strategy, or from conventional tillage to reduced or no-tillage systems), adopters may face increased risks as they learn about these new management and input-use practices (Graff-Zivin & Lipper, 2008). PES programs should recognize that learning takes time, and that production variability – as well as subjective assessments of risk and uncertainty – is likely to be quite high, at least initially. Such information may well lead to the design of an ES program that has higher initial payouts.

2.5 Conclusions

PES programs could play a major role in improving global and local environmental conditions, protecting endangered species, and sustaining biodiversity and wildlife. They also offer the potential for improving the lot of the poor in developing countries as both suppliers and consumers of ES. The analyses in this chapter suggest that successfully establishing these programs is a major challenge, and designing them for dual environmental and social objectives is even more difficult.

Our analysis has indicated that PES can be categorized by the type of environmental externality they address: reducing negative externalities from pollution, correcting dynamic social inefficiencies, and generating positive externalities. These categories have implications for the design of PES programs, as well as for the source of payments.

The discussion on demand in this chapter indicates that there is still considerable research needed on the issue, and this is an important contribution to the creation of demand and willingness to pay for ES. Our analysis suggests that the demand for ES may stem from environmental as well as broader development objectives, and there is a need for better information on the role of PES in both – but particularly the latter. Better information on the potential for ES to contribute to increased agricultural productivity, human health improvement, and risk management are three important areas where further research is merited. Creating demand for ES through the development and dissemination of this type of information and its subsequent incorporation into broad public policymaking are critical for realizing the full potential of this new instrument to contribute to social welfare.

We discuss several supply-side issues that need to be taken into account when designing PES programs for environmental and broader development objectives, including effective targeting, scale and location effects, property rights to land and water, and risk to the suppliers. Our analysis indicates the importance of explicitly defining the objectives of the program, albeit solely environmental or with the incorporation of other social objectives. The discussions in Sects. 2.3.1 and 2.3.2 indicate the significant differences in targeting strategy associated with the variation in program objective. Clearly a major challenge is developing a design that will allow for the targeting of resources in an efficient manner that is both useful for the environment and equitable. There are many issues of design that have to be addressed that will be specific to the ecological and social conditions present, but it is clear that effective ES programs cannot be established without significant interdisciplinary cooperation and organizational entrepreneurship to generate environmental resources that will enhance the economic well-being of the population involved.

Our analysis indicates that designing efficient PES programs requires consideration of traditional issues of market functions – such as market power, as well as non-traditional issues associated specifically with ES provision such as scale and agglomeration effects, and location specificity. For programs designed to address economic development and poverty reduction objectives, consideration of incomplete or problematic property rights to land and water, as well as issues of risk in impacting supply response, is necessary as well.

We recognize that not all PES programs are useful for poverty alleviation; in some cases, they may hurt the poor, and this possibility must be acknowledged and addressed in advance. It is not sufficient to have a good design of payment mechanisms; the actual challenge is development of effective implementation systems that will obtain reliable data on performance, dispense money fairly, and monitor outcomes effectively. Further research on PES should address both some of the major conceptual issues as well as case studies and policy analyses that will provide insight into the realities on the ground.

References

Alix-Garcia J, de Janvry A, Sadoulet E (2005) The role of risk in targeting payments for environmental services. Available at SSRN: http://ssrn.com/abstract=836144.

Anton WRQ, Deltas G, Khanna M (2004) Incentives for environmental self-regulation and implications for environmental performance. J Environ Econ Manag 48(1):632–654.

Arrow KJ, Fisher AC (1974) Environmental preservation, uncertainty, and irreversibility. Q J Int Agric 88(2):312–319.

Baumol WJ, Oates WE (1974) The theory of environmental policy. Prentice-Hall, Englewood Cliffs, NJ.

Bennett MT, Xu J (2005) China's sloping land conversion program: Institutional innovation or business as usual? Paper presented at the ZEF/CIFOR-Bonn workshop on payments for environmental services (PES) – Methods and design in developing and developed countries, Titisee, Germany. Available at http://www.cifor.cgiar.org/pes/_ref/news_events/events/germany/papers.htm.

Buchanan JM, Tullock G (1975) Polluters' profits and political response: Direct control versus taxes. Am Econ Rev 65:139–147.

Clean Development Mechanism (2001) Report of the conference of the parties on its seventh session, held at Marrakesh, from October 29 to November 10, 2001. United Nations Framework Convention on Climate Change. Available at http://unfccc.int/resource/docs/cop7/13a01.pdf#page=3.

Coase RH (1960) The problem of social cost. J Law Econ 3:1–44.

Economic Research Service (2001) Agricultural outlook. September issue. US Department of Agriculture, Washington, DC.

Food and Agriculture Organization of the UN (2007) State of food and agriculture report: Paying farmers for the environment. FAO, Rome.

Fischhendler I, Zilberman D (2005) Packaging policies to reform the water sector: The case of the Central Valley Project Improvement Act. Water Resour Res 41:W07024.

Global Environmental Facility (2007) Ethiopia – A dynamic farmer-based approach to the conservation of the African plant genetic resources. Project Web page available at http://www.gefonline.org/projectDetails.cfm?projID=351.

Graff-Zivin J, Lipper L (2008) Poverty, risk and the supply of soil carbon sequestration. Environ Dev Econ 13(3):353–373.

Grieg-Gran M (2000) Fiscal incentives for biodiversity conservation: The ICMS Ecológico in Brazil. IIER Environ Econ Discussion Paper 00-01, London.

Grieg-Gran M, Porras I, Wunder S (2005) How can market mechanisms for forest environmental services help the poor? Preliminary lessons from Latin America. World Dev 33:1511–1527.

Heal G (2003) Bundling biodiversity. J Eur Econ Assoc 1(2–3):553–560.

Landell-Mills N, Porras I (2002) Silver bullet or fool's gold: A global review of markets for forest environmental services and their impact on the poor. International Institute for Environment and Development, London.

Lipper L, Cavatassi R (2003) Land use change, carbon sequestration, and poverty alleviation. ESA Working Paper No. 03–13. Available at www.fao.org/es/esa.

Livingstone I (1991) Livestock management and "overgrazing" among pastoralists. Ambio 20 (2):80–85.

Mayrand K, Paquin M (2005) Payments for environmental services: A survey and assessment of current schemes. Commission for Environmental Cooperation, Montreal.

Muñoz-Piña C, Guevara A, Torres JM, Braña J (2005) Paying for the hydrological services of Mexico's forests: Analysis, negotiations and results. Instituto Nacional de Ecología, Universidad Iberoamericana, Centro de Investigación y Docencia Económica, Mexico.

Niamir-Fuller M (1999) Managing mobility in African rangelands: The legitimization of transhumance. Intermediate Technology Publications Ltd., London.

Pagiola S, Arcenas A, Platais G (2005) Can payments for environmental services help reduce poverty? An exploration of the issues and the evidence to date from Latin America. World Dev 33(2):237–253.

Parkhurst GM, Shogren JF, Bastian C, Kivi P, Donner J, Smith RBW (2002) Agglomeration bonus: An incentive mechanism to reunite fragmented habitat for biodiversity conservation. Ecol Econ 41(2):305–328.

Pfaff A, Kerr S, Lipper L, Cavatassi R, Davis B, Hendy J, Sanchez-Azofeifa GA (2007) Will buying tropical forest carbon benefit the poor? Evidence from Costa Rica. Land Use Policy 24 (3):600–610.

Pigou AC (1932) The economics of welfare, 4th ed. Macmillan & Co. Ltd., London.

Reto-o-Reto Policy Brief 1 (n.d.) Available at http://www.reto-o-reto.org/downloads/policyinfo-briefs/ PolicyBrief1KitengelaLeaseProgramme.pdf.

Sandford S (1983) Management of pastoral development in the third world. John Wiley and Sons, New York.

Smith J, Scherr SJ (2003) Capturing the value of forest carbon for local livelihoods. World Dev 31 (12):2143–2160.

Smith RBW, Shogren JF (2002) Voluntary incentive design for endangered species protection. J Environ Econ Manag 43(2):169–187.

Swallow B, Leimona B, Yatich T, Velarde SJ, Puttaswamaiah S (2007) The conditions for effective mechanisms of compensation and rewards for environmental services. CES scoping study issue Paper No. 3. ICRAF Working Paper No. 38. World Agroforestry Centre, Nairobi, Kenya.

ten Kate K (2005) Biodiversity offsets: Good for business and biodiversity? Paper presented at the Conference on Payments for Environmental Services and Poverty Reduction. Titisee, Germany.

Tinbergen J (1956) Economic policy: Principles and design. North-Holland, Amsterdam.

U.S. Department of Interior (1994) The impact of federal programs on wetlands. A report by the Secretary of the Interior to Congress. Available at http://www.doi.gov/oepc/wetlands2/index.html.

World Resources Institute (2005) The wealth of the poor: Managing ecosystems to fight poverty. WRI, Washington, DC.

Wu J, Adams RM, Boggess WG (2000a) Cumulative effects and optimal targeting of conservation efforts: Steelhead trout habitat enhancement in Oregon. Am J Agric Econ 82(2):400–413.

Wu J, Adams RM, Zilberman D, Babcock B (2000b) Targeting resource conservation expenditures. Choices (Second Quarter):33–38.

Wu J, Zilberman D, Babcock BA (2001) Environmental and distributional impacts of conservation targeting strategies. J Environ Econ Manag 41(3):333–350.

Zhao J, Zilberman D (1999) Irreversibility and restoration in natural resource development. Oxf Econ Pap 51(3):559–573.

Zhao J, Zilberman D (2001) Fixed costs, efficient resource management, and conservation. Am J Agric Econ 83(4):942–957.

Zilberman D, Lipper L (1998) Sustainability and information. In: D'Souza GE, Gebremedhin TG (eds.). Sustainability in agricultural and rural development. Ashgate, Aldershot, England.

Zilberman D, Lipper L, McCarthy N (2008) When could payments for environmental services benefit the poor? Environ Dev Econ 13:255–278.

Zilberman D, Parker D (1998) Internal water disputes: Causes and solutions. In: Just R, Netanyahu S (eds.). Conflict and cooperation on trans-boundary water resources. Kluwer, Boston, MA.

Chapter 3
Designing Payments for Environmental Services with Weak Property Rights and External Interests

Stefanie Engel and Charles Palmer

Abstract Payments for environmental services (PES) are often promoted as a mechanism for alleviating poverty and providing environmental benefits. This chapter analyzes PES design in a context where actors such as forest-dependent communities have only weak property rights over the forest, and where firms interested in commercial resource exploitation are present. A game-theoretical model of community-firm interactions is applied to the Indonesian setting where communities have been observed to negotiate logging deals with firms. As an alternative, PES design could focus on those communities with the lowest expected payments from logging deals. But these communities may not be able to enforce a PES agreement, while others would conserve the forest anyhow. Most importantly, the introduction of PES may increase a community's expected payoff from a logging deal. A failure to consider this endogeneity in expected payoffs would lead to communities opting for logging deals despite PES, simply allowing communities to negotiate better logging deals. Potential trade-offs are shown to exist between maximizing environmental benefits and poverty alleviation, which implies the need for two policy tools, and not just one.

3.1 Introduction

In this chapter, we discuss the design of payments for environmental services (PES) in a specific setting, namely, when the actors taking the land-use decisions have only weak property rights over the land, and when other actors (e.g., firms) interested in commercial resource exploitation are present. Recent decentralization trends have involved a full or partial transfer of rights over natural resources from the state to local communities or user groups in many countries worldwide (Ribot, 2002). Nevertheless, these rights frequently remain weakly defined in a legal sense and poorly enforced by the government (Engel & López, 2004). Weak property

L. Lipper et al. (eds.), *Payment for Environmental Services in Agricultural Landscapes*, Natural Resource Management and Policy.
doi: 10.1007/978-0-387-72971-8_3;

rights are a recurrent theme in the literature on PES, particularly in development research and with respect to the potential impacts of PES on the rural poor (see, for example, Landell-Mills, 2002). At the same time, globalization has increased the commercial pressure on natural resources, often resulting in the presence of firms interested in their commercial exploitation. The World Resource Institute (WRI) highlights how the livelihoods of the poor have increasingly been impacted by being in “direct conflict with extractive industries such as large-scale fishing, logging and mining” (WRI, 2005, p. 4). Property rights and commercial resource extraction are brought together in this chapter with their implications for PES.

Following Engel and Palmer (2007), we use the example of Indonesia to motivate and illustrate the analysis. The decentralization of Indonesia’s forest sector has resulted in the acknowledgement of communities’ forest rights (Palmer, 2004). Communities have exercised such rights by negotiating logging agreements with timber firms, although formal community rights remain weakly defined and rarely enforced by the government. From these agreements, communities received a relatively small proportion of actual timber rents and environmental damages were common (Palmer, 2006). While logging is not the only threat to Indonesia’s forests, it has been a major factor underlying deforestation in previous decades (FWI/GFW, 2002). Through the capitalization of non-market forest values, PES could potentially provide an alternative source of income to communities while maintaining environmental services (ES) provided by the forest. As this chapter shows, however, designing PES in a context of weak property rights and commercial interests requires an improved understanding of the interactions between resource owners and commercial actors.

Effectiveness and efficiency are both critical in the design of any PES. An effective PES require that it leads to an increase in ES as compared to the situation that would result without such a payment. An efficient scheme implies maximizing ES with a given budget. This would require, for example, an estimation of communities’ opportunity costs, which in the Indonesian example would be the levels of the expected payoffs to communities from logging deals. Fieldwork revealed wide variation in Indonesian communities’ payoffs from such deals (see Engel & Palmer, 2006). The common intuition on PES design would suggest that, for a given ES per hectare, program design should focus on those communities with the lowest opportunity costs, which in our context implies the lowest expected payoffs from logging deals (e.g., Siikamäki & Layton, 2006; Wünscher et al., 2006).

This chapter demonstrates, however, that with weak property rights the issue is far more complex. First, communities with low opportunity costs may be unable to enforce a PES agreement, i.e., they may be unable to prevent logging activities by firms. Second, some communities would conserve the forest regardless; in these cases PES would not induce additionality. Finally, by raising the communities’ reservation utilities in negotiations, the introduction of PES may impact on expected payoffs from a logging deal. A failure to consider this endogeneity in expected payoffs would result in communities opting for logging deals despite PES, by simply allowing communities to negotiate better logging deals. In all of these cases, a PES scheme would be ineffective. We also show that there may be trade-offs between achieving efficiency, and thus environmental gains, and poverty reduction.

In the remainder of the chapter, Sect. 3.2 presents further background on the Indonesian setting and the data collected. The intuition underlying a game-theoretic model, an adaptation of an earlier one developed by Engel et al. (2006) to illustrate the above issues, is presented in Sect. 3.3. It combines conflict and bargaining theory to model the interactions between communities and logging firms. From this conceptual treatment, lessons for the design of PES schemes are drawn in Sect. 3.4. Using data collected in Indonesia, we illustrate how the theoretical results could be used to guide PES design empirically in Sect. 3.5. Section 3.6 concludes.

3.2 Background: The Indonesian Context and Data Collection

This section is derived from the existing literature and data from fieldwork conducted by the authors during 2003–2004 and described in detail in Palmer (2006). In particular, 62 communities in East Kalimantan province were surveyed using community- and household-level questionnaires.[1] All communities were sampled on the basis of having negotiated logging agreements that became operational and ended before the survey began. Data were collected on community characteristics and experiences, and are used throughout this chapter to illustrate how the theoretical model could be empirically implemented.

Tropical forests cover around 40% of Indonesia's land area and support the livelihoods of an estimated 30–40 million people (FWI/GFW, 2002). Over the period 1950–2000, the rapid expansion of commercial logging has at least been partially responsible for a 40% decline in the country's forest coverage, with a rate of loss of ~2 million hectares recorded per year in the late 1990s (FWI/GFW, 2002). Lowland forests are forecast to disappear from Sumatra and Kalimantan by 2010 if this rate of loss continues (Holmes, 2000; cited in FWI/GFW, 2002). The consequences of such a rapid disappearance of Indonesia's forests include an as yet unquantified loss in local ES, carbon sinks, and habitat for the country's disproportionately high share of the world's stock of biodiversity.

From 1997–1998 onwards, Indonesia rapidly decentralized, resulting in changes to the institutions and processes governing the management of natural resources, including the logging concessions system (Barr & Resosudarmo, 2002). These changes empowered forest-dependent communities, which increasingly exerted their property rights over customary (*adat*) forest. As a result, many communities negotiated directly and legitimately with logging firms in exchange for access to financial and in-kind benefits (Casson & Obidzinski, 2002).

---

[1] These were taken from 65 community-level and 687 household interviews, although for direct comparison the sample was reduced to 62 communities (see Palmer, 2006).

Community rights were, however, weakly defined in a legal sense (Wollenberg & Kartodihardjo, 2002). Weak state law enforcement along with endemic corruption in the forestry sector meant that local government rarely, if ever, enforced community-firm logging agreements. In our sample, 84% of communities claimed that the government played no role in contract enforcement (see Palmer, 2006). As a consequence, communities came to depend more on self-enforcement rather than the state to defend their property rights. Community-company conflicts due to firm non-compliance occurred in 50% of cases surveyed. Thus, companies could claim *de facto* property rights over community forests by either making agreements that were not complied with later (Barr et al., 2001) or by simply logging without community consent (Engel et al., 2006). Alternatively, *de facto* property rights could be claimed by communities through blockades (Engel et al., 2006).

In exchange for access to commercially valuable timber on land claimed by the communities, timber firms agreed to pay a fee per m^3 of timber harvested in addition to the provision of in-kind developments. Engel and Palmer (2006) illustrate the variation in actual benefits received by communities (all amounts are per m^3). For example, the mean level of financial payments plus in-kind benefits was around US $3.60 per m^3. The minimum and maximum levels were US $0.30 and US $11.80. By contrast, average domestic timber prices for the common Meranti species were US $30–70 per m^3 over the period 1999–2002 (see Palmer & Obidzinski, 2002).

Logging from these community-firm agreements led to substantial environmental damage. Over 70% of the sampled communities indicated a decline in drinking water quality and over 65% indicated an increase in flooding as a result of their concessions over the period 1998–2003 (Palmer, 2006). Furthermore, Resosudarmo (2004, p. 113) notes that while it might still be too early to draw conclusions about these environmental impacts, the changes so far indicated "a substantial increase in logging with little regard for environmental consequences... This increase is likely to lead to forest deterioration and conversion." In January 2003, the Ministry of Forestry estimated the total area of forest allocated for small-scale concessions by district governments, since the system was established, to be in the order of 2 million hectares (Resosudarmo, 2004).[2]

Thus, the opportunities for Indonesian communities to utilize their forest claims for income generation have thus far concentrated solely on forest timber values, resulting in a decline in the value of the country's forest ES; a cost that will rise given increasing domestic demand for timber products (see Arifin, 2005). An alternative to logging agreements would be for communities to negotiate agreements for maintaining ES in exchange for financial and in-kind benefits. Given the scale of logging in Indonesia, PES may be an important option in the provision of an alternative stream of income to communities deciding on forest use. There are currently no formal PES schemes established in Indonesia, although a number of

[2] Note that data on the proportion of these concessions that were operationalized and the quantity of logs harvested are unavailable.

schemes have been experimenting with the PES concept.[3] For example, Rewarding Upland Poor for Environmental Services (RUPES), a project established by the World Agroforestry Centre (ICRAF),[4] is involved in various schemes across the country (see Arifin, 2005). RUPES orients its "Rewards for Environmental Services" (RES) specifically toward the poorest groups in society.

PES have also been gaining interest in the international development community, including the World Bank and the United Nations, as a potential policy tool for both poverty alleviation and the provision of environmental benefits (WRI, 2005). However, whether considered as a theoretical concept or a practical policy tool, PES programs were not originally designed with poverty reduction as a primary objective (Pagiola et al., 2005). Landell-Mills and Porras (2002) review a large amount of empirical evidence on the potential for PES to simultaneously protect the environment and alleviate poverty. On the basis of relatively limited evidence, they argue that it is too early to reach conclusions about the impacts of PES on poor communities (Landell-Mills & Porras, 2002). Pagiola et al. (2005) suggest, nonetheless, that there can be important synergies when programs are well designed and local conditions are favorable. One key condition for effective PES is that of secure property rights.

3.3 Conceptual Model

In the context of weak property rights and where loggers might be present, the design of an effective and efficient PES scheme as an alternative to logging requires an understanding of the interactions between communities and outside commercial actors. We review the intuition behind a game-theoretic model of these interactions, following Engel et al. (2006).[5] We simplify and apply this model in that we do not consider endogenous policy interventions, and in that we take the logging area as exogenously given. Instead, we focus on community payoffs and analyze the impacts of PES on community-firm interactions.

Conflict and bargaining theory are linked in this model. A community's ability to self-enforce its rights over the forest is critical for understanding its performance in negotiating a logging agreement in a context of weak property rights. Conflict theory models what happens in the absence of negotiations and, hence, sheds light on this ability for self-enforcement. The results are subsequently incorporated into bargaining theory to investigate first why some communities receive higher payoffs from logging deals than others, and second, to predict under what conditions negotiations will succeed or fail. Formal derivations of the simplified and more general theoretical model follow Engel et al. (2006), and can be seen in the Appendix of this chapter.

---

[3] For a recent review, see Suyanto et al. (2005).

[4] For details on RES projects, see http://www.worldagroforestry.org/sea/Networks/RUPES/map-site_indonesia.htm.

[5] A more general version of the model is presented in Engel and López (2004).

3.3.1 Conflict Theory and Property Rights Formation

De facto property rights are modeled as the outcome of a "war of attrition" between a commercial actor such as a logging company (referred to as "the firm" hereafter) and a resource owner ("community"). For simplicity, it is assumed that both actors have perfect information about each other's parameters.[6] Logging requires a specific factor (physical capital) that is only available to the firm.[7] Under the assumption of weak community property rights, each of the actors can in principle obtain *de facto* rights over the forest, e.g., the firm may unilaterally exploit the forest if it can win a war of attrition, or the community may prevent this unilateral exploitation if the power conditions are reversed. The complementarity between the firm and the community in terms of access to the factors of production required for logging is what gives rise to the possibility of bargaining. So, while the firm has access to physical capital, the community may have the ability to control access to the forest.

The community obtains use and non-use values from the undisturbed forest for each period it is able to stop firm operations. If, however, the firm wins the conflict, it receives profits from logging unilaterally. Generally, the war of attrition is won by the actor that is able to stay in a potential conflict longer.[8] First, consider a situation where the costs of setting up a blockade for just one period exceed the present value of benefits obtained from the standing forest forever. Here, the community will never blockade, and the firm will log as long as net profits from doing so are positive. If, on the other hand, the benefits from protecting the forest for a single period already outweigh the costs of blockading in that period, the community will always fight, and the firm, knowing this, will withdraw. For intermediate values of the present value of forest benefits, the boundary condition can be obtained by computing for each actor the maximum time that he can stay in conflict and still receive a non-negative payoff, and then by equating these maximum times. This boundary condition can be seen as a line in Fig. 3.1 – line "*BC 1* (*War of Attrition*)" – and is more formally derived in the Appendix, Eq. (3.1). The figure shows the community's valuation of the standing forest per period and the firm's logging profits on the horizontal and vertical axes, respectively. The location of *BC 1* in the figure generally depends on other parameters, such as logging costs, blockading costs, and both actors' time preferences (discount rates).

The firm is able to stay in conflict longer than the community and thus wins the conflict for all points located above and to the left of *BC 1* (area I in Fig. 3.1). Here,

---

[6] This implies that the player that would lose the conflict withdraws immediately. Actual conflict is possible with imperfect information, although the outcome will generally depend on the same parameters listed here (see Burton, (2004) , for a related model with imperfect information).

[7] This assumption can be justified on the basis that communities are poor (have low savings) and have a disadvantage in the credit market *vis-à-vis* the firm, associated with capital market imperfections (see, e.g., Bose, 1998).

[8] For a more formal treatment of the conflict game, see Engel et al. (2006).

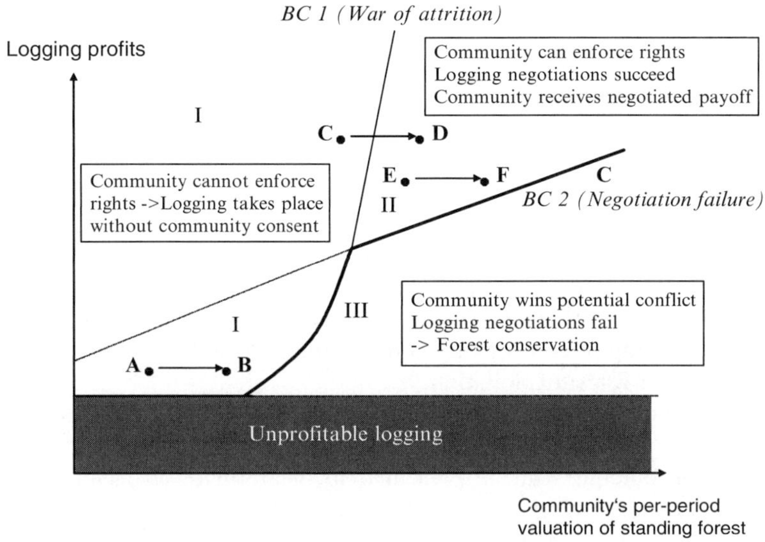

Fig. 3.1 Outcomes of community–firm interactions (Engel & Palmer, 2007)

the firm effectively has access to both physical and natural capital and, hence, is able to exploit the forest unilaterally and without community consent. The implication here is that the community effectively loses its property rights over the forest without any payment from the firm. For all points below and to the right of *BC 1*, the community is able to stay in conflict longer than the firm. The community is therefore able to self-enforce its property rights over the forest, which leads to two possible outcomes: the community may opt to prevent logging altogether (resulting in forest conservation, area III), or it may bargain with the firm over a logging deal (area II).

The outcome of the war of attrition is dependent on the model parameters. In particular, the community is more likely to win a potential conflict with the firm and thus obtain *de facto* rights over the forest if the profitability of logging, blockading costs, and the community's discount rate are low, and/or if the community's valuation of the standing forest, logging costs, and the firm's discount rate are high, see Appendix, Eqs. (3.2) and (3.3). Intuitively, an increase in profits for the firm or benefits from the standing value of the forest for the community enables this actor to stay in conflict longer, and thereby raises the likelihood that he is able to win the conflict. An increase in fighting costs on the other hand (logging costs for the firm, community blockading costs) has the opposite effect. An increase in an actor's discount rate similarly induces him to value the immediate fighting costs more than the long-run benefits from winning, which reduces the maximum length of time this actor would stay in conflict.

3.3.2 Community-Company Bargaining over a Logging Agreement

If the community loses its property rights over the forest (area I in Fig. 3.1), this would imply that the firm has access to both factors of production and, hence, would have little incentive to share logging profits in a negotiated agreement.[9] Thus, negotiations over a logging agreement would not be feasible. We now focus on the case where the community is able to win a potential conflict with the firm and, hence, self-enforce its rights over the forest. The firm, in this case, has access to physical capital while the community effectively controls the forest. Hence, there is scope for bargaining over a logging contract that would allow the actors to pool their resources for forest exploitation.

Negotiations over a logging agreement can be modeled as an alternating-offers bargaining game between the community and the firm. For an interior solution, the negotiation outcome can be presented in the form of an asymmetric Nash Bargaining Solution (NBS). This implies that each actor obtains his reservation utility,[10] with the remaining surplus divided in proportion to bargaining power, Appendix, Eq. (3.4). Thus, community payoffs are increasing in the "size of the cake" (the net profits from logging), the community's bargaining power, and the community's reservation utility, Eq. (3.5). Community payoffs are decreasing in the firm's bargaining power and the firm's reservation utility.

Reservation utilities are the outcomes that result in the absence of negotiations. When the community is able to win a potential conflict, it will self-enforce its property rights over the forest and prevent any logging by the firm. Hence, the community's reservation utility would be the communities' (perceived) present value of the stream of benefits provided by the standing forest over time, while the firm would receive its net profits from using its capital in the next-best alternative activity.

3.3.3 Combining Results from the Two Stages

The results of the two stages are combined and the reduced-form relationship between community payoffs and the model parameters are derived in Engel and Palmer (2006), and reproduced in the Appendix, Eqs. (3.7) and (3.8). Indonesian data on payoffs and on proxies for the parameters are used to test the resulting hypotheses through econometric analysis. In general, Engel and Palmer's (2006)

---

[9] In reality a *pro forma* agreement may still be negotiated and a minimum payment made to the community. This minimum payment is motivated by the idea that firms maintain political capital with the government officials who issued the logging permits (Palmer, 2005). However, this consideration does not affect the qualitative results that are of relevance for our purposes.

[10] More precisely, their inside options (see Appendix).

empirical analysis supports the theoretical model's predictions. Of particular relevance to PES, they find that communities that value the forest more, in particular, those that derive a large proportion of their income from the forest, are more likely to obtain higher payoffs from logging deals. This is both because such communities have a greater ability to self-enforce property rights and because they request more compensation for environmental damages.

With respect to the design of PES, one approach would be to use these results to predict communities' expected payoffs from logging deals, and to use these to proxy for communities' opportunity costs. Thus, data on model parameters for potential PES communities could be used in combination with the results to predict where on the observed range of logging payoffs (US \$0.28–11.81 per m^3) the communities' expected payoffs under a logging deal would be likely to lie. PES should then be at least as large as these expected payoffs. Furthermore, if the objective of the scheme is to maximize ES provision with a given budget, and if ES per hectare are assumed to be approximately equal across communities,[11] it might be most efficient to choose those communities for PES that have the lowest expected payoffs from a potential logging deal, i.e., around US \$0.28 per m^3. These are likely to be the poorest communities, i.e., the ones with high discount rates, which implies that by targeting poor communities, PES could alleviate poverty and maximize environmental benefits at the same time. However, as will be shown in Sect. 3.4, such an approach may neither be effective nor efficient. To show this, it is necessary to consider the conditions under which negotiations fail.

3.3.4 Negotiation Failure

For PES design, when do communities opt for logging agreements and when do they prefer forest conservation, i.e., when do negotiations fail? To analyze the conditions for negotiation failure, we again consider the case where bargaining is feasible due to the community being able to win a potential conflict over *de facto* property rights. Negotiations will fail if the sum of both the firm's and the community's reservation utilities exceeds the "size of the cake" to be divided in a negotiated contract. In this case the "cake" is simply too small to make players better off in a negotiated agreement as compared to their next-best alternative activities. The boundary condition determining the success or failure of negotiations is represented by the line "*BC 2* (*Negotiation failure*)," in Fig. 3.1. This line represents all the points where the sum of both actors' reservation utilities is equal to the "size of the cake."

[11] In practice, of course, ES provided by a hectare of standing forest may differ according to geographic and ecological conditions, an aspect that is ignored here as it is beyond the scope of this chapter.

In summary, the two boundary conditions in Fig. 3.1 yield three potential outcomes of community-firm interactions. First, the firm may effectively control the physical and natural capital, resulting in unilateral logging without community consent and no or low community payoffs (area I in Fig. 3.1). Second, the community may be able to self-enforce its forest rights, which may result in a negotiated logging agreement between the community and the firm (area II). Here, community payoffs from the agreement are increasing in the community's valuation of the standing forest. Third, the community may be able to self-enforce its rights over the forest but its valuation of the standing forest may be so high or logging profits so low that there is no negotiated outcome that both players would agree to. In this case, negotiations would fail and the forest would be conserved (area III).

3.4 Lessons for PES Design

3.4.1 The Introduction of PES

What are the implications of the above results for PES design? In the absence of PES, the community's valuation of the standing forest may include direct uses such as the collection of fuelwood and non-timber forest products. It may also consider local ecological services from the standing forest (e.g., erosion prevention, water retention) as well as non-use values, such as cultural values. The introduction of PES adds an additional value to the standing forest for those communities receiving PES, Appendix, Eq. (3.8). This value may reflect all or part of the benefits from the forest obtained by society at large. For simplicity, it is assumed that PES are simply paid on a constant per-hectare basis for conservation of the standing forest, and not for any specified service such as carbon storage or biodiversity, which may differ across forest plots. Moreover, PES are assumed to be made per period and on condition of actual forest conservation.

The introduction of PES raises the communities' per-period valuation of the standing forest (a horizontal shift to the right in Fig. 3.1). This has two implications for the model. First, the community's reservation utility in negotiation rises, see Appendix, Eqs. (3.9) and (3.10). Intuitively, the introduction of PES implies that the community, if able to enforce its rights over the forest, has a better alternative to logging, as compared to the situation without PES. Second, by raising the community's benefits from protecting the forest, PES increase the community's willingness and ability to stay in a potential conflict, because the community's benefits from fighting rise. Thus, PES strengthen the community's ability to win a potential conflict over *de facto* property rights to the forest.

3.4.2 Conditions for Effective PES

There are three conditions that need to be satisfied for PES to be effective in inducing forest conservation (Engel & Palmer, 2007). First, the community has to be able to effectively enforce its property rights over the forest. To see this, consider the case where a community is initially situated at point A in Fig. 3.1. PES, by raising the community's valuation of the standing forest, induce a horizontal shift to the right, say, to point B. While this community would prefer conservation with PES over a logging agreement (since point B is situated to the right of *BC 2*), it is unable to win a potential conflict with the firm and, hence, cannot enforce the PES contract. Therefore, the firm logs unilaterally, and PES are ineffective. Thus, in the absence of external enforcement, this implies that PES need to induce a shift large enough to take the community to the right of boundary condition *BC 1* as well as *BC 2*, i.e., where it is able to self-enforce property rights over the forest and prevent the firm from logging unilaterally. Note that the introduction of PES in situations of weak property rights need not be futile. Rather, where PES raise the community's value of the standing forest sufficiently, it can help induce it in successfully enforcing its property rights, i.e., property rights are endogenous to PES.

The second condition is that the PES amount has to be large enough to induce negotiations with logging firms to fail. When PES are introduced, the community either has the choice to accept this and, thus, conserve the forest, or to negotiate a logging agreement with the firm. For PES to be effective in achieving forest conservation, it is required that communities, given the PES offer, would prefer PES and forest conservation over the logging agreement. In other words, PES have to induce a shift into area III in Fig. 3.1. First, consider a community located initially at point C in Fig. 3.1 and for which PES induce a shift to point D. Before PES, this community was unable to enforce its property rights over the forest. PES in this case raise the community's ability to enforce its rights, but the community would opt to use this ability to negotiate a logging contract and share in logging profits rather than opt for conservation and receiving PES. Thus, PES fail to induce forest conservation. Consider a second community for which PES imply a shift from point E to point F in Fig. 3.1. In this case PES merely raise the community's reservation utility in bargaining, allowing it to negotiate a better logging deal. In other words, by putting external forest values on the table, the introduction of PES simply results in the firm offering a better deal than it would have done in the absence of PES. Here again, PES are ineffective in achieving forest conservation. Note, however, that in both cases and despite the failure to induce forest conservation, the community will be financially better off with PES than compared to the case without.

The third condition for effective PES is that it should achieve additionality in ES. First, logging needs to be profitable otherwise the forest would be conserved anyway, i.e., where logging is unprofitable (anywhere in the grey area of Fig. 3.1), PES would not provide additionality. Moreover, the same holds for communities initially located in area III. In the absence of PES, these communities already value the forest so highly that they would have turned down any feasible logging agreement anyway. To achieve additionality, ES buyers should thus focus on communities that, prior to the PES intervention, are located in areas I or II, i.e., in the absence of PES they would opt for a logging agreement. In summary, the conditions for effective PES are:

The community, given PES, needs to be able to win a potential conflict with the firm (*BC 1*). Otherwise the firm can log despite PES agreement.

PES need to induce a breakdown of any potential logging agreement – (*BC 2*). Otherwise PES would only raise community payoffs from logging.

PES need to focus on communities where logging is likely to occur in the absence of PES (i.e., communities initially located in areas I or II). Otherwise PES fail to induce additionality.

These conditions are also given formally in the Appendix, Eq. (3.12). Condition (2) implies that the level of PES required for effectiveness is not the expected payment according to what is observed in terms of actual logging payments (with an observed average of US \$3.60 per m^3). Again, this is because the firm, realizing that PES have improved the community's reservation utility, will also raise their offer to the community, so long as the firm can still retain enough logging profit to be better off than under its next-best activity. Thus, orienting PES amounts toward currently observed logging fees may only allow communities to negotiate better logging deals and the attempt to use PES to compensate the community for the lost payoffs from a potential logging deal with the firm is futile. To be effective, the present value of PES over time (together with the community's other values of the standing forest) needs to be large enough to outbid the highest potential offer from the firm; the community needs to be better off under PES than under the most favorable potential logging deal. This maximum possible offer by the firm is likely to be unobserved and may substantially exceed the maximum payment observed (US \$11.81 per m^3). Therefore, effective PES might have to be much higher than initially expected.

3.4.3 PES and Poverty Alleviation

3.4.3.1 Efficiency and Pro-Poor Targeting of PES

We now consider how a PES scheme can maximize ES provision with a given budget. Again, assume for simplicity that the ES provided by 1 hectare of forest protected are similar for all communities.

Inclusion in a PES scheme will shift the location of a particular community in Fig. 3.1 to the right by the amount of the periodic payment made because it raises the community's per-period valuation of the standing forest. Moreover, to be effective, PES should focus on communities in areas I or II and be large enough to induce a shift into area III. It is easy to see from the figure that such a shift could be achieved at lowest cost for communities near to, but to the left of the thick line in Fig. 3.1. This line reflects the binding condition among conditions (i) and (ii) for effective PES. Note, however, that the communities close to and to the left of this line are not the ones with the lowest expected payoffs from a negotiated deal. Rather, within area II, these are the communities with the highest expected payoffs prior to PES. Communities located in area I reflect those with the lowest expected payoffs but, unless located close to the thick line, a low payment would fail to induce effective PES as communities would be unable to enforce forest conservation.[12]

The location of *BC 1* and *2* depends on the model parameters. An increase in the community's discount rate shifts the two boundary conditions to the right. Intuitively, communities with higher discount rates are less likely to win a potential conflict with the firm because they value the immediate costs of fighting more than the longer run benefits from winning the conflict (*BC 1*). Moreover, negotiation failure is also less likely for communities with higher discount rates (*BC 2*). This is because the community's reservation utility is the present value of the stream of benefits from the standing forest as perceived by the community. The higher the discount rate, the lower this present value. Hence, *ceteris paribus*, a community with higher discount rates is less likely to be located close to the thick line in Fig. 3.1. This implies that the maximization of environmental benefits would not stipulate the targeting of PES to the poorest communities, which tend to have high discount rates. First, the poorest communities are also the ones least likely to be able to enforce a PES contract, even if they agreed to it in the first instance. Second, poor communities – due to higher discount rates – put comparatively more weight than richer communities on the immediate benefits from a logging deal than on the periodic payments from a PES contract. Therefore, for a given level of PES, these communities are more likely than richer ones to opt for a logging deal rather than PES, thus implying a poverty-environment trade-off in PES design in our setting.

3.4.3.2 Property Rights

To what degree do our results rely on the assumption of weak property rights? Moreover, how would the provision of better-defined and government-enforced property rights for communities affect environmental outcomes and poverty? These issues are important because the improvement of government provision and enforcement of property rights is a frequently discussed policy option (see WRI, 2005). In addition, third-party organizations wanting to implement PES could also help communities in property right enforcement, e.g., by lowering blockading costs.

Property rights are clearly a crucial determinant of community payoffs from a logging contract. The issue of self-enforcement, represented by the line *BC 1* in Fig. 3.1, would become irrelevant if community property rights to the forest are well defined and government enforced. A firm could no longer log unilaterally without

[12] Note that if PES is are ineffective in inducing forest conservation for a low pay-off community, then it may not receive any payment, i.e., the PES program would have achieved zero environmental gains but at zero cost. Thus, where PES is are ineffective, it is not necessarily inefficient. In reality of course, upfront payments may be made before any conservation outcome is observed and there may be high transactions costs from negotiating the original PES agreement.

community consent, leading to an increase in community payoffs for those communities that are unable to self-enforce rights (area I in Fig. 3.1). Poor communities (with high discount rates) are more likely to fall into this group. Thus, the provision of more secure property rights would assist in poverty alleviation as it prevents poor communities from being exploited by commercial actors. With respect to environmental outcomes, the only relevant line in Fig. 3.1 would then be *BC 2*. Communities located above and to the left of this line would still make logging deals, while those below this line would opt for forest conservation. The provision of secure property rights would induce forest protection in all communities located between *BC 1* and *BC 2* in the lower-left of Fig. 3.1, i.e., those that would have been unable to self-enforce property rights, although their valuation of the forest is high enough (or logging profits in the area low enough) to make them prefer conservation over a feasible logging deal. Thus, providing more secure property rights would lead to environmental improvements in these communities.

With secure property rights, PES design also becomes less complex. Condition (1) would be satisfied by default. For effective PES, conditions (2) and (3) would remain. From condition (3), solving for the minimum effective payment results in an equation, Eq. (3.13) in the Appendix, which implies that the minimum per-period PES amount required to effectively induce forest conservation is likely to be increasing in logging profits and the community's discount rate. It is also decreasing in logging costs, the firm's profits in the next best activity, and the community's valuation of the standing forest in the absence of PES. For very high valuations, no PES are required; the community would opt for forest conservation anyhow (area III). These final points imply that efficiency could be enhanced by focusing on communities with intermediate, rather than low valuations of the standing forest in the absence of PES.

Similarly, communities with very low discount rates may fall in area III (no additionality). Together with the result that poorer communities have higher discount rates, the objective of efficiency would imply focusing on communities with medium discount rates rather than the poorest communities. Again, these results imply that there may be a potential trade-off in PES design between the objectives of maximizing ES provision and poverty alleviation. This result is independent of whether property rights are weak or secure. The provision of more secure property rights in itself is a policy that could achieve both environmental and welfare improvements.

3.5 Empirical Application

PES design in our setting would require obtaining information about the "location" of communities in Fig. 3.1. In particular, an estimate of the starting location of a potential PES candidate relative to *BC 1* and *BC 2* is required. While the Indonesian data described in Sect. 3.2 were originally collected for a different purpose, we illustrate how the theoretical results could be combined with empirical analysis to guide PES design.

Palmer (2006) used data on the logging benefits received by the communities surveyed to analyze the probability of these communities being located in areas I and II. A minimum payment of US $1.70 per m^3 of log production was established as the cutoff point on the basis of this being the lowest negotiated payment in the sample. This is interpreted as the minimum acceptable payment. Actual payments frequently differed from the negotiated value. Any actual payment falling below this threshold indicated that the community was unable to self-enforce property rights. Nineteen communities (31% of the sample) received a fee level that came below this threshold, while the remainder received US $1.70 or more per m^3. A sensitivity analysis varying the threshold level was undertaken to test the robustness of the results to this assumption. Given the discrete nature of the dependent variable, a logit model was run on a combination of proxies for the theoretically relevant parameters with the dependent variable equal to one where the community received a payment above the US $1.70 threshold (assumed to fall into area II) and equal to zero otherwise (area I). The econometric results are shown in Table 3.1.

The results generally confirm the theoretical hypotheses presented earlier in this chapter. Community blockading costs were proxied by the proportion of households containing members of dominant ethnic groupings, the proportion of households participating in community organizations (both variables proxies for social capital), distance to the market and the proportion of households with government

Table 3.1 Econometric results on probability of community being able to enforce property rights

Variable (parameter proxied)	Coefficient	Standard error	*t*-stat.	p[|*T*| > *t*]
Constant	−3.2680	2.6455	−1.235	0.2167
Average percentage of household incomes derived from sale of forest products (*b*)	0.1587	0.2272	0.698	0.4850
Forest quality: area logged before by commercial operation (Yes = 1, No = 0) (*v*)	2.1276	1.1978	1.776	0.0757*
Actual area logged (hectare) (*v*)	−0.009559	0.004437	−2.154	0.0312**
Percentage of households containing at least one government employee (*s*)	−0.7598	0.5081	−1.495	0.1348
Percentage of households that participate in community organizations (*s*)	0.7434	0.3270	2.273	0.0230**
Percentage of households containing members of dominant ethnic grouping (*s*)	0.3034	0.2534	1.198	0.2311
Distance to nearest market (km) (*s*)	−0.2566	0.1570	−1.635	0.1021
Percentage of households holding savings before agreement (r^C)	0.1267	0.5515	2.298	0.0216**
No. of observations	62			
Restricted log likelihood	−38.2			
Chi-squared	37.8			
Percentage of outcomes predicted correctly	71.0			

Note: *Significant at 0.10 level; **Significant at 0.05 level. All results corrected for heteroskedasticity

Source: Palmer (2006)

employees (both proxies for opportunity costs).[13] From Table 3.1, the directions of the effects follow the predictions of the theoretical model, i.e., the probability of a community being able to enforce its rights over the forest is decreasing in the community's blockading costs. Only market distance and household participation in community organizations are statistically significant. Community discount rates were proxied by the proportion of households holding savings before the onset of negotiations, with poorer communities expected to have higher discount rates. The results indicate a significant positive effect on the ability for communities to enforce property rights, supporting the theoretical hypothesis that lower discount community being able to enforce its rights over the forest is decreasing in the community's blockading costs. Only market distance and household participation in community organizations are statistically significant. Community discount rates were proxied by the proportion of households holding savings before the onset of negotiations, with poorer communities expected to have higher discount rates. The results indicate a significant positive effect on the ability for communities to enforce property rights, supporting the theoretical hypothesis that lower discount rates are associated with a better ability to fight. Average household incomes derived from forest products proxied for community valuation of the standing forest. As expected, the effect was positive, although insignificant. This may be because this variable reflects a percentage of total income rather than the absolute value of the income from forest products. Finally, the probability of a community being able to enforce property rights significantly increases if the community's forest was logged before and is significantly decreasing in the size of the area logged. Both of these factors are proxies for logging profitability. Over 70% of observations are predicted accurately by the model. The results can be used to assess the location of a particular community relative to *BC 1* by using estimates of the explanatory variables[14] to calculate the predicted probability of it being able to enforce its property rights. If this probability is less than 0.5, the community would be expected to be located in area I, otherwise it would be expected to be located in area II.

While this analysis illustrates an empirical estimation for *BC 1*, some words of caution are in order. Since the original research focus required that all communities sampled in this survey had involved in negotiations and made agreements with firms, it is likely that area I communities were undersampled and additional data would be needed to improve the analysis. Moreover, the proxies used were often not ideal and, for firm discount rates and logging costs, were missing altogether. The latter could potentially introduce omitted-variable bias, although it is not

---

[13] In the Indonesian context, where all households regularly go to the market to sell surplus produce for cash income regardless of distance, and where other employment opportunities are negligible, opportunity costs are likely to increase with distance to the market (Palmer, 2006). This is in contrast to other contexts where household participation is elastic to distance from market and greater distance implies lower opportunity costs.

[14] If estimates only on some of the variables are available, then the average observed sample value could be used for the other variables for simplicity.

unreasonable to assume that in this particular setting firm parameters hardly varied across the sample.

Regarding *BC 2*, the research focus on negotiated agreements meant that data were not available on communities falling into area III. There is anecdotal evidence, however, for at least two communities that had declined all offers for logging deals, opting for forest conservation instead (for one of these cases, see Iwan, 2004). Similar to the analysis on areas I and II, yielding *BC 1*, the additional collection of data on communities located in area III would allow for the estimation of the predicted probability for negotiation failure. The dependent variable in this case would be a dummy indicating whether a community opted for forest conservation or a logging deal. The results in Sect. 3.3.3 indicate that relevant explanatory variables should include for example, proxies for net logging profits and the community's present value of the standing forest (which itself depends on the community's per-period valuation of the standing forest and its discount rate).

The collection of more complete data would permit an adequate estimation of the two boundary conditions, which could be used in PES design. Data would need to be collected on the empirically relevant independent variables for all communities that could be potentially included in a PES scheme. Alternatively, communities could self-report these characteristics when applying for PES.[15] The econometric results could then be used to estimate each community's predicted probabilities of (a) winning a war of attrition, and (b) opting for forest conservation. Those communities that satisfy the following conditions could then be considered for PES. First, they should have one of the two predicted probabilities greater than 0.5 (indicating that they lie on the right of one of the boundary conditions). Second, the other predicted probability should be below, but close to 0.5 (indicating that the community lies close to and to the left of the other boundary condition). In summary, these conditions imply that PES candidate communities lie to the left, but close to the thick line in Fig. 3.1.

Estimating the payment necessary to induce selected communities to opt for PES and forest conservation is an even more complex task. The results indicate that this minimum payment depends on the model parameters. In practice, there are obvious logistical and financial constraints in collecting all the data required for the design of a PES scheme. Perhaps more promisingly, the approach described here could be used to identify communities predicted to lie in area III (namely those with predicted values greater than 0.5 in both regressions). These communities should not be considered in order to assure additionality. Auction or contract design could be used to elicit the opportunity costs of the remaining communities under consideration.[16]

[15] Of course, self-reporting may induce problems of asymmetric information, an important issue that is beyond the scope of this chapter.

[16] Auction design would, however, require that communities be aware of the firm's potential to raise its payments.

3.6 Conclusions

This chapter analyzed PES design in a context where community property rights over the forest are weak and logging firms seeking to commercially exploit the forest are present. We began with the common intuition that PES should compensate resource owners for their opportunity costs induced by land-use change, and that an efficient PES design should focus on those communities with the lowest opportunity costs. In the Indonesian setting, opportunity costs are the potential payoffs to be made in a logging agreement. Based on fieldwork observations that actual payoffs vary greatly among communities involved in logging agreements, a game-theoretic model of community-firm interactions was presented that tries to explain the causes of this variation.

In Sect. 3.4, the model's implications for effective and efficient PES design were analyzed. The results indicate that the conventional wisdom in this regard is misleading for two reasons. First, communities with very low expected payoffs from negotiations also tend to be those that are unable to self-enforce property rights and prevent unilateral forest exploitation by firms. Therefore, PES contracts with these communities may be ineffective. Second, the introduction of PES may increase a community's valuation of the standing forest, thus impacting on its ability to self-enforce its property rights as well as its expected payoff from a negotiated logging agreement. If this endogeneity of community payoffs is ignored, then PES implementation may only result in better logging deals for communities, without achieving forest conservation. That said, improved logging deals may have positive welfare impacts on communities even after taking the environmental costs of logging into account. This endogeneity also implies that the relevant logging payment to be considered should be the best possible offer by the firm, which is unobserved, and that effective PES may have to be significantly larger than expected. However, it should be noted that this problem is not unique to PES and that any conservation intervention using positive incentives should consider the value of foregone logging fees.

PES are often promoted as a mechanism that can potentially both alleviate poverty and provide ES. The results presented here, however, indicate a potential trade-off in PES design between the objectives of maximum ES provision (efficiency) and poverty reduction. This is not to say that poverty reduction *per se* is not beneficial to the environment. Rather, the result here is that PES may not be "the one stone to kill the two birds" of poverty alleviation and maximum environmental benefits. Targeting PES to maximize the provision of ES will usually not be consistent with targeting the poorest communities. First, the poorest communities are also the least likely to be able to enforce a PES contract. Second, the poorest communities tend to have higher discount rates and thus will value the immediate benefits from a logging agreement more than periodic PES. Therefore, for a given level of PES, poor communities are more likely to opt for logging than PES. Nevertheless, it is interesting to note that introducing PES, even if ineffective in achieving conservation, may still help poor communities to obtain greater benefits through improved logging deals. Given the high transaction costs of PES implementation, however, it is likely that there are more effective and efficient poverty alleviation strategies than PES. In summary, achieving maximum environmental gains and poverty alleviation is likely to require two policy tools, and not just one.

This result is independent of whether property rights are weak or secure; it is sufficient that external actors seeking to commercially exploit the forest are present. Since property rights are endogenous, the introduction of PES in a situation of weak property rights, by raising the value of natural resources to communities may actually enhance their ability to enforce their rights. The provision of more secure property rights, on the other hand, could achieve both environmental and welfare gains. In particular, for those poor communities with high enough valuations of the forest, but who are unable to self-enforce forest rights and, hence, prevent exploitation of the forest by commercial actors. While the potential of this policy with respect to environmental outcomes remains limited by the fact that communities are unlikely to consider external values of the forest, the provision of more secure property rights helps to reduce the complexities of PES design. This implies that there may be strong synergies between the two policy options.

Acknowledgment Funding for this research was provided by the Robert Bosch Foundation, Germany (Grant No. 32.5.8041.0003.0).

Appendix

Let b denote per period values from the undisturbed forest to the community. If the firm wins the conflict, it receives profits v from logging unilaterally. The discount rates of the community and the firm are, respectively, r^C and r^F. Let c denote the firm's fixed logging costs in each period it attempts to log unilaterally. The community can blockade firm activities at a cost of s per period. Where $s > b/r^C$ (in which $\int_{t=0}^{\infty} be^{-rt}dt = b/r_C$), the community never fights and the firm logs as long as $v - c > 0$. Where $s < b$, the community always fights, and the firm, knowing this, withdraws. For intermediate values of b ($sr^C < b < s$), the boundary condition is found by equating the time that each actor can stay in conflict and still obtain non-negative net benefits:

$$v = \Omega, \textit{where} \ \Omega \equiv \frac{c}{r^F}\left[\left(\frac{s(1-r^C)}{s-b}\right)^{\frac{\ln(1-r^F)}{\ln(1-r^C)}} - 1 + r^F\right]. \tag{3.1}$$

The probability that the community is able to establish *de facto* property rights (PR) can thus be written as a function of model parameters:

$$PR = g(v, b, r_C, r_F, c, s). \tag{3.2}$$

As argued in Engel et al. (2006), inspection of $v = \Omega$ indicates that

$$\frac{\partial PR}{\partial v} \leq 0, \ \frac{\partial PR}{\partial b} \geq 0, \ \frac{\partial PR}{\partial r_C} \leq 0, \ \frac{\partial PR}{\partial r_F} \geq 0, \ \frac{\partial PR}{\partial c} \geq 0, \ \frac{\partial PR}{\partial s} \leq 0. \tag{3.3}$$

As explained in the text, bargaining leading to a logging agreement is possible only when $v < \Omega$. We distinguish two types of reservation utilities. Inside options are the payoffs obtained by each player while parties temporarily disagree and negotiations are ongoing (Bulow & Rogoff, 1989), denoted by d^C and d^F, for the community and the firm, respectively. Outside options, denoted by R^C and R^F, are the parties' payoffs available when bargaining fails permanently (Binmore, 1985).

For an interior solution, the negotiation outcome in an alternating-offers bargaining game is given by the NBS (Muthoo, 1999). Thus, the negotiated payments to the community (Π^C) and firm (Π^F) are obtained by solving the following Nash bargaining problem

$$\max_{\Pi^C, \Pi^F} \ \left[\Pi^C - d^C\right]^{\tau} \left[\Pi^F - d^F\right]^{1-\tau} \quad \text{s.t.} \quad \Pi^C + \Pi^F = v - c,^{17} \tag{3.4}$$

where τ is the community's bargaining power vis-á-vis the firm ($0 \leq \tau <$). The solution to Eq. (3.4) for the community's payment is

$$\tilde{\Pi}^C = d^C + \tau\left(v - c - d^C - d^F\right). \tag{3.5}$$

The total "cake" to be divided in negotiations is $v - c$. The community's inside option in the case where the community is able to self-enforce its property rights is given by the present value of the standing forest to the community. Thus, $d^C = \frac{b}{r_C}$, and:

$$\tilde{\Pi}^C = \frac{b}{r_C} + \tau\left(v - c - \frac{b}{r_C} - d^F\right). \tag{3.6}$$

Equation (3.6) implies that community payoffs from a logging agreement are increasing in b and decreasing in r_C. Engel and López (2004) show that the community's payment can be written as

[17] Following Engel and Palmer (2008), we abstract here from possible differences in the logging area between unilateral logging and a negotiated agreement.

$$\Pi^C = h\left(v, c, b, \tau, d^F, PR\right) \quad \text{with}$$
$$\frac{\partial \Pi^C}{\partial v} \geq 0,\ \frac{\partial \Pi^C}{\partial c} \leq 0,\ \frac{\partial \Pi^C}{\partial b} \geq 0,\ \frac{\partial \Pi^C}{\partial \tau} \geq 0,\ \frac{\partial \Pi^C}{\partial d^F} \leq 0,\ \frac{\partial \Pi^C}{\partial PR} \geq 0. \tag{3.7}$$

Using the expressions for PR, given in Eqs. (3.2) and (3.3), and considering that τ is itself increasing (decreasing) in the firm's (community's) discount rate, and increasing in other factors associated with higher bargaining power (p), Engel and Palmer (2006) obtain the following reduced-form relationship between Π^C and the model parameters:

$$\Pi^C = \tilde{h}\left(v, c, b, r_F, r_C, \bar{d}^F, s, p\right) \quad \text{with}$$
$$\frac{\partial \Pi^C}{\partial v} \gtrless 0,\ \frac{\partial \Pi^C}{\partial c} \gtrless 0,\ \frac{\partial \Pi^C}{\partial b} \geq 0,\ \frac{\partial \Pi^C}{\partial r_F} \geq 0,\ \frac{\partial \Pi^C}{\partial r_C} \leq 0,\ \frac{\partial \Pi^C}{\partial \bar{d}^F} \leq 0, \tag{3.8}$$
$$\frac{\partial \Pi^C}{\partial s} \leq 0,\ \frac{\partial \Pi^C}{\partial p} \geq 0.$$

As shown by Muthoo (1999), the interior solution in Eq. (3.5) holds only if $\tilde{\Pi}^C > R^C$. Otherwise, the community would receive R^C in negotiations. Moreover, if $R^C + R^F > v - c$, then negotiations will fail and the actors obtain their outside options.

Let P denote the per-period payment made under PES, which is conditional on the conservation of the forest, and b_0 is the community's per-period valuation of the standing forest in the absence of PES. Thus, the community's total valuation of the standing forest is

$$b = b_0 + P. \tag{3.9}$$

The community's outside option increases with PES:

$$R^C = \frac{b_0 + P}{r_C}. \tag{3.10}$$

By raising b_0 + P, PES also raise the community's willingness and ability to stay in a potential conflict. In summary, community payoffs from interacting with the logging firm are:

$$\Pi^C = \begin{cases} 0 & \text{if} \quad v > \Omega \\ \tilde{\Pi}^C & \text{if} \quad v \leq \Omega \quad \text{and} \quad \tilde{\Pi}^C > R^C \\ R^C & \text{if} \quad v \leq \Omega \quad \text{and} \quad \tilde{\Pi}^C \leq R^C. \end{cases} \tag{3.11}$$

Moreover, if $R^C + R^F > v - c$, where R^C is given by Eq. (3.10), and $v < \Omega$, negotiations would fail and forest protection would result (Engel et al., 2006). Because P affects Ω, R^C, and possibly $\tilde{\Pi}^C$ (the latter if and only if inside options are also affected by PES), the community's expected payoff from a logging agreement is also affected by PES. More specifically, Eqs. (3.1), (3.6), (3.9), (3.10), and (3.11), together imply that Π^C are (weakly) increasing in P.

The conditions for effective PES can be summarized as follows:

Proposition: *An effective PES scheme with per-period payment P conditional on resource conservation requires*

1. $v < \Omega(\mathrm{P})$ *(Community can enforce resource conservation),*
2. $\frac{b_0}{r_C} + R^F < v - c$ and $v > c$ *(Additionality, i.e., no negotiation failure in the absence of PES and logging is profitable), and*
3. $(b_0 + P)/r_C + R^F > v - c$ *(PES induce a breakdown of any potential resource extraction agreement).* (3.12)

With secure property rights, condition (1) in the proposition would be satisfied by default. Effective PES then have to satisfy conditions (2) and (3). Solving for the minimum effective payment (from condition 3), $P_{\min}$,

$$P_{\min} = r_C\left[v - c - R^F\right] - b_0. \tag{3.13}$$

References

Arifin B (2005) Institutional constraints and opportunities in developing environmental service markets: Lessons from institutional studies on RUPES in Indonesia. World Agroforestry Centre (ICRAF), Bogor, Indonesia.

Barr C, Resosudarmo IAP (2002) Decentralization of forest administration in Indonesia: Community livelihoods and economic development. Center for International Forestry Research, Bogor, Indonesia.

Barr C, Wollenberg E, Limberg G, Anau N, Iwan R, Sudana M, Moeliono M, Djogo T (2001) The impacts of decentralisation on forests and forest-dependent communities in Malinau district, East Kalimantan, Decentralisation and forests in Indonesia series: Case study 3. Center for International Forestry Research (CIFOR), Bogor, Indonesia.

Binmore K (1985) Bargaining and coalitions. In: Roth A (ed.) Game-theoretic models of bargaining. Cambridge University Press, Cambridge.

Bose P (1998) Formal-informal sector interaction in rural credit markets. J Dev Econ 56:265–280.

Bulow J, Rogoff K (1989) A constant recontracting model of sovereign debt. J Politi Econ 97:155–178.

Burton PS (2004) Hugging trees: Claiming *de facto* property rights by blockading resource use. Environ Resour Econ 27:135–163.

Casson A, Obidzinski K (2002) From new order to regional autonomy: Shifting dynamics of “illegal” logging in Kalimantan, Indonesia. World Dev 30:2133–2151.

Engel S, López R (2004) Exploiting common resources with capital-intensive technologies: The role of external forces. ZEF-Discussion Papers on Development Policy 90. Center for Development Research, Bonn (ZEF).

Engel S, Palmer C (2006) Who owns the right? The determinants of community benefits from logging in Indonesia. For Policy Econ 8:434–446.

Engel S, Palmer C (2007) Payments for environmental services and poverty alleviation: The case of weak property rights and commercial interests, Chair of environmental policy and economics working paper, ETH Zurich, Switzerland.

Engel S, Palmer C (2008) Payments for environmental services as an alternative to logging under weak property rights: The case of Indonesia. Ecol Econ 65(4):799–809.

Engel S, López R, Palmer C (2006) Community-industry contracting over natural resource use in a context of weak property rights: The case of Indonesia. Environ Resour Econ 33:73–93.

Forest Watch Indonesia/Global Forest Watch (FWI/GFW) (2002) The state of the forest: Indonesia. Forest Watch Indonesia, Washington DC. Global Forest Watch, Bogor, Indonesia.

Iwan R (2004) Mobilizing community conservation. A community initiative to protect its forest against logging in Indonesia. Paper prepared for the 10th Biennial Conference of the International Association for the Study of Common Property (IASCP), Oaxaca, Mexico.

Landell-Mills N (2002) Developing markets for forest environmental services: An opportunity for promoting equity while securing efficiency? Philos Transact A Math Phys Eng Sci 360:1817–1825.

Landell-Mills N, Porras I (2002) Silver bullet or fools' gold? A global review of markets for forest environmental services and their impact on the poor. International Institute for Environment and Development (IIED), London.

Muthoo A (1999) Bargaining theory with applications. Cambridge University Press, Cambridge.

Pagiola S, Arcenas A, Platais G (2005) Can payments for environmental services help reduce poverty? An exploration of the issues and the evidence to date from Latin America. World Dev 33:237–253.

Palmer C (2004) The role of collective action in determining the benefits from IPPK logging concessions: A case study from Sekatak, East Kalimantan, Working Paper No. 28. Center for International Forestry Research (CIFOR), Bogor, Indonesia.

Palmer C (2006) The outcomes and their determinants from community-company contracting over forest use in post-decentralization Indonesia. Dev Econ Policy 52, Peter Lang, Frankfurt.

Palmer C, Obidzinski K (2002) Indonesia case study prepared for the report: Higher international standards or rent-seeking race to the bottom? The impacts of forest trade liberalisation of forest governance. Food and Agriculture Organisation (FAO), Project GCP/INT/775/JPN. Mimeo, Center for Development Research, Bonn, Germany.

Resosudarmo, IAP (2004) Closer to people and trees: Will decentralisation work for the people and the forests of Indonesia? Eur J Dev Res 16:110–132.

Ribot J (2002) Democratic decentralization of natural resources, institutionalizing popular participation. World Resources Institute, Washington, DC.

Siikamäki J, Layton DF (2006) Potential cost-effectiveness of incentive payment programs for biological conservation. Discussion Paper No. 06-27, Resources for the Future, Washington, DC.

Suyanto S, Leimona B, Permana RP, Chandler FJC (2005) Review of the development of an environmental services market in Indonesia. RUPES Working Paper. World Agroforestry Centre, South East Regional Office.

Wollenberg E, Kartodihardjo H (2002) Devolution and Indonesia's new forestry law. In: Colfer CJP, Resosudarmo P (eds.) Which way forward? People, forests and policymaking in Indonesia, pp. 81-109. Resources for the Future, Center for International Forestry Research (CIFOR) and Institute for Southeast Asian Studies (ISEAS).

World Resources Institute (WRI) (2005) World resources 2005: The wealth of the poor – managing ecosystems to fight poverty. United Nations Development Programme, United Nations Environment Programme, World Bank, World Resources Institute, Washington, DC.

Wünscher T, Engel S, Wunder S (2006) Payments for environmental services in Costa Rica: Increasing efficiency through spatial differentiation. Q J Int Agric 45(4):319–337.

Chapter 4
Marketing Environmental Services

Amir Heiman, Yanhong Jin, and David Zilberman

Abstract Environmental services (ES) are diverse both in terms of the benefits that they provide, and in terms of potential buyers. This suggests a mixture of institutional arrangements and strategies required to create demand for ES. In some cases, there is an advantage to commoditizing ES and marketing them through large exchanges. In other cases, the ES are unique, and special efforts and patience for finding the appropriate buyers are needed. In all cases, increasing consumer awareness of availability, value, and benefit of ES is important and leads to increased demand. Furthermore, in all cases, buyers have to be assured of product reliability, which requires explicit mechanisms for monitoring, enforcement of contracts, and insurance. The demand from ES can come from governments and industry that may use it as a least-cost approach to solving environmental problems or a source of revenue and consumers who may use ES as a source of direct consumption benefits or altruism of pro-social behavior. Marketing strategies should be targeted for the specific characteristic of needs of various market segments.

4.1 Introduction

Payments for environmental services (PES) are major tools for improving environmental quality. They include government programs, such as the Conservation Reserve Program (CRP) in the United States, forest conservation for debt arrangements reached between rich countries in the north and tropical countries in the south, and purchases of lands by The Nature Conservancy to become nature reserves.

While there are plenty of offers and schema to supply environmental services (ES), the emergence of demand is less obvious. Some of the prominent PES arrangements are also transfer payments by governments to constituencies like farmers. While there may be much stated willingness to pay for ES, some of which is probably cheap talk, actual willingness to pay is significantly smaller. To make PES a major environmental policy tool requires establishing procedures to create and utilize the demand for ES.

L. Lipper et al. (eds.), *Payment for Environmental Services in Agricultural Landscapes*, Natural Resource Management and Policy.
doi: 10.1007/978-0-387-72971-8_4;

The objective of this chapter is to develop a framework to study the challenges of marketing ES. Marketing efforts should consist of identifying the various market segments and their social and environmental needs and developing products that balance demand and social and environmental concerns for the segments that evaluate the environmental concept. Marketing efforts should be directed to changing the structure of demand to include environmental concerns by enhancing the self- and social image of those individuals who are environmental conscious and de-marketing the segments, activities, and firms that do not comply with environmental concerns. Marketing efforts should be directed to convincing managers to sponsor social and environmental organization and activities. US firms spend about $12.1 billion in 2005 on affinitive activities, but only 9% of it was directed to social cause.[1] The low share of sponsorship of social activities does not correspond with results of many studies showing that consumers "will respond with more favorable rating and higher likelihood of choice to a brand that has a certain social-cause affiliation" (Bloom et al., 2006, p. 51). Following the introductory section where we present the various categories of ES, we will identify the different categories of potential buyers of ES, the factors enhancing their valuation of ES and willingness to pay for them, and the product design and marketing strategies necessary to establish significant effective demand.

4.2 The Diversity of ES

ES have a wide range of functions and their associated goods and services serve various purposes. To make comparative ecological economics possible, de Groot et al. (2002) classify ES into four primary categories[2]: (a) regulation functions that provide many services that have direct and indirect benefits to humans (e.g., clean air); (b) habitat functions that contribute to the conservation of biological and genetic diversity and evolutionary process by providing refuge and reproduction habitat to plants and animals; (c) production functions which consist of soil formation, crop pollination, natural pest control, and so forth and serve as inputs to marketed services (e.g., food, fuel); and (d) information functions related to the benefits from ES through recreation, cognitive development, relaxation, and spiritual reflection (heritage value of natural ecosystems and features).

PES are used as an incentive for pollution control, resource conservation, and provision of environmental amenities. Notable examples of pollution control are arrangements through which water utilities pay dairy farmers not to use grazing practices in the watershed of their reservoir. Conservation is achieved, for example, by payments to forest communities to control deforestation (Alix-Garcia et al., 2004). Wetlands and wildlife refuges are being created to provide recreational

[1] IEG Sponsorship report December 24, 2004 as reported by Bloom et al. (2006).

[2] Table 1 in de Groot et al. (2002) provides functions, goods and services of ES corresponding to discussions of ecosystem processes, and components in details.

opportunities for hunters and bird watchers, and to provide water purification as well as buffering for flood control for nearby cities.

Control of pollution can be achieved by several means. Command and control have been heavily used by regulators, but there is vast evidence showing that it may lead to inefficient outcomes. The polluter pays principle, as embodied by a Pigovian tax, has been long advocated by economists and environmental activists alike as an efficient and effective policy. However, as Buchanan and Tullock (1975) argue, polluters may use their political muscle to prevent imposition of these policies. Cap and trade is another mechanism that can lead to efficient outcomes that have been widely used in recent years, for example, in the Kyoto Protocol. It may be more accepted by polluters because, unlike pollution taxation, it doesn't withdraw resources (tax payments) outside of the polluting sector. Yet, sometimes polluters may prefer not to spend resources on pollution prevention, or they may not be able to afford to do so. PES can be a subsidy to reduce pollution, and it may occur when polluters are politically strong and have rights to pollute, or when they are very poor and do not have the resources to pay for pollution prevention. Similarly, payments for conservation may occur in situations where the self-interests of the resource owners may conflict with those of a third party wanting the resource preserved. Forest communities in developing countries may have high discount rates and are likely to perceive large gains from conversion of forest resources to rangeland and lumber, while environmentalists and natural resource agencies may prefer to see the forests preserved. Growing demand for eco-tourism, for example, safaris, bird-watching tours, or hunting, may lead to payments to landowners and developers to preserve or create the environment that provides these specific amenities coveted by consumers and recreationists.

Another way to distinguish between various ES is to separate those providing consumptive use from those providing non-consumptive use. The use of natural systems for ecotourism, outdoor sports, school excursions, and scientific research represents consumptive functions. Both the case of a soft drink company paying farmers to divert waste disposal to improve water quality and the case of a recreational club paying to preserve hunting grounds are examples of payments for consumptive use. Yet, many individuals may pay for non-consumptive use as well. Some PES represent the existence value of knowing that a rare species will survive or a pristine environment (e.g., heritage value of natural ecosystem and features) will remain unchanged.

Many ES, for example, sequestered carbon in the soil, trees, or other media, are becoming standard commodities treated by large markets. Others are more unique, and more resistant to commodification. In some cases, PES are a one-of-a-kind experience or for the continued existence of a unique natural phenomenon.

The same land or water resource may provide more than one ES. Diversion of land from intensive farming to forest may both improve air quality and protect against soil degradation. A wetland may provide both water purification and wildlife habitats. In these cases, both complementary activities can provide income that will allow for greener activities. In some cases, ES obtained from a resource may be substitutes. A piece of land may be conserved to provide forest services or may be diverted to become a wetland.

Like many assets, both timing and location affect the use and valuation of ES. A well-maintained vineyard may have a much higher value as a source of recreation and aesthetic beauty in the urban fringe than it would have in the agricultural heartland. A water reservoir carries a much higher value during a drought than during the rainy season. The opposite is true for a wetland acting as a flood control buffer, which may be especially valuable during periods of heavy rain.

The spatial dimensions of resources providing ES may vary necessarily, often depending on biological considerations. Some environmental amenities exhibit increasing returns to scale. For example, a critical mass of land resources may be needed to sustain certain wildlife species. In other cases, there may be gains from maintaining land or water resources spread over separate locations for protection and diversification. Some protected species may be spread over a larger area but must have open pathways to move from place to place.

The multiple typologies of ES and their benefits are useful in designing marketing strategies and identifying potential buyers, sales channels, mechanisms for exchange, and information awareness and promotion activities.

4.3 The Nature of Demand for ES

The development of a marketing strategy for ES is derived from understanding the demand for it. As we have seen, ES are diverse and are likely to be purchased by diverse buyers. The consideration affecting some of the buyers is analyzed below.

4.3.1 Governments

Most of the spending on ES thus far has been done by government agencies at several levels – local, regional, national, and international. As Rausser (1982) notes, some government policies aim to address market failure and others are distributional. That holds true for ES programs. The CRP in the United States has served for a large part as a farm subsidy program, and the policy debate surrounding its design is about aligning environmental criteria with political power of affected parties (Babcock et al., 1996). The environmental quality improvement program (EQIP) is paying farmers to reduce pesticide use and livestock producers to reduce animal waste. It combines pollution control with transfer payment. The payment for forest conservation in Costa Rica is another example where the ES are combining transfer of resources with attaining environmental objectives. The global environmental facility is an example of a mechanism for provision of funding for ES by "global" government.

The allocation of funding for ES can be modeled by a cooperative game framework similar to the one proposed by Zusman (1976). Let i be an indicator of ES projects, assuming that the funding of I projects is considered, where $i = 1 \dots I$, and let x_i be the budget allocated to the ith ES project, where $X = \sum_{i=1}^{I} x_i$. Assume that there are K groups affected by these projects (positively and negatively), and the benefit the kth group obtains from the ith projects is $B(x_i, k)$. Each of the groups affects the political process, and the political weight of the kth group is denoted by $w(k)$. Let's also assume that there is a political cost (or cost of alternative usage for a constant governmental budget) associated with spending given by $c(x)$. With these definitions, the budget will be allocated to the various ES projects to maximize a weighted sum of benefits minus the total political cost of the budgetary expense. The optimization problem, solving for the various x_i, is thus

$$\underset{x_1,x_2}{\text{Max}}, \cdots, \sum_{k=1}^{K} \sum_{i-1}^{I} w_k B(x_i, k) - c\left(\sum_{i-1}^{I} x_i\right). \tag{4.1}$$

$$\underset{\{x_i\}}{\text{Max}} \sum_{k=1}^{K} \sum_{i-1}^{I} w_k B(x_i, k) - c\left(\sum_{i-1}^{I} x_i\right). \tag{4.2}$$

The expenditure on the ith ES is determined by equating the sum of the politically weighted marginal benefits of that ES with the political marginal cost of expenditure. Groups with more political muscle, or groups that are ready to sacrifice more political capital on garnering ES projects that benefit them, will bias the allocation of ES money to favor their projects. Thus, agricultural groups in the United States and forest owners in Costa Rica may obtain significant support to ES that benefits them disproportionately because of their political investment. The optimality condition also suggests that when the government finances are in better shape, i.e., lower shadow price of the budget leading to lower cost, more will be spent on ES. This modeling approach assumes a cooperative game. There are other forms of solution, such as non-cooperative game in the case of environmental concerns. The non-cooperative concept may be more realistic in situations where the environmentalists have very little political power.

Governments can also induce demand for ES indirectly. Stricter water quality standards, restrictions on chemical use, and tougher land-use regulation are activities that may lead to a PES scheme by affected firms. Higher water quality standards may induce water utilities to pay landowners in their water catchment areas to modify their activities. The imposition of constraints on wetlands conversion in the United States led to the evolution of wetland banking. The inclusion of carbon sequestration activities as a means to obtain carbon credits consistent with the Kyoto Protocol is creating new ES-generating activities. The political process has been a major avenue for creating demand for ES, and indeed environmental groups have been working through the political process to introduce policies that directly or indirectly lead to ES programs. Such political activities may require establishing alliances with various groups that sometimes may be adversarial to environmental causes. Indeed, farmers or forest landowners are likely to work together to garner payments for conservation activities, despite differences on major issues. Marketing of new PES requires political action, persuasion, and coalition building to obtain the political support for policies that directly or indirectly induce the emergence of either public or private demand for ES.

4.3.2 *Industry*

Most private firms will view PES primarily from a profitability perspective. The weighting of risks, short-term versus long-term earnings, and the goodwill generated because of environmental stewardship may vary among firms and affect their demand for ES; but, fundamentally, the demand for ES is derived from the economics of their main business activities. Some firms may pay for ES to improve the quality of production inputs, other firms will pay for ES to reduce risks, and still others may pay for ES to help them improve their image, thus enhancing demand for their final product.

Consider first the demand of the bottled water (or soft drink) industry for ES that will enhance the quality of water it provides. For simplicity, we assume that the industry is competitive, but firms collaborate to induce major socially desirable and privately beneficial activity through a producer's association. The analysis can be easily modified for a monopolistic or oligopolistic one without substantially changing the implications.

Suppose that one can have either high quality or low quality water. The quality of the water is affected by agricultural production activities. One way to improve water quality is to pay farmers for modifying their production technology. We denote the derived demand of the product by $D(p, m)$, where p is the price and m is an indicator of water quality. Hence, the derived demand with high quality water is $D(p, l)$ in Fig. 4.1. The marginal production cost of the product with water of either quality is MC. If the higher quality water is utilized, the price is p_h, the quantity consumed is q_h, and the industry's profit is the area ACP_h. If the lower quality water is used, the price of the product is p_l, the quantity consumed is q_l, and the profit is the area ABP_l. The increase in the industry's gross profits from the provision of ES

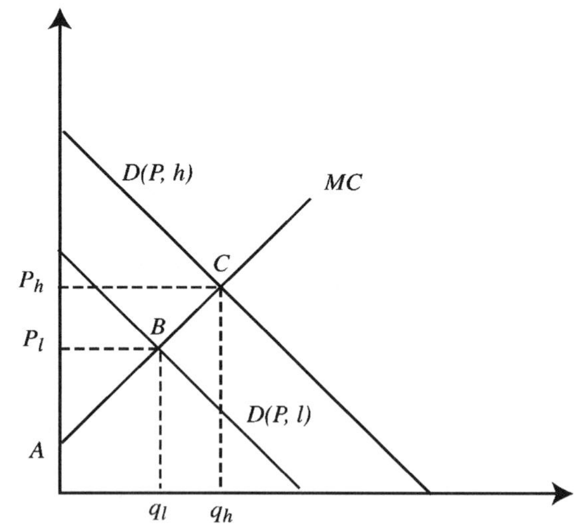

Fig. 4.1 An example of PES of acquiring high quality water. Amir Heiman, Yanhong Jin, and David Zilberman

that will increase water quality if the consumers are aware of the quality issue is denoted by MG and is represented by the area P_lBCP_h in Fig. 4.1.

This area is the maximum amount that the industry (represented by a producer association) may be willing to pay farmers for the ES. Of course, the industry would like to keep some of the extra profits, and the final distribution of profits will be the result of negotiation. Let the payment for ES be denoted by PES. If the cost to the farmers for modification of their production system is $FC \leq PES \leq MG$. If the negotiation results in a fair solution (Rabin, 1997) where the gain is shared, then $PES = (MG - FC)/2$, but that is one among many possible outcomes.

Consider the case where an insurance company is exposed to flood risk. Let i be an indicator of state of nature among I possible states, and W is the size of the flood buffer zone (wetland) in the flood region. Let $q_1(W)$ denote the probability of a flood and $D_1(W)$ be the total cost of the damage caused by a flood. Both probability and damage depend on the size of the buffer zone (wetland) and the farm production. Let $c(W)$ denote the cost to implement and maintain the wetland. A risk-neutral firm will determine W to minimize expected total cost:

$$\min_{W} \left\{ \sum_{i=1}^{I} q_1(W) D_1(W) + c(W) \right\}. \tag{4.3}$$

The optimal size of the wetland will equate the expected value of the marginal savings with the marginal cost of additional wetland capacity. Changes in expectation with respect to liabilities may lead to revised calculations and increased willingness to pay for ES (wetlands). For example, suppose a wetland is near a city. As the city grows, the damages from flood are likely to increase. Thus, we may expect insurance companies to increase their investment in risk-reducing ES over time. The growth of the city leads to scarcity of land increasing the cost of maintaining the wetland $c(W)$. The increasing cost may offset the benefit of preventing the expected future disaster. The analysis should be expanded to cases where the insurers are risk averse, and where the levels of liability, risk premium, and wetland capacity are determined simultaneously. It would be useful to investigate situations where investment in ES may reduce risk premiums and increase profitability for insurers simultaneously.

Frequently, insurance is competitive and several insurers bear exposure within a given flood zone. Thus, raising funds to support wetlands or other ES to reduce risk may entail high transaction costs associated with collective action among the companies or with multi-party negotiations with the agency associated with wetlands construction and preservation. In these cases, designing a marketing campaign to raise awareness about the potential gains from ES, as well as building the goodwill to allow financing the project, is a major challenge. Goodwill is built if consumers internalize, as a result of the marketing activities, that keeping the environment is an important personal goal. In some cases, re-insurers could be better prospects for financing risk-reducing ES, because they are small in number and may be the ultimate bearers of the risk (in the case of large catastrophic losses).

Since the government acts as the insurer of the last resort for flood risk in the United States and other countries, it would be worthwhile to examine the role of governments as a purchaser of ES to mitigate this risk, or to consider a public-private partnership to this end.

Firms may use PES as a mechanism to enhance public image. This is likely to happen when consumer preferences, and therefore demand, depend on the environmental record of the firm. Let the demand for a firm's product x be affected by ES expenditures, so if x is quantity of output, and p is price, then demand will be $x = D(p, ES)$, and the revenue of the firm B is $R(x, ES)$. Let us suppose that the cost of production is $c(x)$, and $v(ES)$ is the cost of ES. Then the firm has to decide about optimal output and optimal spending on ES. The optimal spending on ES is determined where the marginal increase in revenues because of ES is equal to the marginal cost of ES:

$$\frac{\partial R(x^*, ES^*)}{\partial ES} = \frac{\partial v(ES^*)}{\partial ES}. \tag{4.4}$$

Revenues will increase from ES spending when synergies exist between the product and ES. For example, buyers of outdoor recreation gear will value more a company that actively supports the preservation of pristine wilderness. In some cases, companies with a bad environmental reputation may use PES to soften their image and recapture some environmentally conscious consumers. In these cases, the benefit from PES can go beyond increased sales, and may reduce the fervor that leads to expensive lawsuits or restrictive legislation.

4.3.3 Consumers

Consumer demand for ES programs will be expressed by paying for the output of these programs, for example, entrance fees for a wetland or a park, purchasing hunting or fishing licenses, or buying products that are produced by indigenous tropical communities living in an ecologically sustainable manner within the forest. Another form of consumer demand that may be more intriguing is contributions to non-governmental organizations (NGOs) or other organizations for funds generating ES and preserving forests, wildlife habitats, wetlands, etc. Indeed "angel investors" that provide significant sums to environmental causes have played important roles in financing green activities that include provision of ES.[3] Our analysis will concentrate mostly on contributions that fund ES programs.

[3] See, for example, Investor's Circle (www.investorscircle.net) a group that coordinates the efforts of environmentally conscious angel investors for sustainable business.

The motivation to support the environment can be classified into three clusters: direct motivation (e.g., hunters), altruism and pro-social behavior (Schwartz, 1977; Price et al., 1995), and demonstrating concerns about the state of the world. The third motivation is more general and is related to a general satisfaction from donating and building a self-image of being superior in values. The latter is enhanced if it is accompanied by social recognition. Marketing efforts can aim at building such a mechanism of social recognition and providing cover to such activities. Consumers' pro-social activities are moderated by social norms (Schwartz, 1977; Osterhus, 1997), and therefore the first marketing step is to make environmental protection a prioritized personal goal. This can be done by education and advertisement campaigns.

Individuals will pay for ES because they either directly benefit or indirectly gain from existence value. Hunters and bird watchers will contribute to support the establishment of a local wetland that will enhance the value of the local ES they consume. However, some individuals may support an ES program that is farther away because it generates valuable outputs that are global public goods, it contributes to biodiversity preservation, or they value the existence of an amenity. Another element exists as well – consumers may contribute for their reputation effects or to gain social status as donors. Many public goods such as symphonies, theaters, and museums thrive on donations from individuals who derive social standing or out of a sense of social obligation or citizenship. Individuals may appreciate the heritage value of natural ecosystems and features (de Groot et al., 2002). The combination of these incentives should drive marketing strategies that aim to raise funds for ES projects.

Let an individual's willingness to contribute to an ES fund be denoted by $V(W, A, S, t)$, where W is wealth, A is awareness and experience, S is social pressure, and t is incentives. Contributions to causes are likely to be luxury goods, and the marginal contribution is likely to be positive and increasing with wealth ($\partial V/\partial W > 0$ and $\partial^2 V/\partial W^2 > 0$). However, wealth by itself is not a sufficient condition for contributing to an ES fund. People are more likely to contribute when they are familiar with the causes and care about them ($\partial V/\partial A > 0$). Moreover, people who care about the causes and can afford to pay have the best potential as valuable contributors ($\partial^2 V/\partial A \partial W > 0$). Contributing to a cause may generate social externalities and reputation effects, implying that people are more likely to contribute to causes supported by their peers or social group ($\partial V/\partial S > 0$). Finally, contributions to ES programs and other causes can be induced by incentives such as tax exemptions ($\partial V/\partial t > 0$) and, with a progressive income tax, the tax savings are likely to be more significant for wealthier individuals ($\partial^2 V/\partial W \partial t > 0$). Furthermore, naming opportunities are valuable incentives, especially when they are appreciated by the donor's peer group ($\partial^2 V/\partial S \partial t > 0$).

Marketing efforts aimed to increase consumers' contribution to ES funds have to first target a group of potential donors who combine income and have the potential of becoming interested in marketing activities. It is important to recognize the heterogeneity among potential donors and enable the marketer to be more efficient in spending marketing resources when targeting consumers. Proper targeting will assure that the marketer will not spend marketing efforts on groups that have low probability of donating, starting with the group with the highest potential and moving to the next group, using the first group as a social reference group. Within the target group, marketers of ES can direct willingness to contribute to their products and can create the willingness to contribute. The former, directing existing motives to the environmental product, is less complex than the latter generating needs. Consumers who are willing to donate have already successfully moved through the first three stages of product adoption – having a need, being aware of activity and alternatives, and attitude (preferences) to ES providers/organizations. The marketer needs in this case to first build a differentiated offer, which will have the potential of having prominent advantage over competing activities and then generating awareness to the marketing offer.

Differentiating the marketing offer is perhaps the most complex step at preparing the strategic marketing plan. It has to recognize customer needs and motives that are often hidden, explore the maximum that the potential donor may donate, and analyze the competition. For example, rich individuals who want to donate to satisfy their egos, desire to be portrayed as philanthropists, or wish to be memorialized by their donation may consider donating to hospitals, school systems, museums, the local philharmonic orchestra, and environmental activities. The alternatives differ in their perceived value to society; their contribution to humans, nature, and culture; and in their tangibility. Contributions to hospitals are more tangible and visible than contributions to education or the environment. Supporting the preservation of a rare species in the middle of a remote forest will get much less attention than financing an ambulance displaying a nameplate of the donor. Understanding that the lack of tangibility is a barrier to donation and internalizing that competitive alternatives require large investment created two successful marketing plans to promote environmental activities. The first was the London Zoo's plan to offer donors the opportunity to adopt a specific animal. The price tag of an animal depended on its size and the derived cost to feed it. A donor with high willingness and ability to invest in this activity could adopt an elephant or a lion and provide for the animal's lifetime well-being, while a donor with limited resources could adopt a beetle. The second concept was Israel's Jewish National Fund's (JNF) 30-year offer to plant any number of trees in honor of someone and pay per tree. The donor selects the number of trees, the location the trees will be planted, and the names of the people being commemorated. To guarantee visibility and tangibility, donors receive a document specifying that a tree(s) was (were) planted in their honor, in honor of someone they wish to acknowledge, or in memory of a loved one. This system allows individuals who may not be wealthy to participate in an important environmental activity – replanting the historic bible country and converting a semi-arid area into a subtropical area. The disadvantages of the JNF were converted into marketing advantages. If competitors target the big money, The JNF targeted the segment with less income and low ability to donate but with the same willingness to do something for Israel. A simple offer is a key requirement in convincing a potential donor to donate, but it may be not enough. Social norms are often moderated by trust (Osterhus, 1997), and donors may be uncertain about the success of the project. In this case guarantees (such as money-back guarantees) if the project fails may reduce the uncertainty.

In some cases there is need to generate the necessity (and understanding) to participate in environmental activities. This is done by relating the basic needs of any individuals to be recognized as a generous person (or the alternative of not been considered as stingy) or by adopting the alternative of marketing through building guilt feelings. Interest and awareness can be generated by various informational policies, including advertisements, brochures, and newsletters, and events to expose potential donors to environmental causes and their benefits. The Sierra Club organizes outings both to generate awareness of environmental causes and to build social networks that are likely to enhance ES fund contributions. The Sierra Club, Duck Unlimited, and other environmental groups provide products that bundle fun with efforts to support a cause. This synergetic product has wide appeal. Awareness of environmental causes can also be enhanced by improving environmental education in schools and in colleges, and by "teaching the teachers." Education and awareness of ES are not only beneficial for fundraising, but they are also beneficial for building political support for public provision of ES. Finally, garnering public support for ES programs requires establishing an incentive structure that includes tax deductions, as well as various forms of recognition. Universities may name buildings after important donors, and cities may name streets and public plazas after politicians. Similarly, a contribution to establish ES could lead to other naming opportunities.

The heterogeneity of the public suggests that efforts to create public demand and contribution should adjust according to the situation (Kotler & Levy, 1969). Some people who may not be familiar with a problem or concept behind the ES should be targeted by basic educational programs and other awareness-raising efforts. Individuals who are more familiar with the issue should be targeted more intensively as potential donors. Of course, some especially promising prospects may deserve special attention. Past donations reveal preferences for the ES, thus a record of donations in the past can be used as a basis for pursuit of future contributions.

Contribution to the environment is not limited to donation. A successful marketing activity would result in a state where consumers would prefer to purchase from a firm that contributes more to the environment. This is an analogy to preferences of some individuals to purchase from firms that are members of the fair-trade organizations (Becchetti & Rosati, 2007), preferring local grown flowers over imported flowers because they save transportation costs and preferring organic flowers since they save the environment and are pesticide-free.

There is a need to recognize that most individuals do not see social activities as part of their personal goals; therefore, their participation in social and community activities is very limited (Dawes & Thaler, 1988). The low contribution can be modified by relating the social benefit to personal goals (e.g., health, setting an example to your children) or by generating the feeling that environmental support is a consensus, and not supporting it will be interpreted as antisocial behavior. The number of individuals who contribute to the environment is a key factor in creating the state where environmental protection is a norm. If it is not the case, i.e., the majority does not contribute, then feelings of unfair burden may hamper the goodwill and willingness to cooperate in multi-player activities with the nature of public good outcomes (Akerlof, 1982; Kahneman et al., 1986; Rabin, 1993; Li, 2008). Advertisement and public relations may help in reducing the feelings of unfair burden and promoting the importance of personal contribution.

4.3.4 NGOs

Frequently, NGOs are formed to articulate the preferences of economic agent groups engaged in activities on their behalf. When society consists of heterogeneous consumers, there may be a group of individuals in society that will have stronger preferences than the average citizens for a public good, say, preservation of certain ecosystems. The members of the group may perceive underinvestment by the government in the preservation of the ecosystem and will raise the funds to both invest in the preservation directly and to lobby for enhanced public support. The World Wildlife Fund initiates and supports activities that preserve and protect wildlife using donations from individuals who have strong preferences for wildlife protection and improved well-being. NGOs may pay for ES as part of their activities or use their political muscle to enhance government PES. The Nature Conservancy has adopted PES as a major element of its mode of operation. Other environmental NGOs may emphasize the role of penalties and direct control to achieve environmental goals. That may be because of either lack of capacity to raise funds for PES, lack of expertise in their design and use, and objection to using these forms of financial incentives. PES is a relatively new institutional innovation, and it has been only partially adopted by organizations that can benefit from it, in particular, NGOs. As the adoption models suggest (Feder et al., 1985), the adoption of PES may increase over time as a result of imitation among NGOs, education and training of NGO personnel, and accumulation of experience that will lead to reducing the cost and increasing the effectiveness of PES programs.

As we analyze demand for ES, it is useful to differentiate between mechanisms and institutions that are responsible for funding ES, and the institutions that are responsible for the actual purchase transaction with the seller. NGOs tend to be organizations that buy ES, but the funding of the purchases is generally supplied by consumers. Industries may, for the most part, be direct buyers of ES (even if they make direct donations to NGOs), and governments in general are direct buyers of ES.

4.4 Institutional Arrangements for Funding and Purchasing ES

Introduction of PES requires establishment of an institutional setup that will allow sellers to discover buyers, provides an environment for price negotiations and the details of the ES, and offers mechanisms for protection against various risks. When the ultimate buyer is different from the provider of funding, then there is a need for institutions to raise the funds. The exact setup depends on the number and identity of buyers and sellers and the features of the ES.

A *market* embodied in an *exchange* is likely to emerge as a mechanism where ES are bought and sold in cases where the number of buyers and sellers is large, and the ES can be standardized and commodified. The markets for carbon emission rights are obvious examples as sequestration or emission of 1 ton of carbon in a well-defined product, and in this case location of emission or sequestration does not matter. However, some means of sequestration are obviously more certain and stable than others, and implementation of these markets requires a system of certification. When there are many buyers and sellers who are concerned with issues of liquidity and certainty, in addition to the spot markets in the ES, markets for options as well as futures markets may emerge.

When there is one buyer (or a small number of buyers) and many sellers, then the purchase of ES will be managed through *bidding*. For example, in the case of the CRP program in the United States, the government asks farmers to provide a bid regarding the amount of land that they will offer to the program, the annual amount of money they will ask, and the environmental benefits of the land. The government will determine the sellers of ES based on a formula that weighs the costs and benefits of different proposals. A similar system of bidding exists in government programs in Mexico and Costa Rica that pay for ES produced by forests.

Sometimes, when there are many buyers and sellers but the products may be very diverse and nonstandard, many of the transactions will be done through a *bulletin board*, where buyers describe their product, possibly showing a picture and providing contact information or, alternatively, where the sellers define what they want and provide contact information. In this case, the negotiations between buyers and sellers establish the price and the specifics of the transaction. A bulletin board is very popular in many important markets – for example, used cars, dating, and jobs. A special example with many similarities to ES markets is the real estate market, where every house has its own unique attributes. There is an electronic water market for the California Central Valley, where buyers and sellers use the Internet to meet and arrange for transactions (Olmstead et al., 1997). Trading through a bulletin board requires higher transaction costs than trading through an exchange, and market information within a bulletin board system is not as transparent, but these extra costs are appropriate if buyers and sellers need to adjust for the uniqueness of products.

A seller of ES may choose to open a *storefront*, namely, they may provide a specific address or location where people can purchase ES. Storefronts are opened when there are relatively many buyers and a small number of sellers, and the buyers need the opportunity to examine and compare a variety of products in order to find the one that fits their needs. If we compare wheat versus clothing, wheat is a standardized product traded in an exchange, while buyers of clothing require the opportunity to examine the product in advance, necessitating the existence of storefronts. In the case of wetland banking, specialized organizations have an inventory of wetlands that can provide the appropriate services to developers who need wetlands services credit in order to develop their own land. A storefront can be a physical location, or it can be a virtual storefront online. When there are differences in the quality and nature of the ES provided by different vendors, a potential buyer would look for the opportunity to compare prior to purchase. For example, because of distinct experiences that exist in different recreational ES, there are distinct providers, and individuals or organizations may shop virtually through catalogs or brochures, or through the Internet (e.g., comparison of websites for national parks or hunting grounds). When storefronts exist and buyers of ES have a choice between different types of products, then the role of marketing tools like advertisements and pre-purchase demonstrations becomes particularly important. Mechanisms to reduce buyer uncertainty about the product may increase demand and may increase sales.

When it comes to raising funds for ES, a major mechanism is soliciting through various forms of fundraising. The potential donors are heterogeneous; they vary in their preferences and in their ability to pay. Furthermore, the fundraising organization may have information about some subset and may be uncertain about another. Some forms of solicitation through advertisements and general media are used both to identify individuals who may have interest in becoming donors for the purchase of ES, and also to compel them to contribute. A letter campaign may target individuals who are known to have preferences for certain types of ES (members of certain groups or associations). For richer individuals, the organization can afford to perform solicitations in person.

4.5 Enhancing the Demand for ES

The diversity of situations that give rise to ES leads to the use of different channels and mechanisms to fund and sell ES. The demand for all the diverse categories for ES is likely to increase if several basic principles are recognized. First, consumers are concerned about risk and, in particular, risk about product reliability and fit (Heiman et al., 2001). Buyers are not likely to purchase a product that has a high likelihood of failure, and mechanisms like warranties and dealer backups have been used to improve the performance of a product and reduce the cost of failure for the consumer, thus increasing demand. Similarly, consumers are less likely to buy a product that may not fit their particular needs. Mechanisms such as product demonstration, money-back guarantees, and secondary markets have been used to reduce fit uncertainty.

In the case of ES, reliability is of utmost importance. When a utility in the North buys carbon credit from farmers or forest communities that are to engage in carbon sequestration, the utility is concerned about the possibility of the other parties failing to fulfill their commitment. Similarly, when a water utility pays livestock operators to control animal waste runoff, they may be concerned about violations. The concern with risk of violations or failure to fulfill contracts may reduce the willingness to pay for ES in both cases. Thus, one solution is to develop effective mechanisms for monitoring the behavior of providers of ES, and for enforcement of contracts. This is quite a challenge, especially when many small farmers are providing carbon sequestration, or are committed to reducing certain forms of polluting behavior. Thus, creative mechanisms that reduce the cost of monitoring and reduce the likelihood of violations are of significant value. One example is the use of collective punishments as a threat, and another is reduced penalties as compensation for self-reporting when accidents occur, in order to reduce the damage.

Some ES programs pay the provider annual rent for access to services. For example, the CRP of the United States, as well as the forest protection programs in several developing countries, pay landowners on an annual basis for the choices that provide the ES. In these situations, the providers for ES are paid what can be perceived as annual rents. In other situations, the buyer purchases an asset, which is assumed to provide the ES for each period during its lifetime. For example, a program to purchase rainforest acreage pays for the ES associated with that acreage indefinitely into the future. Similarly, when a farmer is paid for sequestration of a ton of carbon, his behavior is intended to reduce the stock of carbon for a long period of time, if not forever. In particular, the farmer may be paid for using no-tillage or other technologies to reduce carbon emissions, and for not reversing these choices in the future. It is much easier to assure delivery of ES when paid on an annual basis than to assure delivery of future ES from a purchased asset. Therefore, establishing reliable mechanisms of monitoring and enforcement, as well as insurance, is a bigger challenge when one plans the long-term sale of ES-generating assets. Providers of ES may offer payments of varying durations, and adjust prices for uncertainty costs of insurance and the discount rates.

Buyers of ES may be subjected to fit risk as well. For example, a utility may pay farmers or invest in a wetland protection to improve water quality, and then realize that these activities did not solve the water quality problems. Contracts that provide the buyers some compensation or refund in the event of the ES failing to serve its intended purpose are likely to increase the expected net earning from the provision of the ES, as buyers will be willing to pay more for the ES when their fit risks are smaller. In addition to the buyers' preferences for lower risks, they also value that the buying process is less costly in terms of time and money. Buyers need to spend less time purchasing a product if there are fewer uncertainties about its performance, use, and price, and when the purchasing process is streamlined and simple. One avenue to reduce the costs and risks of purchasing is selling standard commodities. One of the challenges in marketing ES is commodification, namely, established uniform product standards that allow large-scale exchange. When ES that remove pollution and waste can be sold as commodities, they can be incorporated into cap-and-trade arrangements. That is the case with carbon sequestration activities that incorporated the trading of market emission rights. Wetlands that reduce the nitrates loading of a body of water through bioremediation may be paid for these ES once pricing for removal of the nitrates in the water is established, through cap-and-trade arrangements, or pollution penalties. When an activity is providing a mix of ES, if each ES has a market or policy-determined price, the activity will be paid by the sum of values it creates. In other cases, weights will be given to the various ES resulting from the activity, and markets or other mechanisms will determine the price of "standard" units of ES that will be used as a benchmark for assessment. This approach has been used in assessing bids or determining payments for participation in several governmental ES programs in the United States, Europe, and Mexico.

Buyers of ES, especially private buyers, may be concerned about liquidity. They would like to be able to re-sell the product if their preferences or financial situations change. A related concern of buyers and sellers of products is price variability and uncertainty. Therefore, futures and options markets have evolved, allowing hedging of prices into the future. Active exchanges for well-defined commodities serve to enhance liquidity, and are especially effective if futures and options markets are included because it reduces the risk of market participation. Indeed, there are some active environmental and resource markets that have futures and options; they include the water bank in California, which serves as a mechanism to protect water districts and farmers against drought. The exchanges for tradable permits for air pollutants (including the emerging market for carbon emission rights) also combine spot, futures, and options markets.

The pursuit of lower transaction costs and uniformity has to be balanced against the gains of specialization. Buyers of ES are diverse in their preferences and ability to pay. Providers of ES should aim to establish quantitative standards that will allow for easy value assessment and trading of ES, yet at the same time allow for differentiation among them. Differences in terms of size, quality, and other dimensions that can be easily monitored allow buyers the opportunity to be selective and enable the seller to take advantage of various categories of buyers interested in the ES. A forest community may gain from payments for improved water quality by a nearby utility, earnings from carbon credit sold to a global carbon exchange, and premiums for wildlife protection paid by a conservation group.

The gains from differentiation are not limited to the sale of actual ES; they can also be present in fundraising by environmental groups planning to purchase ES. Many donors would like recognition by having a location named after them, receiving special acknowledgment (e.g., in a newsletter or other publication), or being taken for a personal visit to exotic sites. Differences in giving capacity and interests among donors can be accommodated by offering a variety of naming opportunities, establishing different categories of donations that are related to contribution size, or by organizing special and exclusive environmental tourisms or adventure in nature programs. This type of entrepreneurship will enable environmental groups to obtain more funding for its activities. An alternative approach to raising funds and support for ES is through appropriate development projects. Tourists who came to a resort may appreciate nature adventures around it, and thus tourists visiting recreational facilities may provide the funds that support provision or maintenance of ES in the vicinity of these facilities. Again, price discrimination that increases the surplus taken by the facility, and therefore by the environmental group for purchase of ES, can be a viable strategy. For example, having luxury and standard rooms in a hotel located near an animal reservation will generate higher earnings than having just standard rooms and, if a given fraction of the income ends up as PES, the higher earnings of the hotel with differentiated rooms will result in higher ES.

The environmental sector could learn from the arts and leisure sectors about how to enhance resource availability by conferring social status and recognition on the basis of the size of contributions, which will allow certain consumers to differentiate themselves.

4.6 Conclusion

The literature on ES has been, to a large extent, one sided. It emphasizes mechanisms that lead to the provision and supply of ES, but without much attention to the creation of demand for ES. In order for PES to play a major role in improving environmental quality and reducing poverty, both supply and demand aspects of the problem must be addressed.

ES are diverse both in terms of the benefits that they provide, and in terms of potential buyers. This suggests a mixture of institutional arrangements and strategies required to create demand for ES. In some cases, there is an advantage to commoditizing ES and marketing them through large exchanges. In other cases, the ES are unique, and special efforts and patience for finding the appropriate buyers are needed. In all cases, increasing consumer awareness of availability, value, and benefit of ES is important and leads to increased demand. Furthermore, in all cases, buyers have to be assured of product reliability, which requires explicit mechanisms for monitoring, enforcement of contracts, and insurance.

A resource manager needs to be creative in garnering PES. The diversity of ES that may be provided by individual resources should create opportunities to sell different types of ES to different buyers. The PES may sometimes be modest, but the combined payments may allow for sustainable management of natural resources to provide valuable environmental amenities.

References

Akerlof G (1982) Labor contracts as a partial gift exchange. Q J Econ 97:543–569.

Alix-Garcia A, de Janvry A, Sadoulet E (2004) Payments for environmental services: To whom, for what, and how much? Department of Agricultural and Resource Economics, Working Paper, University of California, Berkeley.

Babcock BA, Lakshminarayan PG, Wu J, Zilberman D (1996) The economics of a public fund for environmental amenities: A study of CRP contracts. Am J Agric Econ 78(4):961–971.

Becchetti L, Rosati F (2007) Global social preferences and the demand for socially responsible products: Empirical evidence from a pilot study on fair trade consumers. CEIS Tor Vergata - Research Paper Series, 31(91). Available at http://papers.ssrn.com/paper.taf?abstract_id=962488.

Bloom PN, Hoeffler S, Keller KL, Basurto Meza CE (2006) How social-cause marketing affects consumer perceptions. Sloan Manag Rev 47(2):49–55.

Buchanan JM, Tullock G (1975) Polluters' profits and political response: Direct controls versus taxes. Am Econ Rev 65(1):139–147.

Dawes RM, Thaler RH (1988) Anomalies: Cooperation. J Econ Perspect 2:187–198.
de Groot RS, Wilson MA, Booumans RMJ (2002) A typology for the classification, description, and valuation of ecosystem functions, goods, and services. Ecol Econ 41:393–408.
Feder G, Just RE, Zilberman D (1985) Adoption of agricultural innovations in developing countries: A survey. Econ Dev Cult Chang 32(2):255–298.
Heiman A, McWilliams B, Zilberman D (2001) Demonstrations and money-back guarantees: Market mechanisms to reduce uncertainty. J Bus Res 54:71–84.
Kahneman D, Knetsch JL, Thaler RH (1986) Fairness as a constraint on profit seeking: Entitlements in the market. Am Econ Rev 76:728–741.
Kotler P, Levy SJ (1969) Broadening the concept of marketing. J Mark 33(1):10–15.
Li J (2008) The power of convention: A theory of social preferences. J Econ Behav Organ 65:489–505.
Olmstead J, Sunding D, Parker D, Howitt R, Zilberman D (1997) Water marketing in the 90s: Entering the electronic age. Choices (3rd quarter):24–28.
Osterhus TL (1997) Pro-social consumer influence strategies: When and how do they work? J Mark 61(4):16–29.
Price LL, Feick LF, Guskey A (1995) Everyday market helping behavior. J Public Policy Mark 14:255–266.
Rabin M (1997) Psychology and economics. Department of Economics, Working Paper No. 97–251, University of California, Berkeley.
Rabin M (1993) Incorporating fairness into game theory and economics. Am Econ Rev 83(5): 1281–1302.
Rausser GC (1982) Political economic markets: PERTs and PESTs in food and agriculture. Am J Agric Econ 64(5):821–833.
Schwartz SH (1977) Normative influences on altruism. In: Berkowitz L (ed.) Advances in experimental social psychology, Vol. 10. Academic Press, New York.
Zusman P (1976) The incorporation and measurement of social power in economic models. Int Econ Rev 17(2):447–462.

Chapter 5
Economics of Carbon Sequestration Projects Involving Smallholders

Oscar Cacho

Abstract Afforestation and reforestation projects have the potential to help mitigate global warming by acting as sinks for CO_2. However, participation in carbon-sink projects may be constrained by high costs. This problem may be particularly severe for projects involving smallholders in developing countries. Of particular concern are the transaction costs incurred in developing projects and measuring, certifying, and selling the carbon-sequestration services generated by such projects. This chapter addresses these issues by analyzing the implications of transaction and abatement costs in carbon-sequestration projects. A typology of transaction costs is presented, and estimates of the five cost types are derived based on a review of existing projects. The influences of project design on abatement costs and transaction costs are explored, and the critical values of a set of three project-design variables (farm price, number of participating farms, and minimum farm area) are identified for any given combination of transaction costs.

5.1 Introduction

Interest in payments for environmental services provided by landholders and rural communities has increased considerably in recent years as shown elsewhere in this book. The services receiving the most attention have been biodiversity conservation, clean water supply, and climate mitigation. Climate mitigation services related to land use can be generated through conservation of forests, to avoid CO_2 emissions, or through afforestation and reforestation (AR), to capture CO_2 from the atmosphere. This chapter concentrates specifically on the latter type of activity when undertaken as small-scale forestry on farms.

The Kyoto Protocol has provided the context under which much of the debate on mitigation of global warming has occurred, although, as discussed later, this is by no means the only source of demand for climate mitigation services. Under the Clean Development Mechanism (CDM) of the Kyoto Protocol, developing countries have the opportunity to engage in activities that offset greenhouse gas emissions. Emission

L. Lipper et al. (eds.), *Payment for Environmental Services in Agricultural Landscapes*, Natural Resource Management and Policy.
doi: 10.1007/978-0-387-72971-8_5;

offsets are measured in tons of CO_2 equivalents (CO_2e) and are traded as Certified Emission Reductions (CERs). Energy-efficiency projects, large-scale forestry projects, and small-scale agroforestry projects may all supply CERs. Most of the literature on the cost of participating in the CDM has focused on the energy sector and the role of transaction costs on project developers (Fitchner et al., 2003; Michaelowa & Jotzo, 2005; Michaelowa et al., 2003; Yamada & Fujimori, 2003). AR projects face additional challenges due to the *permanence* problem, which arises because AR projects tend to be temporary in nature, since CO_2 captured during forest growth is released upon harvest.[1] In contrast, projects in the energy sector that reduce emissions are permanent, in the sense that an avoided emission will never reach the atmosphere. Cacho, Hean and Wise (2003) discuss this problem and present alternative accounting procedures; Marland et al. (2001) argue for a rental market to avoid arbitrary and complex accounting methods. This chapter contributes to the literature by focusing on AR projects and considering the transaction costs faced by smallholders as well as by project developers.

Some AR projects in tropical countries are competitive in terms of abatement costs[2] per ton of CO_2 captured, but much of the land in the tropics is managed by semi-subsistence farmers and shifting cultivators, so their willingness to participate in mitigation is important (de Jong et al., 2004). Obstacles faced by smallholders in remote areas may be lessened in carbon markets relative to markets for agricultural commodities for two reasons: the "product" does not need to be transported in order to be sold, and a ton of CO_2 removed from the atmosphere has the same effect independently of where it resides. So the problems often faced by smallholders in not being able to obtain transportation to markets for their perishable goods, or to achieve the quality required by international markets, do not apply in the carbon market. Notwithstanding the fact that smallholders may prove to be efficient providers of climate mitigation services in terms of abatement costs, transaction costs[3] may constrain their participation in the carbon market. Transaction costs arise because of the need to ensure that the service being purchased actually exists. Monitoring costs are a major component of transaction costs, but there is also an important fixed-cost component.

We will refer to the project developer as the "Buyer" of the carbon sequestration service and to the farmers as the "Sellers" of the service. Although the Buyer is also a seller in the international market, we will not deal with this segment of the market, but will assume that the Buyer simply faces the market price of CERs. The Buyer could be a non-governmental organization (NGO), a government agency, or a

[1] Depending on the final use of the wood, this release can take a long time, such as when construction timber decomposes.

[2] Abatement costs are the actual costs of "producing" one emission reduction, usually measured as an opportunity cost.

[3] Transaction costs are the costs of participating in the CER market, or converting an emission reduction into a CER. For AR activities, transaction costs include the costs of monitoring and certifying carbon sequestration rates by measuring biomass accumulation over time.

private entity. The Buyer will have to absorb transaction costs, at least temporarily, and be exposed to risks. Presumably the Buyer would expect the sale of CERs to at least cover its costs.

The CDM requires sustainable development goals to be met as well as greenhouse mitigation goals. The host country determines whether a project contributes to its sustainable development objectives, hence AR projects involving large plantations may qualify under the CDM if the employment they generate for local people contributes towards these objectives. Designing projects that benefit smallholders engaged in forestry activities, therefore, requires strategies that reduce the transaction costs of dealing with a large number of farmers relative to the costs of dealing with a single large forestry company.

This chapter presents an overview of the abatement costs and transaction costs that are likely to be experienced by a smallholder forestry project. A simple model is developed to identify the incentives required by sellers and a buyer to engage in carbon-sequestration activities under the CDM. The model is used to identify the critical values of three project-design variables: the farm price of carbon, the number of participating farms, and the minimum size of these farms. The critical values are those at which the project becomes feasible for a given set of transaction costs.

5.2 A Model of Project Participation

Consider a project composed of one buyer and many sellers. The Buyer is an NGO (the project developer) and the Sellers are smallholders. The Sellers are paid for adopting forestry land uses that sequester carbon above a baseline. The Buyer purchases these carbon offsets and sells them in the CER market. So the Buyer acts as an intermediary between the landholders and the international carbon market.

Let Sellers be identified by an index $j = 1, 2, \ldots n$. A Seller j will participate in the project if the reward received for carbon sequestration (v_{Cj}) is larger than the opportunity cost of switching land uses (the abatement cost, v_{Aj}) plus the transaction cost of participating in the project (v_{Tj}). The condition for seller participation is

$$v_{Cj} > v_{Aj} + v_{Tj}. \tag{5.1}$$

The three variables are measured in terms of present value. The present value of carbon payments received by Seller j is:

$$v_{Cj} = a_j \sum_t p_F (C_{jt} - C_{0t})(1 + \delta_s)^{-t}, \tag{5.2}$$

where a_j is the area of land converted to forestry in year zero; $(C_{jt} - C_{0t})$ represents the stock of carbon above the baseline per hectare of land in year t; p_F is the farm price of carbon, and δ_S is the Seller's discount rate. The abatement cost to Seller j is:

$$v_{Aj} = a_j \sum_t (r_{0t} - r_{jt})(1 + \delta_s)^{-t}, \tag{5.3}$$

where r_{0t} and r_{jt} represent the net revenues per hectare in year t for the baseline and the proposed land use, respectively. The transaction cost experienced by Seller j is the discounted sum of a stream of annual transaction costs (q_{jt}):

$$v_{Tj} = \sum_t q_{jt}(1 + \delta_s)^{-t}. \tag{5.4}$$

Now consider the Buyer. The Buyer will implement a project if the present value of carbon payments received in the CER market (V_C) is at least equal to the present value of payments to smallholders (the abatement cost to the Buyer, V_A) plus the transaction costs of designing and implementing the project (V_T). The condition for Buyer participation is:

$$V_C \geq V_A + V_T. \tag{5.5}$$

V_C is the discounted sum of payments obtained by accumulating the carbon offsets produced by all landholders in the project, certifying them and selling them in the CER market:

$$V_C = \sum_j a_j \left[\sum_t p_C \left(C_{jt} - C_{0t} \right) \, (1 + \delta_B)^{-t} \right], \tag{5.6}$$

where p_C is the rental price per ton of carbon and δ_B is the Buyer's discount rate. The abatement and transaction costs for the Buyer are, respectively:

$$V_A = \sum_j a_j \left[\sum_t p_F \left(C_{jt} - C_{0t} \right) \, (1 + \delta_B)^{-t} \right] \tag{5.7}$$

$$V_T = \sum_t Q_t (1 + \delta_B)^{-t}. \tag{5.8}$$

The Buyer must set the farm price of carbon (p_F) at a level that satisfies conditions (5.1) and (5.5), but this decision is influenced by the size of the project. To see why, let the total area of the project be $A = \sum a_j$, assume that prices and the number of participating farms remain constant through time, and express condition (5.5) as $V_C - V_A \geq V_T$, substituting Eqs. (5.6) and (5.7) into this expression and simplifying yields:

$$A(p_C - p_F) \sum_t \left(C_{jt} - C_{0t}\right) (1 + \delta_B)^{-t} \geq V_T. \tag{5.9}$$

This implies that, if the transaction costs faced by the Buyer are relatively fixed, increasing the total project area (A) will make it possible to pay a higher p_F, and therefore provide a stronger incentive for landholders to participate in the project. Although there is evidence that a large proportion of transaction costs is fixed, there is an important variable-cost component associated with monitoring. This variable cost would attenuate the complementary relationship between A and p_F implied in Eq. (5.9), but it would not invalidate it because V_T is likely to increase slower than $A(p_C - p_F)$ as A increases. To implement this model for empirical analysis and gain an understanding of the project-design parameters that most influence project feasibility, it is necessary to obtain estimates of the transaction costs and abatement costs experienced by buyers and sellers.

5.2.1 *Carbon Prices*

The feasible range of farm prices (p_F) is influenced by the market price of carbon (p_C). Here we express both these variables as annual rental prices per unit of biomass carbon stored in trees. This avoids the need of dealing with the permanence problem by imposing arbitrary carbon accounting procedures or constraining the duration of temporary CERs.[4] The use of rental prices is a departure from the official rules of the CDM, but it simplifies the analysis and provides more flexibility than the current rules.

To estimate rental prices, consider the present value (PV) of an asset that yields a perpetual stream of annual payments Y discounted at rate i:

$$PV = \frac{Y}{1 - e^{-i}}. \tag{5.10}$$

In a perfect market the ratio Y/PV is equivalent to the rental price of the asset expressed as a proportion of the asset's value. If we let the asset be a CER (expressed as a ton of CO_2e) valued at price p_{CER}, and consider that the process of photosynthesis converts 3.67 units of CO_2 into one unit of biomass carbon, then the rental price of biomass carbon is:

$$p_C = 3.67 \left(1 - e^{-i}\right) p_{CER}. \tag{5.11}$$

[4] A temporary CER or '*t*CER is a CER issued for an AR project activity which expires at the end of the commitment period following the one during which it was issued (United Nations Framework Convention on Climate Change (UNFCCC) document FCCC/CP/2003/6/Add.2).

Clearly, the CER price places an upper limit on the feasible farm price because the Buyer would set $p_F \leq p_C$ even in the absence of transaction costs. Estimates of CER prices in the literature vary widely, but generally fall within the range of \$5–50/t of CO_2, with lower values being more common because of the risk of investing in developing countries that may have weak institutions. Lecocq and Capoor (2003) in a review of carbon markets state that prices for emission reductions from "small projects with a strong sustainable development contribution command premiums in the marketplace, with prices ranging from US\$5–12/t CO_2e." They also point out that "Retailers report a marked preference by customers for community-based agro-forestry and other forestry deals."

5.3 Transaction Costs

Williamson (1985) distinguishes the costs of contracting as ex ante and ex post transaction costs, corresponding with activities undertaken in achieving an agreement and then coordinating implementation of the agreement, respectively. In the context of this chapter, transaction costs are the costs of participating in the carbon market, which include the costs incurred in the process of obtaining approval for the project. To date, the demand side of project-based carbon transactions has been dominated by the Government of the Netherlands and the Prototype Carbon Fund (PCF), which accounted for 30% and 26% of market transactions, respectively, in terms of volume in 2002–2003 (Lecocq & Capoor, 2003). Other buyers include the Community Development Carbon Fund (CDCF), the BioCarbon Fund, and Japanese entities and private firms that have been increasing their profile in recent years (Lecocq & Capoor, 2003). Some of these projects might exhibit lower transaction costs than projects under the CDM, which has stringent and perhaps cumbersome validation and certification requirements.

Cacho et al. (2003, 2005) present a typology of transaction costs applicable to carbon-sink projects, largely based on Dudek and Wiener (1996). Below we aggregate the seven categories of Cacho et al. (2005) into five and distinguish between the costs borne by buyers and sellers (Table 5.1). The transaction costs experienced by buyers and sellers in time period t are, respectively:

$$Q_t = W_{St} + W_{At} + W_{Pt} + W_{Mt} + W_{Et} \tag{5.12}$$

$$q_{jt} = w_{Sjt} + w_{Ajt} + w_{Pjt} + w_{Mjt} + w_{Ejt}, \tag{5.13}$$

where the subscripts represent search and negotiation (S), approval (A), project management (P), monitoring (M), and enforcement and insurance (E). The CDM project cycle (Fig. 5.1) can help identify these costs.

The CDM project cycle starts with the preparation of a Project Design Document (PDD). This requires the project developer to identify a suitable region; gather agricultural, social, and economic information about the region to develop the baseline; identify suitable land uses and estimate their carbon sequestration potential;

Table 5.1 Classifications of transaction costs in AR projects for carbon sequestration

Cost type	Buyer (Q)	Seller (q)
Search and negotiation	ex ante	
	W_S	w_S
	Find sites, establish contact, organize information sessions, draft contracts, provide training, promotion; establish baseline for region; estimate potential C stocks and flows of project; design individual farm plans; produce PDD	Attend information sessions; undertake training; design farm plan
Approval	ex ante	
	W_A	w_A
	Approval by host country (DNA); validate the project proposal (DOE); submit to CER Board	Obtain permit
Project management	ex ante	
	W_P	w_P
	Buy computers and software, establish office establish permanent sampling plots	Purchase tape and equipment for measuring trees and sampling soil
	ex post	
	Maintain database and administer payments; coordinate field crews, pay salaries; distribute payments to landholders; interest costs	Attend regular project meetings
Monitoring	ex post	
	W_M	w_M
	Enter data from farmer sheets, calculate C payments Process soil C samples; measure random sample of plots to check farmer estimates; verification and certification of carbon (DOE)	Measure trees, fill in form, and deliver to project office Sample soil C
Enforcement and insurance	ex post	
	W_E	w_E
	Maintain buffer of C; purchase liability insurance; settle disputes	Protect plot from poachers and fire; participate in dispute settlement

Source: Oscar Cacho

contact and establish relationships with the local people; negotiate the terms of the project and the schedule of payments for carbon-sequestration services; and possibly undertake environmental and social impact studies. These activities are included within *search and negotiation* costs in Table 5.1. Estimates of these costs in the literature vary widely depending on the nature of the activities within the project, the scale of the project, assumptions regarding the presence of local NGOs and farmer groups that may facilitate the process of contacting local people, and the availability of local experts to design the monitoring strategy and prepare the PDD.

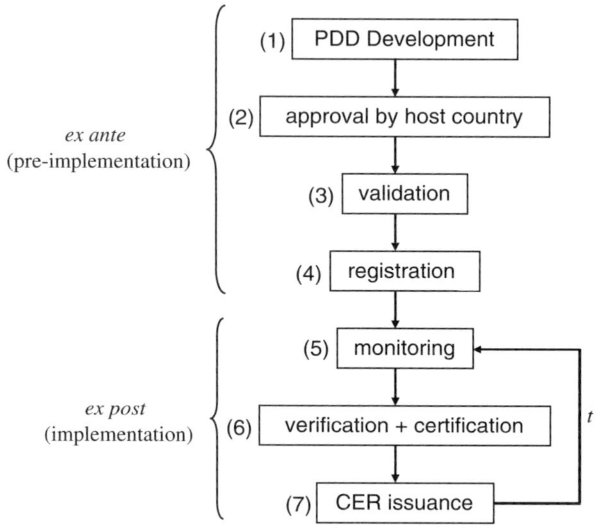

Fig. 5.1 The CDM project cycle used as a base for estimating transaction costs. Oscar Cacho

Steps 2, 3, and 4 of the CDM cycle (Fig. 5.1) fall within the *approval costs* category. They include approval by the Designated National Authority (DNA) of the host country; validation of the PDD by a Designated Operational Entity (DOE) accredited by the CDM Executive Board; and registration of the project when submitted to the Executive Board. The costs of these activities depend on several factors, including the institutional infrastructure of the host country and the availability of a local DOE that can validate the PDD as a cheaper alternative to an international consultant.

Steps 5, 6, and 7 of the CDM cycle (Fig. 5.1) fall within the *monitoring costs* category of Table 5.1. These are the costs of measuring the CO_2 abatement actually achieved by the project, including certification and verification by a DOE. Once the CDM Executive Board issues the appropriate number of CERs, the project developer (the Buyer) becomes a seller in the international carbon market. Any additional transaction costs that may be associated with the latter step are not accounted for below. It is assumed that the project developer can access the full price per CER, although it is a simple matter to reduce the price by a brokerage fee if applicable. Monitoring costs are recurrent, as they are incurred every time a new batch of carbon is submitted for CER crediting.

Two types of transaction costs listed in Table 5.1 do not fit neatly within the CDM project cycle; nonetheless, they are necessary for the approval and operation of the project.

Project management costs include the cost of keeping records of project participants and administration of payments to Sellers, as well as salaries and transportation costs of project employees. Ex ante project management activities include the establishment of a local project office and the training of staff. Project management costs are not normally recognized explicitly in the literature on transaction costs of Kyoto mechanisms, but they are expenses incurred in buying and selling carbon-sequestration services, so they should be considered.

Enforcement and insurance costs arise from the risk of project failure or underperformance, which might be caused by fire, slow tree growth, or leakage.[5] Enforcement costs may be incurred in the form of litigation and dispute-resolution expenses. Insurance options may include purchase of an insurance policy, deduction of a risk premium from the price of carbon, and maintenance of buffer carbon stocks that are not sold. These activities form part of the risk-management strategy required within the PDD.

5.3.1 Estimates of Transaction Costs

A selection of CDM transaction-cost estimates published in the literature is presented in Table 5.2. The search and negotiation costs (W_S) range between \$22,000 and \$160,000; the approval costs (W_A) range between \$12,000 and \$120,000; and

Table 5.2 A selection of transaction costs estimates of CDM projects

	Source					
	A	B	C		D	
			Low	High	Low	High
Search	15,000		19,000	29,000	5,000	20,000
Negotiation	25,000		10,500	10,500	20,000	25,000
Baseline determination	35,000	18,000				
Preparation of PDD		3,618	6,500	120,000	25,000	40,000
W_S total	75,000	21,618	36,000	159,500	50,000	85,000
Approval	40,000		1,000	10,000		
Validation	15,000	28,000	6,000	80,000	10,000	15,000
Registration	10,000	4,000	5,000	30,000	10,000	10,000
W_A total	65,000	32,000	12,000	120,000	20,000	25,000
Monitoring	10,000	750	6,550	6,550		
Verification + cert.	8,000	20,500	10,112	50,559	3,000	15,000
Adaptation fee			10,193	212,349	2,100	21,000
W_M total	18,000	21,250	26,855	269,458	5,100	36,000
Risk mitigation (% of CERs)					1%	3%
W_E total					1,050	10,500

Sources: (A) Michaelowa et al. (2003) – low-cost scenarios; (B) de Gouvello and Coto (2003) – hydroelectric project in Guatemala; (C) Krey (2004) – survey of projects in India; (D) EcoSecurities and UNDP (2003) – biomass power generation

Values for A are expressed in Euros (€) and for B, C, and D are expressed in dollars (\$)

[5] Leakage refers to the possible increase in emissions outside the project boundary caused, for example, by displaced people clearing forest elsewhere.

the monitoring costs (W_M) range between \$5,000 and \$270,000. Only one source in Table 5.2 presents risk-mitigation costs, which are classified under enforcement and insurance (W_E); these values were calculated based on the assumed price of \$3.00 per CER used in the original source. The wide range of values in all categories illustrates the fact that transaction costs are highly sensitive to the type and size of project assumed.

The CDM user's guide (EcoSecurities and United Nations Development Programme, 2003) assumes a biomass power plant with a 20-year lifetime. The low and high estimates for this source correspond to a small power plant (35,000 t CO_2/year) and a large plant (350,000 t CO_2/year), respectively. Their *feasibility assessment* values were classified under *search* in Table 5.2, and their *legal fees* estimates under *negotiation*. In addition to verification and certification, monitoring costs also include an adaptation fee that goes to a fund established by the United Nations Framework Convention on Climate Change (UNFCCC) to help vulnerable countries adapt to the effects of climate change. The cost estimates in Table 5.2 are largely based on energy projects (including biomass energy) rather than on AR projects, and do not involve negotiation with a large number of smallholders.

Useful additional information regarding transaction costs of projects involving smallholders is provided by the *Scolel Te* project in Southern Mexico, which has developed a useful management system called "Plan Vivo." De Jong et al. (2004) outline the transaction costs associated with designing the Plan Vivo Management System. Under the *search and negotiation* category, we could include the costs of undertaking the feasibility study, the carbon inventories, the land-use analysis, and the development of the regional baseline. The total cost of these activities was ~\$830,000. In addition, the system requires the design of individual farm plans.

Arifin (2005) presents estimates of the transaction costs incurred by community-based forestry management groups in Sumber Jaya, Indonesia. These groups are participating in the RUPES[6] project. Activities cited by Arifin included under *search and negotiation* are obtaining information and joining farmer groups; under *approval* is the cost of obtaining a permit to participate; under *project management* is the cost of attending meetings; and under *enforcement and insurance* are the costs of guarding crops and participating in dispute settlement. Arifin calculated these costs as the per-household time allocated to perform activities multiplied by the wage rate.

5.4 Abatement Costs and Carbon Payments

Abatement costs in an AR project are the costs of producing one unit of uncertified carbon sequestration services. For Sellers, abatement costs are measured as the opportunity cost of not undertaking the most profitable land-use activity as a result

[6] RUPES stands for Rewarding the Upland Poor for the Environmental Services they provide.

of adopting a prescribed AR activity that stores additional carbon. This opportunity cost is the present value of the stream of net revenues foregone as a result of participating in the project, Eq. (5.3). The abatement cost to the Buyer is the present value of the stream of payments to landholders for carbon-sequestration services, Eq. (5.7). In order to implement these equations, we require information on carbon sequestration rates and net revenue streams for the baseline and the forestry activity.

Carbon trajectories for the baseline and the project activity, $C_0(t)$ and $C_j(t)$, are normally calculated using either single-equation models of tree growth or more complex models, such as CO_2Fix (i.e., de Jong et al., 2004), WANULCAS (i.e., Wise & Cacho, 2005a), and SCUAF (i.e., Wise & Cacho, 2005b), that include soil as well a biomass carbon. Below we use single-equation models and do not consider soil carbon, which is more expensive to measure. For projects that are known to have non-decreasing effects on soil carbon, it may not be necessary to measure this pool after the baseline is established. As a general rule, reforestation projects in agricultural lands tend to increase soil carbon and, if the marginal cost of measuring this carbon pool is greater than the marginal benefit of the carbon credits obtained, the project developer would prefer not to measure this pool (see Cacho et al., 2004).

Net revenue streams for the baseline and the project activity, $r_0(t)$ and $r_j(t)$, can be calculated on a spreadsheet based on inputs (labor, fertilizer, and seedlings, etc.) and outputs (grain, fruit, timber, resin, etc.), preferably obtained through surveys in the region of interest.

Table 5.3 presents a baseline (cassava) and three alternative land uses for Sumatra, Indonesia. The carbon stock of the baseline is assumed to be zero because biomass is harvested every year and soil carbon is not accounted for. The three alternative land uses presented in Table5.3 include a traditional system popular among smallholders (rubber), a complex agroforest with high biodiversity (damar), and a fast-growing tree used for pulp and timber (*Acacia mangium*).

Table 5.3 A selection of possible land uses for AR projects in Sumatra, Indonesia

	System			
	Cassava (baseline)	*Rubber*	*Damar*	*Acacia Mangium*
Mean carbon stock (t C/ha)	0.0	42.4	102.7	129.6
NPV ($/ha)	2,705	122	1,317	2,367
Years to positive cash flow	–	13	6	8
Establishment cost ($)	–	898	1,855	642
Return to labor ($/day)	4.65	1.71	2.20	9.59
Labor requirement (d/ha/year)	112	105	116	62
Opportunity cost ($/ha)	–	2,583	1,388	338
Carbon cost ($/t)	–	60.93	13.51	2.61
CO_2 cost ($/t)	–	16.60	3.68	0.71

Source: Ginoga et al. (2002) and data collected as part of ACIAR research project ASEM 2002/066

Rubber production has a long history in Sumatra and various production systems exist, ranging from jungle rubber in community land to large commercial plantations (Budidarsono et al., 2001; Tomich et al., 2002). The rubber system represented in Table 5.3 assumes a relatively fast-growing tree that can be tapped for latex starting in year six after planting. The damar system is a complex agroforest developed by the Krui people of Lampung, southern Sumatra. The system consists of a sequence of crops building up to a "climax that mimics mature natural forest" (ASB, 2001). The main tree species is damar (*Shorea javanica*), a source of resin that provides a flow of income. Other outputs include fruits, pepper, and firewood. *A. mangium* is a fast-growing nitrogen-fixing tree used for furniture, firewood, and pulp. This species is very popular in Indonesia and represents one of the main plantation trees in Sumatra.

The net present values (NPV) were calculated for a 60-year period to allow time for the damar system to mature. Establishment cost was calculated as the NPV of the stream of costs incurred until the cash flow becomes positive, and the return to labor was calculated as the wage rate that makes NPV = 0. Returns to labor range from \$1.71/day for rubber to \$9.59/day for *A. mangium*. Compared to an average wage rate of \$1.50/day in south Sumatra, all four systems provide attractive returns to labor. Average annual labor requirements are considerably lower for *A. mangium* than for the other three systems. The amount of carbon sequestered in aboveground biomass for each system (Fig. 5.2) was estimated using Gompertz functions:

$$C_t = \beta_1 \left(\frac{\beta_2}{\beta_1}\right)^{\exp(\beta_3 t)}. \tag{5.14}$$

The parameter values (β_1, β_2, β_3) were set to (81.7, 0.933, 0.138) for rubber; (198.1, 2.696, 0.068) for damar; and (282.3, 1.216, 0.517) for *A. mangium*. The mean carbon stock (Table 5.3) was calculated by dividing the area under the corresponding curve (Fig. 5.2) by 60 years. This is an estimate of the "permanent" increase in carbon stocks, assuming that the land use will not change and land productivity will not decrease with subsequent production cycles.

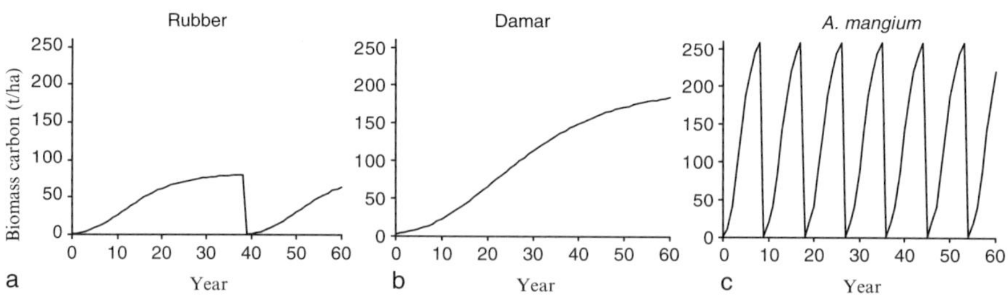

Fig. 5.2 Biomass carbon trajectories for three land-use systems in Sumatra, Indonesia. Oscar Cacho

5.5 Assumptions and Experiments

The CDM rules are not followed in the numerical analysis undertaken below in two major respects: (1) it is assumed that the project is eligible to continue receiving CERs as long as the trees planted by participating farmers continue capturing CO_2, so the first commitment period of the Kyoto Protocol (2008–2012) is ignored; (2) temporary CERs are not considered, rather an annual rent, Eq. (5.11), is assumed.

A series of computer experiments are performed on a hypothetical project. The project is assumed to consist of n identical farms each consisting of a hectares. The baseline is an annual cassava crop and the project activity is an *A. mangium* plantation harvested every 8 years, which is the land use with the highest carbon stock and NPV of the land uses in Table 5.3. The mean carbon stock of this system (130 t C/ha) is of similar magnitude to the systems used in the *Scolel Te* project (128 $\pm$ 25.6 t C/ha) reported by de Jong et al. (2004). The project developer establishes individual contracts whereby farmers agree to change their land use from cropping to forestry and receive payments for the carbon captured in their trees. In designing the project the Buyer decides on the number of participants (n), the carbon price paid to farmers (p_F), and other features such as monitoring and risk-mitigation strategies.

Transaction cost assumptions are presented in Table 5.4. These values are based on the estimates from the literature reviewed earlier. The units of measurement of these costs vary. In the case of the Buyer, costs can be ex ante fixed costs ($), annual fixed costs ($/year), or variable costs dependent on the number of participating farms ($/farm) or on the size of the project ($/ha/year). In the case of the Seller, costs are expressed in terms of labor. In the numerical analysis, they are later converted to dollars based on the wage rate p_L. The original five transaction-cost categories are disaggregated to account for variation in the units of measurement. The expanded classification is presented under "Cost type" (column 1, Table 5.4), where number subscripts denote the different cost types. For example, there are three types of monitoring costs; W_{M1} ($/ha/year), W_{M2} ($/year), and W_{M3} (CER/year).

Monitoring costs of AR projects can be high, and designing the right monitoring strategy involves consideration of sampling principles. The intensity and frequency of sampling will affect the accuracy of biomass carbon estimates and, therefore, determine the amount of CERs that can be claimed (Cacho et al., 2004). Involving smallholders in the monitoring effort can help achieve the higher accuracy of intensive sampling at a lower cost. Hairiah et al. (2001) noted that farmers in Sumatra can assess the volume of wood in their trees by sight, and Delaney and Roshetko (1999) found that a crew could learn carbon measurement methods in forest plots in 2 days. These facts suggest that farmers could be taught to monitor their plots at low cost. The farmers would deliver their completed sampling sheets to the project office, and the project office would enter these data into a database and estimate carbon stocks. The project office would also perform random checks to confirm or adjust farmers' estimates. This method is assumed to cost $8/ha/year to the Buyer and require 3 days of a farmer's time (Table 5.4). Monitoring also

Table 5.4 Transaction cost assumptions for base case

Cost type	Activity	Cost	Units
	Buyer (project manager)		
W_{S1}	Consultation and negotiation	20,000	$
W_{S1}	Establish baseline and C flows of project for region	6,500	$
W_{S1}	Design monitoring plan, establish PS plots	50,000	$
W_{S1}	Prepare PDD	8,000	$
W_{S2}	Design individual farm plans	500	$/farm
W_A	Approval by host government	1,000	$
W_A	Validate the project proposal (DOE)	6,000	$
W_A	Submit to CER Board (registration fee)	[a]	$
W_{P1}	Purchase IT infrastructure, establish local office	20,000	$
W_{P2}	Maintain database/software and administer payments	10,000	$/year
W_{P2}	Coordinate field crews, pay salaries	60,000	$/year
W_{M1}	Randomly check C stocks reported by farmers	8	$/ha/year
W_{M2}	Verification and certification of carbon by DOE	15,000	$/year
W_{M3}	Adaptation fee	0.02	CERs/year
W_{E1}	Maintain buffer of C	0.10	CERs/year
W_{E2}	Settle disputes	100	$/farm/year
	Sellers (farmers)		
w_S	Attend information sessions	6	days
w_S	Undertake training	10	days
w_S	Design farm plan	4	days
w_A	Obtain permission to participate in project	4	days
w_P	Attend regular project meetings	5	days/year
w_M	Measure trees, fill in form, and deliver to project office	3	days/ha/year
w_E	Protect plot from poachers and fire	10	days/year
w_E	Participate in dispute resolution	2	days/year

Source: Oscar Cacho

[a]Registration fees vary with project size <15,000 CERs = $5,000; 15,000 to <50,000 CERs = $10,000; 50,000 to <100,000 CERs = $15,000; 100,000 to <200,000 CERs = $20,000; >200,000 CERs = $30,000

involves verification and certification of carbon stocks by a DOE. This is assumed to cost $15,000/year (Table 5.4), but the cost could be higher if international experts are required or the project sites are scattered over a large area.

Designing individual farm plans (W_{S2}) involves a technician visiting each farm and drawing a land-use change plan in consultation with the farmer. This is assumed to cost $500 per farm to the Buyer, which would include 1 day or 2 days of a local technician's time plus travel expenses, and take 4 days of the Seller's time (Table 5.4).

Enforcement and insurance are assumed to involve maintaining a buffer of 10% of biomass carbon not sold as CERs, plus an average cost of $100 per farm per year to settle disputes (Table 5.4); this expense would include any legal fees involved. The buffer is a risk-mitigation strategy to account for leakage or the possible loss of trees.

Table 5.5 Other assumptions for base case

Variable	Value	Description
p_{CER}	20	Price of CERs (\$/t CO_2e)
p_C	4.28	Farm price of carbon (\$/t C)
p_L	1.5	Price of labor (\$/day)
n	500	Number of farms in project
a	1.5	Average area of farm (ha)
δ_B	0.06	Buyer discount rate
δ_S	0.15	Seller discount rate
i	$\ln(1 + \delta_B)$	Discount rate in carbon rental market

Source: Oscar Cacho

Using the expanded notation introduced in Table 5.4, transaction costs can now be calculated as:

$$V_T = W_{S1} + W_A + W_{P1} + nW_{S2}$$
$$\sum_t \begin{bmatrix} W_{P2} + W_{M2} + n(W_{E2} + aW_{M1}) \\ + (W_{M3} + W_{E1})(C_{jt} - C_{0t})p_C \end{bmatrix} (1 + \delta_B)^{-t} \tag{5.15}$$

$$v_T = \left[w_S + w_A + \sum_t [w_P + w_E + a\, w_M](1 + \delta_S)^{-t} \right] p_L. \tag{5.16}$$

Assumptions regarding prices and discount rates are presented in Table 5.5. The price of CERs is set initially at a high value (\$20/t CO_2) to ensure the project is feasible.

Replacing Eqs. (5.4) and (5.8) with (5.15) and (5.16), respectively, and inserting parameter values in the appropriate equations, we can now solve the model and determine under what conditions the project is feasible. Based on conditions for project participation in Eqs. (5.1) and (5.5), dropping the j subscripts and including functional relationships, the project is feasible if the following two conditions are satisfied:

$$v_C(a, p_F, C(t), \delta_S) - v_A(a, p_L, r(t), \delta_S) \geq v_T(p_L, \delta_S) \tag{5.17}$$

$$V_C(a, p_C, C(t), \delta_B) - V_A(a, p_F, C(t), \delta_B) \geq V_T(a, n, p_C, C(t), \delta_B), \tag{5.18}$$

where $C(t)$ is the eligible carbon trajectory ($C_{jt} - C_{0t}$). The expressions on the left of the inequalities are the carbon margins. Experiments consist of solving the model for different values of the arguments (in particular, p_C, p_F, a, and n) and determining when both conditions (5.17) and (5.18) are satisfied.

5.6 Analysis of Project Design

5.6.1 *Farm Price*

The first step in the numerical analysis is to determine bounds for the farm price. This involves finding the minimum price acceptable to the average Seller[7] and the maximum price the Buyer is willing to pay. First, p_F is set such that $v_C - v_A = v_T$ and the resulting value is called p_S; then p_F is set such that $V_C - V_A = V_T$ and the resulting value is called p_B. The project is feasible only if $p_B \geq p_S$, and the farm price falls within the range $p_S \leq p_F \leq p_B$. The actual value of p_F depends on the market power of the participants, the objectives of the Buyer, and the outcome of negotiations.

The carbon margin for the Seller ($v_C - v_A$ in Fig. 5.3a) increases linearly with p_F, whereas the carbon margin for the Buyer ($V_C - V_A$ in Fig. 5.3b) decreases linearly with p_F. The intersections of the carbon margin curves with their respective transaction cost curves indicate the price bounds (p_S, p_B). Given the assumptions in Tables 5.4 and 5.5, the feasible farm price ranges between \$0.61/t C and \$1.80/t C (Fig. 5.3). For simplicity, we set $p_F = (p_S + p_B)/2$ as the base price to determine the effects of other project design variables; therefore, $p_F =$ \$1.21/t C in the base case.

5.6.2 *Minimum Farm Size*

The assumptions in Table 5.5 imply that the project covers 750 ha (500 farms of 1.5 ha each) and increases the biomass carbon stock by 97,200 t C above the base-line (129.6 t C/ha × 750 ha); corresponding to a total of 356,724 CERs (97,200 t C × 3.67 t CO_2/t C). Given that we are dealing with smallholders, it is important to determine to what extent the size of participating farms affects the feasibility of the project. To answer this question, we solve the model for a range of values of a, while simultaneously adjusting n to keep project size constant at 750 ha (or 356,724 CERs). This operation does not affect the carbon margin, but it has a significant effect on transaction costs for the Buyer (Fig. 5.4).

As farm size increases, the Buyer's transaction costs decrease at a decreasing rate and become relatively flat at farm sizes beyond 5 ha or so. Reducing farm size below 1 ha causes transaction costs to increase exponentially. The minimum farm size for the given parameters is 0.86 ha (Fig. 5.4), which would require 875 participating farms to maintain total project area at 750 ha. At this point the Buyer's transaction costs would be ~\$4.19M, which translates to \$11.76/CER. By comparison, for a project with 5-ha farms (requiring 150 farms to maintain the project area at 750 ha), the Buyer's transaction costs would be \$2.65M, or \$7.43 per CER, still a

[7] For simplicity, we will refer to the average Seller and assume all Sellers have identical farms.

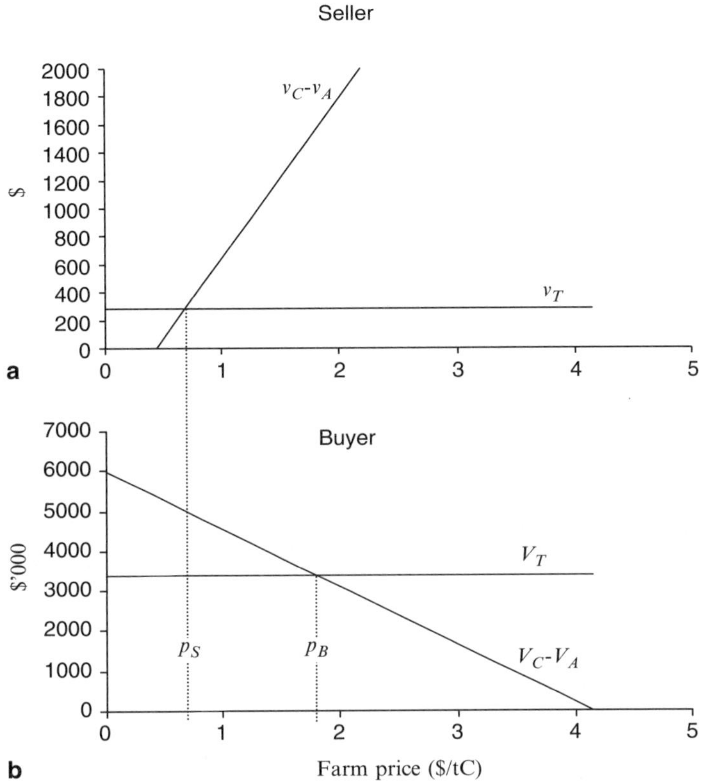

Fig. 5.3 The feasible range of farm prices within which both the Sellers and the Buyer would participate in the project is derived by finding the minimum price acceptable to Sellers in **(a)** and the maximum price acceptable to the Buyer in **(b)**. Oscar Cacho

high value but considerably lower than with the smaller farms. In Indonesia many farmers have areas of 1 ha or less, and they would be excluded from our hypothetical project unless they could contract with the project as a group offering a larger area of land. De Jong et al. (2004) report that families participating in the *Scolel Te* project are able to initiate reforestation activities on 0.5–1.5 ha without a significant drain in the labor resource. According to our results only the larger farms in this range would be acceptable in the project, and only at a high CER price of $20/t CO_2.

5.6.3 Minimum Number of Farms

Now assume that farm size remains constant at 1.5 ha, but it is possible to change the total project area by regulating the number of contracts with farmers (Fig. 5.5). As the number of participating farms increases, both the carbon margin ($V_C - V_A$) and transaction costs (V_T) for the Buyer increase linearly, but the former increases faster. Under this scenario, a minimum of 327 farms is required for a feasible

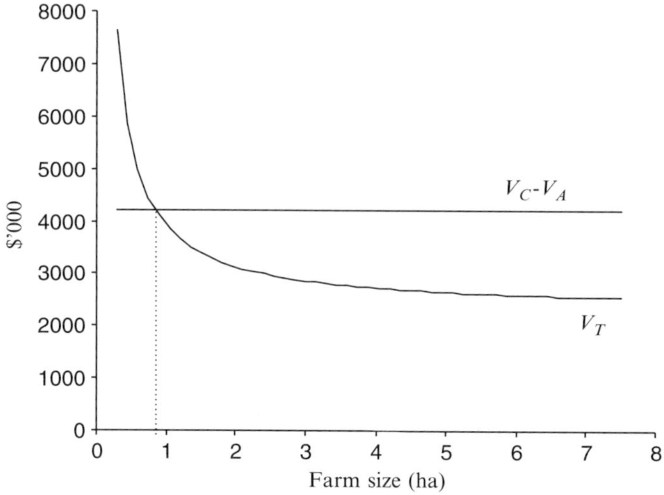

Fig. 5.4 Minimum feasible farm size for a project area of 750 ha and a farm price of \$1.21/t C is indicated by the dotted line at the intersection of the carbon margin ($V_C - V_A$) and the transaction costs (V_T) for the Buyer. Note: the number of farms decreases as farm size increases to keep the project size constant. Oscar Cacho

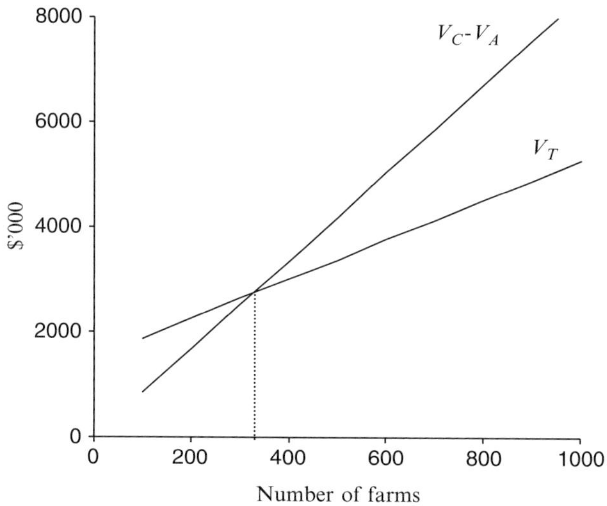

Fig. 5.5 Minimum number of farms required for a feasible project with farm size of 1.5 ha and a farm price of \$1.21/t C is indicated by the dotted line. Note: the total project area increases as the number of farms increases because the farm size is fixed. Oscar Cacho

project with p_F fixed at \$1.21/t C; this will result in transaction costs of ~\$2.74M for a total project area of 490 ha capturing 233,000 t CO_2; this results in transaction costs of about \$11.78 per CER, which totally offsets the carbon margin. With 500 farms, transaction costs represent about 80% of the carbon margin, whereas with 1,000 farms, this decreases to 63% (Fig. 5.5).

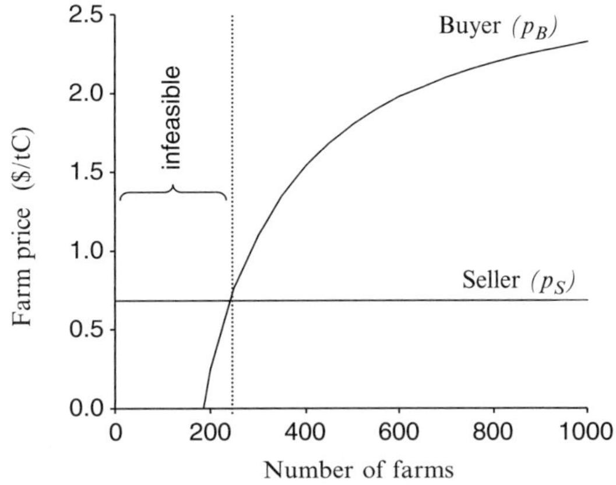

Fig. 5.6 The breakeven number of farms, indicated by the dotted line, is calculated as the point at which the maximum price the Buyer is willing to pay (p_B) equals the minimum price the seller is willing to accept (p_S). At this point there are no profits for either party. The Buyer's curve was derived by finding the values of p_F that make $V_C - V_A = V_T$ for a range of farm numbers. Oscar Cacho

Corbera (2005) reports that the Fondo Bioclimatico, the organization that runs the *Scolel Te* project in Mexico, increased the number of contracts between 1997 and 2004 from 43 farmers in 6 communities to 650 farmers in 33 communities. During the same period, the reforestation area increased from 78 ha to 845 ha. This implies that the average farm area has decreased from 1.8 ha to 1.3 ha and may indicate that it has been feasible to accept smaller farms into the project as the initial infrastructure has been established and learning costs have been covered.

Now consider the implication noted earlier in connection with Eq. (5.9) that as the total project area increases, the maximum farm price the Buyer would be prepared to pay may also increase. To confirm this, we solve the model for different values of n, while holding farm size constant at 1.5 and adjusting p_F until all carbon profits are dissipated. That is, for any given value of n, we solve for the Buyer's breakeven value of p_F that makes $V_C - V_A = V_T$. Results of this analysis are presented in Fig. 5.6. The Buyer's breakeven farm price increases at a decreasing rate, from \$0.68/t C to \$2.33/t C as the number of farms under contract increases from 242 to 1,000; and total project area increases from 363 ha to 1,500 ha. In Fig. 5.6, the minimum number of farms (242) is that at which the breakeven price for the Buyer is the same as the minimum price acceptable to the Seller ($p_B = p_S$). Note that this number of farms differs from that associated with Fig. 5.5 above, because here p_F is endogenously determined as a breakeven price for any given value of n, whereas above p_F was exogenously set at \$1.21 as the average between p_B and p_S.

5.6.4 Effects of CER Price

The CER price used above (\$20/t CO_2e) is rather high, so it is important to determine how a lower price will affect project feasibility. In particular, it is of interest to evaluate how the CER price affects the critical values of p_S, p_B, n, and a identified above. Essentially, this involves changing p_{CER} and repeating the above analysis to identify the points at which the Buyer's carbon margin ($V_C - V_A$) equals

Table 5.6 Effect of CER price on critical values of project-design variables

	Price of CERs (\$/t CO_2e)		
	20	15	12
Seller minimum carbon price (\$/t C), p_S	0.69	0.69	0.69
Buyer maximum farm price (\$/t C), p_B	1.79	0.88	0.33
Farm price (\$/t C), p_F	1.24	0.78	0.69
(A) *With project area constant (750 ha):*			
Minimum farm area (ha)	0.86	1.33	2.91
Corresponding number of farms	868	564	258
CER Project (t CO_2e)	354,343	354,343	354,343
(B) *With farm size constant (1.5 ha):*			
Minimum number of farms	329	459	756
Corresponding project area (ha)	493	688	1,134
CER Project (t CO_2e)	232,879	325,133	535,586
(C) *With minimum farm price* ($p_F = p_S$):			
Breakeven number of farms	244	425	756
Corresponding project area (ha)	367	638	1,134
CER Project (t CO_2e)	173,153	301,415	535,824

Source: Oscar Cacho

the transaction cost (V_T). Results are presented in Table 5.6. The first column of results shows the base case already discussed, the other two columns are the results with p_{CER} values of \$15 and \$12. Given the transaction costs assumed and the default number of farms (500) and farm size (1.5 ha), a p_{CER} of \$12 is not feasible. At this CER price, the Buyer's price (p_B = 0.33) is below the Seller price (p_S = 0.69). Setting the farm price p_F at its lowest feasible value of \$0.69/t C, we find that the minimum farm area with constant project size (750 ha) is 2.91 ha, and the minimum number of farms at constant farm area (1.5 ha) is 756. The former result (block A in Table 5.6) is represented by downward shift of the $V_C - V_A$ line in Fig. 5.4 as the CER price decreases, causing the new intersection with V_T to occur at a larger farm size. The latter result (block B in Table 5.6) is represented by downward shift of the $V_C - V_A$ line in Fig. 5.5 causing the new intersection with V_T to occur at a larger number of farms.

The last three rows of Table 5.6 (the block labeled C) are the most interesting, because they show the absolute minimum possible project size (when $p_F = p_S$), or the breakeven project size; rather than the minimum project size with p_F arbitrarily set at the mean between Buyer's and Seller's prices. The breakeven number of farms increases from 244 at a p_{CER} of \$20 to 756 at a p_{CER} of \$12. This shift represents a tripling in project area from 367 ha to 1,134 ha and is equivalent to an increase in project size (in terms of CERs) from 173.1 kt CO_2e to 535.8 kt CO_2e.

To put our results in perspective consider that, in May 2006, there were 176 CDM projects registered, claiming to reduce emissions by an average of 301,633 t CO_2e/year. Classified by size, there were 71 large-scale projects with average emission reductions of 638,133 t CO_2e/year and 78 small-scale projects claiming 29,554 t CO_2e/year. To convert our results from stocks to flows of carbon and compare them to existing projects, note that the biomass-carbon stock of *A. mangium* is assumed to increase from 0 t C/ha to 259 t C/ha in 8 years (Fig. 5.2c); this represents an annual CO_2 reduction of 118.0 t (3.67 × 259/8); multiplying this value by the breakeven project areas in Table 5.6, we obtain 43,489 t CO_2/year, 75,705 t CO_2/year and 134,572 t CO_2/year for CER prices of $20, $15, and $12, respectively. So our hypothetical project may fit between the medium- and large-scale categories depending on the CER price.

5.7 Introducing Biomass Energy (Fuel Switching)

Now assume that the biomass harvested every 8 years is used to produce biofuel, so it is possible to claim CERs on the reduced emissions caused by replacing a fossil fuel, such as diesel or coal, with biomass fuel. The biomass fuel emissions receive carbon credits because they represent CO_2 recently absorbed from the atmosphere, rather than CO_2 absorbed millions of years ago as is the case with fossil fuels. Therefore, biomass fuel emissions are said to be greenhouse neutral. The number of CERs that can be claimed by the project depend on the fuel that is being replaced. The following assumptions are made:

- Wood biomass contains 50% carbon.
- Only 70% of biomass is usable as a fuel.
- The net calorific values (NCV) of wood, diesel, and coal are as shown in Table 5.7.
- The carbon content factors (CCF) of diesel and coal are as shown in Table 5.7.

Table 5.7 Energy and carbon content of alternative fuels

Variable	Fuel			
	Wood	Diesel	Coal	Units
NCV – Net calorific value	13.8[a]	43[b]	28.2[b]	MJ/kg fuel
CCF – Carbon content factor[b]		0.0201	0.0258	kg C/MJ
Calculations				
(a) Carbon produced by burning fuel = NCV × CCF		0.864	0.728	kg C/kg fuel
(b) Wood required to replace fuel = NCV wood/NCV fuel		3.116	2.043	kg wood/kg fuel
(c) Equivalent carbon produced by wood = a/b		0.277	0.356	kg C/kg wood
(d) CO_2 replaced by wood = 0.7 × day/0.5		1.018	1.307	kg CO_2/kg wood
(e) CO_2 emissions replaced by biomass carbon $(\gamma_k)^c$ = 0.7 × day/0.5		1.425	1.829	kg CO_2/kg biomass C

[a]FAO (2004)
[b]Kazunari (2005) coal values are for coking coal
[c]Assumes biomass contains 0.5 carbon, and only 0.7 of biomass is usable for fuel

Table 5.8 Results for project with a biofuel component

Critical variable	Price of CERs ($/t CO_2e)		
	15	10	5
Diesel replacement			
Breakeven number of farms	100	175	703
Corresponding project area (ha)	150	263	1,055
Project size (CER/year)	24,727	43,400	174,045
Coal replacement			
Breakeven number of farms	84	141	469
Corresponding project area (ha)	126	211	703
Project size (CER/year)	22,362	37,638	125,145

Source: Oscar Cacho

To estimate the number of CERs that can be claimed by the project for fuel substitution, we need to obtain a wood-replacement factor for the fossil fuel in question. This factor (*t* of CO_2 fossil-fuel emissions avoided per *t* of carbon harvested from trees) is $\gamma_1 = 1.425$ for diesel and $\gamma_2 = 1.829$ for coal. The required calculations are shown in Table 5.7. The present value of the fuel-substitution activity can then be calculated as:

$$V_k = \sum_{\tau \in t_H} \gamma_k H_\tau p_{CER} (1 + \delta_B)^{-\tau}, \tag{5.19}$$

where H_τ is the amount of carbon harvested in year τ, t_H is the set of harvest years, and the subscript k represents either diesel (1) or coal (2). The value of V_k is then added to the present value of rental carbon (V_C) to obtain the total value of carbon when a biofuel component is included in the project. The critical project-design variables can then be calculated following the same process as before. The results of this analysis are shown in Table 5.8.

The first thing to notice is that now the project is feasible at lower CER prices. Whereas before a price of less than \$12/t CO_2e made the project infeasible, adding a biofuel component makes the project feasible at a CER price as low as \$5/t CO_2e. The introduction of biofuel in the project complicates the measurement of CERs because now there are two components: (1) annual rental payments on carbon stocks above the land-use baseline (cassava) and (2) purchase payments on the flows of CO_2 emissions replaced every harvest year relative to a baseline given by the fossil fuel being replaced (diesel or coal). Thus, it is now more appropriate to report the project size in terms of carbon flows rather than stocks. Average CER flows per unit area (t CO_2/ha/year) are calculated for every 8-year rotation as consisting of 118 t CO_2/ha/year during tree growth plus either 46.1 t CO_2/ha/year or 59.2 t CO_2/ha/year for diesel or coal, based on the replacement factors γ_k. CER flows per unit area are then multiplied by the project area to calculate the (CER) size of the project in terms of t CO_2/year (Table 5.8). Minimum project size ranges between 24.7 kt CO_2/year and 174 kt CO_2/year for diesel substitution, and between 22.4 kt CO_2/year and 125.1 kt CO_2/year for coal substitution (Table 5.8). The effect

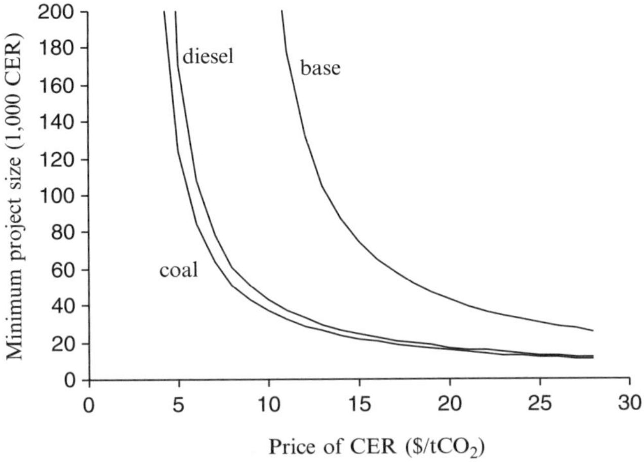

Fig. 5.7 The breakeven number of CERs as a function of CER price for the base project and two alternative biofuel projects replacing either diesel or coal. Oscar Cacho

of CER price (p_{CER}) on minimum project size is nonlinear (Fig. 5.7). Minimum project size decreases rapidly at low values of p_{CER}, and this decrease flattens out at p_{CER} values beyond about \$10 or \$20/t CO_2e depending on whether the project includes a biofuel component. There is a considerable gap between the base case and the biofuel cases, but the actual gap may not be as large if the costs of establishing a biofuel plant have to be covered by the project.

The analysis illustrated in Fig. 5.7 implicitly assumes either that there is an existing plant in the region, which can take the harvested wood and produce a fuel such as biodiesel or ethanol, or that local power demand can switch from fossil fuel to woodfuel at no cost. This replacement, however, may require an investment in new equipment (e.g., to replace a diesel generator with a wood-fired generator). Although it is out of the scope of this chapter to estimate the size of the required investment, it is straightforward to estimate the minimum project size required to cover a given investment cost (I). This cost becomes part of the Buyer's abatement cost, so we replace Eq. (5.7) with:

$$V_A = I + \sum_j a_j \left[\sum_t p_F \left(C_{jt} - C_{0t} \right) \left(1 + \delta_B \right)^{-t} \right]. \tag{5.20}$$

The abatement cost now includes the investment cost (in present-value terms) plus the payments to farmers. On its own, I will shift the carbon margin ($V_C - V_A$) curve down in Fig. 5.3b, resulting in a lower Buyer price (p_B) than without the investment. Of course, the investment also increases V_C by the present value of CERs produced by the biofuel component and this will shift ($V_C - V_A$) up. Equation (5.6) is thus replaced by:

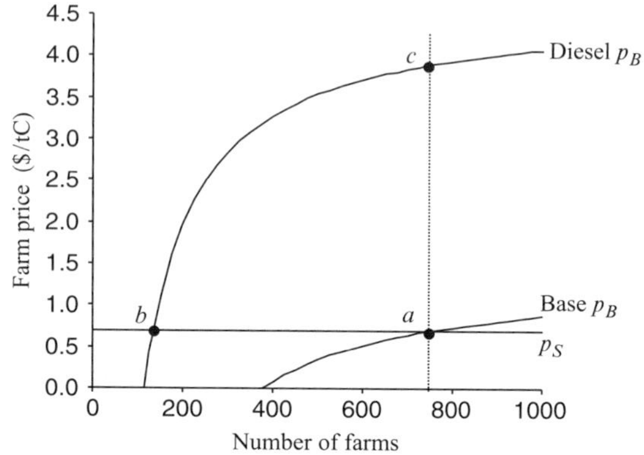

Fig. 5.8 Minimum project sizes in base case **(a)** and biofuel (diesel) case **(b)**, with p_{CER} = \$12/t CO_2e. Oscar Cacho

$$V_C = \sum_j a_j \left[\sum_t p_C \left(C_{jt} - C_{0t} \right) \, (1 + \delta_B)^{-t} + \sum_{\tau \in t_H} \gamma_k H_\tau p_{CER} \, (1 + \delta_B)^{-\tau} \right], \quad (5.21)$$

where the new term within the square brackets is V_k from Eq. (5.19). Thus the final result of introducing biofuels relative to the base case depends on the relative size of the changes in V_A and V_C when the biofuel is introduced. It is also possible that the introduction of a biofuel component will change transaction costs and have a further effect on minimum project size (an increase in V_T in Fig. 5.3b would decrease p_B). Although an upward shift in V_T is the intuitive result of complicating the project by introducing biofuels, it is interesting to question whether I may actually decrease V_T by, for example, reducing monitoring and enforcement costs because farmers have to supply their harvest to the biofuel plant, thus providing a cheap audit on carbon outputs.

Figure 5.8 compares the base project and the biofuel (diesel) project in terms of feasible farm prices, assuming I=0 and V_T is not affected by the introduction of biofuel. Thus, only the shift in V_C, Eq. (5.21) is considered. Points a and b indicate the breakeven number of farms for the base case (756) and the diesel case (134); point c indicates the maximum farm price that would result from designing the diesel project to be the same size as the base project. The difference in $c-a$ is about \$3.20 in Fig. 5.8. This surplus can be used to cover investment costs or to provide additional incentives to farmers, by increasing p_F above the minimum value imposed by p_S. In this example (with p_{CER}=\$12/t CO_2e), the surplus at point c is about \$7.4M. If the investment required to introduce the biofuel is greater than this figure, the biofuel component is unprofitable.

5.8 Summary and Conclusions

This chapter was motivated by a growing interest in environmental services in general and greenhouse gases in particular. Project-based carbon sequestration projects were analyzed based on a model of project participation. The conditions for a buyer (projects developer) and a group of sellers (farmers) to participate in a farm-forestry project were derived. The model accounts for abatement costs, transaction costs, and carbon payments as functions of a set of variables that include discount rates, price of carbon, and the number and size of participating farms. Three important project-design variables were identified: the farm price of carbon, the number of participating farmers, and the area of their farms. These variables are under the control of the project developer, subject to international carbon prices and the availability of enough farms in the area able to change land use from the baseline to the project activity. The model allows a project developer to identify the critical values of the project-design variables that will make a project feasible. Economies of scale were shown to be an important factor, with costs per ton of carbon sequestered dropping considerably as the area covered by the project increases. This result was largely due to the high proportion of transaction costs that are fixed. The model was extended to include biomass-energy, in addition to carbon sequestration, as a component within the project. It was shown that this component has the potential to make smaller projects feasible.

Acknowledgments This chapter was prepared under contract with FAO. The research was largely funded by the Australian Centre for International Agricultural Research (ACIAR) under project ASEM 2002/066. The author is grateful to Robyn Hean for valuable comments on previous drafts, to Russell Wise for research assistance, and to colleagues in CESERF and ICRAF, Indonesia, for providing data and information.

References

Arifin B (2005) Institutional perspectives of lifescape co-management: Lessons learned from RUPES sites in Sumatra, Indonesia. In: Murdiyarso D, Herawati H. (eds.). Carbon forestry: Who will benefit? Workshop on Carbon Sequestration and Sustainable Livelihoods, CIFOR, Bogor, Indonesia.

ASB (2001) The Krui agroforests: A model of sustainable community-based management. Policy Briefs, Alternatives to slash and burn. (http://www.asb.cgiar.org).

Budidarsono S, delos Angeles MS, Wibawa G (2001) A profitability assessment of smallholder rubber agroforest systems in Jambi, Sumatra, Indonesia. Manuscript prepared for the workshop: Complex Agroforests: Farmer Knowledge, Management, Profitability and Conservation, Jambi, Sumatra.

Cacho OJ, Hean RL, Wise R (2003) Carbon-accounting methods and reforestation incentives. Aust. J Agric Resour Econ 47:153–179.

Cacho OJ, Marshall GR, Milne M (2003) Smallholder agroforestry projects: Potential for carbon sequestration and poverty alleviation. ESA Working Paper No. 03-06, Food and Agriculture Organization of the UN.

Cacho OJ, Marshall GR, Milne M (2005) Transaction and abatement costs of carbon-sink projects in developing countries. Environ Dev Econ 10(5):1–18.

Cacho OJ, Wise R, MacDicken K (2004) Carbon monitoring costs and their effect on incentives to sequester carbon through forestry. Mitig Adapt Strateg Glob Chang 9(3):273–293.

Corbera E (2005) Bringing development into carbon forestry markets: Challenges and outcomes of small-scale carbon forestry activities in Mexico. In: Murdiyarso D, Herawati H (eds.). Carbon forestry: Who will benefit? Workshop on Carbon Sequestration and Sustainable Livelihoods, CIFOR, Bogor, Indonesia.

de Gouvello C, Coto O (2003) Transaction costs and carbon finance impact on small-scale CDM projects. PCF Plus, Washington, DC.

de Jong BHJ, Gaona SO, Montalvo SQ, Bazan EE, Hernandez NP (2004) Economics of agroforestry carbon sequestration. In: Alavalapati JRR, Mercer DE (eds.). Valuing agroforestry systems: Methods and applications. Kluwer, Dordrecht.

Delaney M, Roshetko J (1999) Field tests of carbon monitoring methods for home gardens in Indonesia. In: Field tests of carbon monitoring methods in forestry projects. Forest Carbon Monitoring Program, Winrock International, Arlington, VA.

Dudek DJ, Wiener JB (1996) Joint implementation, transaction costs, and climate change (Issue 96). Organisation for Economic Co-operation and Development, Paris, p. 173.

EcoSecurities and United Nations Development Programme (2003) Transaction costs, efficiency and supportive governance. In: The clean development mechanism: A user's guide, Chap. 5. UN Development Programme, New York, pp. 55–68.

FAO (2004) Unified bioenergy terminology (UBET). FAO Forestry Department, Wood Energy Programme, Food and Agriculture Organization of the UN.

Fitchner W, Graehl S, Rentz O (2003) The impact of private investor's transaction costs on the cost effectiveness of project-based Kyoto mechanisms. Clim Policy 3:249–259.

Ginoga K, Cacho O, Erwidodo, Lugina M, Djaenudin D (2002) Economic performance of common agroforestry systems in Southern Sumatra, Indonesia: Implications for carbon sequestration services. Working Paper No. CC03, ACIAR Project ASEM 1999/093. (http://www.une.edu.au/carbon/wpapers.php).

Hairiah K, Sitompul SM, van Noordwijk M, Palm C (2001) Carbon stocks of tropical land use systems as part of the global C balance: Effects of forest conversion and options for "clean development" activities. ASB Lecture Note 4A, ICRAF, Bogor. (http://www.icraf.cgiar.org/sea).

Kazunari K (2005) Revision of default net calorific value, carbon content factor and carbon oxidization factor for various fuels in 2006 IPCC GHG inventory guideline. REITI, IAI, Government of Japan.

Krey M (2004) Transaction costs of CDM project sin India - An empirical survey. HWWA Report No. 238. Hamburg Institute of International Economics.

Lecocq F, Capoor K (2003) State and trends in the carbon market. The World Bank Document.

Marland G, Fruit K, Sedjo R (2001) Accounting for sequestered carbon: The question of permanence. Environ Sci Policy 4:259–268.

Michaelowa A, Jotzo F (2005) Transaction costs, institutional rigidities and the size of the clean development mechanism. Energy Policy 33(4):511–523.

Michaelowa A, Stronzik M, Eckermann F, Hunt A (2003) Transaction costs of the Kyoto mechanisms. Clim Policy 3:261–278.

Tomich TP, de Foresta H, Dennis R, Ketterings Q, Murdiyarso D, Palm C, Stolle F, Suyanto, van Noordwijk M (2002) Carbon offsets for conservation and development in Indonesia? Am J Altern Agric 17(3):125–137.

Williamson OE (1985) The economic institutions of capitalism. Free Press, New York.

Wise RM, Cacho OJ (2005a) Bioeconomic analysis of carbon sequestration in farm forestry: A simulation study of *Gliricidia sepium*. Agrofor Syst 64:237–250.

Wise RM, Cacho OJ (2005b) Tree-crop interactions and their environmental and economic implications in the presence of carbon-sequestration payments. Environ Model Softw 20 (9):1139–1148.

Yamada K, Fujimori M (2003) Technical procedures for CDM project design. Mitig Adapt Strateg Glob Chang 8:221–235.

Chapter 6
Conservation Payments to Reduce Wildlife Habitat Fragmentation and Disease Risks

Richard D. Horan, Jason F. Shogren, and Benjamin M. Gramig

Abstract We investigate the challenges of using payments for environmental services (PES) to protect endangered species given habitat fragmentation in conjunction with disease risks from neighboring livestock. Using a bioeconomic model, we show how greater connectivity of habitat creates an endogenous tradeoff. More connectedness both (1) increases growth of endangered species populations, while (2) simultaneously increasing the likelihood diseases will spread more quickly. We examine payments for habitat connectedness, livestock vaccination, and reduced movement of infected livestock. We find the cost-effective policy to first use subsidies to promote habitat contiguousness. Once habitat is sufficiently connected, disease risks increase to the point where disease-related subsidies become worthwhile. Highly connected habitat requires nearly all the government budget be devoted to disease prevention and control. The conservation payments result in significantly increased wildlife abundance, increased livestock health and abundance, and increased development opportunities.

6.1 Introduction[1]

Habitat fragmentation and loss and infectious diseases are global threats to wildlife conservation. Habitat fragmentation and loss is the most important current cause of extinction among terrestrial vertebrates (MEA,[2] 2005, p. 45), and the conservation risks posed by infectious diseases are significant and escalating (MEA, 2005). Pathogen introductions may achieve a status similar to invasive species, the second most important cause of extinction (Daszak et al., 2000). Moreover, the impacts of these forces are not independent; habitat fragmentation and diseases are considered important joint drivers of extinction, and concern over this joint threat is growing

[1] This chapter draws largely from Horan et al. (forthcoming).

[2] Millennium Ecosystem Assessment.

L. Lipper et al. (eds.), *Payment for Environmental Services in Agricultural Landscapes*, Natural Resource Management and Policy.
doi: 10.1007/978-0-387-72971-8_6;

(MEA, 2005, p. 57; Daszak et al., 2000; Hess, 1996; Gog et al., 2002; McCallum & Dobson, 2002; Simonetti, 1995; McCallum & Dobson, 1995).

Habitat fragmentation increases disease risks by stressing wildlife populations, increasing animal densities in fragmented areas, and allowing greater contact with encroaching human or livestock populations that can serve as disease reservoirs (Daszak et al., 2000; Hess, 1996; Gog et al., 2002; McCallum & Dobson, 2002; Simonetti, 1995; McCallum & Dobson, 1995). But managing disease problems involving endangered species on fragmented habitats may be more complicated than simply reducing the degree of fragmentation, particularly when other disease hosts are present (e.g., Hess, 1996; Gog et al., 2002; McCallum & Dobson, 2002).

For instance, the ability of a disease to establish in a single population depends on the population's density because infectious contacts are more likely in a denser population. If a species is threatened or endangered (i.e., low numbers) or if sub-populations are isolated due to fragmentation, there may be too few infectious contacts for the disease to establish (though an endangered species on a diminutive area might be at most risk; McCallum & Dobson, 2002). This suggests some degree of fragmentation may actually be beneficial in the single-species case. The outlook is potentially quite different, however, when a host reservoir is also present. McCallum and Dobson (2002, p. 2042) define a host reservoir as a species "in which the pathogen is maintained with a less detrimental impact than that on the 'target' endangered host species . . . so that the force of infection can be maintained on the endangered host species as its density declines towards extinction." That is, diseases can more easily establish within an endangered population when a reservoir host is present because the reservoir provides a sustained risk of exposure (McCallum & Dobson, 2002; Dobson, 2004). Moreover, disease mortality can soar among the non-reservoir species (McCallum & Dobson, 1995, 2002; Dobson, 2004). Fragmentation may not halt infectious contacts and the disease could put an endangered species at significant risk.

Yet reducing fragmentation might not always be the answer. While reducing fragmentation leads to greater productivity of healthy members of endangered populations, this can also increase the number of infectious contacts – both among endangered species and host reservoirs such as domestic livestock. For instance, although habitat corridors, national parks, and wildlife reserves can mediate the impacts of fragmentation, many public lands are situated near livestock grazing lands or may be used for grazing themselves (Simonetti, 1995). Reduced fragmentation here increases the chance of contact with livestock, which serves as a reservoir for many diseases harmful to wildlife (Simonetti, 1995; Gaydos & Gildardi, 2004; Peterson et al., 1991; McCallum & Dobson, 1995, 2002).

Box 1 describes some examples of livestock reservoirs infecting threatened and endangered species. These examples come from low-income countries. Habitat fragmentation and disease threats, though they exist globally, can be negatively correlated with economic development (Ceballos et al., 2005). The developing world contains a major share of global biodiversity (ReNED, 2005) – terrestrial species are at most risk in high-density, moderately converted tropical biomes (MEA, 2005, p. 45) – which tend to be located in developing regions. Moreover, conservation resources and legislation are lacking in many developing nations – including measures to encourage the identification and control of diseases in domestic animals (ReNED, 2005; Ceballos et al., 2005).

Ecology and economics are both important for managing disease problems in the context of fragmentation and a host reservoir. Both disciplines matter here because both economic and ecological trade-offs arise in response to ecological investment options that could improve the situation, and because the incentives facing private landowners and ranchers may differ from those of a conservation authority. Incentives are particularly important because the risks of transmission between domestic and wild animals imply that private conservation efforts also matter for success, not just public investments (Povilitis, 1998; Simonetti, 1995). In general, we find that the role of economics as applied to wildlife disease risk given conservation and habitat fragmentation has been underexplored.

Box 1 – Motivating examples

Here we provide three examples of endangered species threatened by habitat fragmentation and a domestic disease reservoir. Control recommendations are indicated in Table 6.1.

Endangered Andean Deer in Chile

The Andean deer (*Hippocamelus bisulcus*), known commonly as huemul in South America, is a cultural symbol of Chile, appearing on its national coat-of-arms alongside the threatened Andean condor (*Vultur gryphus*). There is widespread interest in preserving and restoring the species in Chile. Huemul is found in greatest abundance (about 1,500 individuals) in Argentinian Patagonia (Povilitis, 1983; Smith-Fleuck & Fleuck, 1995). The northern part of the species' historic range has been reduced to a confined area of Central Chile called Nevados de Chillán and was estimated to have a population of about 60 individuals in 1997 (Povilitis, 1998).

Connectivity of habitats and disease risks are both areas of concern for Andean deer conservation. Huemul (and deer in general) are susceptible to *Cysticercus tenuicollis*, a bladderworm that is the larval stage of the canine and feline tapeworm *Taenia hydatigena*. *C. tenuicollis* is a common parasite of deer and other ruminant species (Pybus, 1990; Michigan Department of Natural Resources, 2002), and has been identified as a source of mortality for huemul (Texera, 1974; Simonetti, 1995; Povilitis 1998). *C. tenuicollis*, along with habitat fragmentation, is viewed as a major obstacle to population recovery (Simonetti 1995; McCallum & Dobson, 2002; Povilitis 1998). Experts are concerned with the cross-species transmission of *C. tenuicollis* to huemul from livestock (or other infected wildlife). Free-ranging domestic cattle and goats are known carriers of *C. tenuicollis* and several other pathogens that can be transmitted to huemul (Povilitis, 1998).

The Ethiopian Wolf

The Ethiopian wolf (*Canis simensis*) is found in Afroalpine landscapes throughout the Ethiopian highlands, with about half its total population in the

(*continued*)

Bale Mountains (Marino et al., 2006). This endangered canid feeds on an abundant supply of rodents in open meadows and grasslands in its optimal habitat. Following rainy seasons, pastoralists graze their livestock in the grasslands and meadows. While livestock may impact the quality of wolf habitat, the major threat from this human-wildlife system interaction is disease from domestic dogs used by livestock herders.

Canine parvovirus, canine adenovirus, rabies, and canine distemper virus are common diseases that threaten wild canid species and have as reservoir hosts domestic dogs (Laurenson et al., 1998). Sympatric domestic dogs that range across areas of high wolf density are "expected to increase the likelihood and potential severity of a disease epizootic in Bale" (Haydon et al., 2002; Laurenson et al., 1997; Marino et al., 2006, Sillero-Zubiri et al., 1996). There have been two separate outbreaks of rabies among Ethiopian wolves since observation of the packs in Bale Mountain National Park began, one in the Sanetti plateau and another in Web Valley. While the source of the Sanetti outbreak remains uncertain, the Web Valley outbreak was confirmed to be the same rabies serotype found in domestic dogs. Both outbreaks are suspected to have originated in domestic dogs that came into contact with wolves (Sillero-Zubiri et al., 1996).

Hirola and Others in Kenya and Somalia

The hirola (*Damaliscus hunteri*), commonly known as Hunter's hartebeest and Hunter's antelope, are found in dry grasslands over a limited area of southeast Kenya and contiguous lands in southernmost Somalia. Once common throughout the United Republic of Tanzania, Kenya, and Somalia, only an estimated 400 remain.

Rinderpest (German for "cattle plague") is arguably the epizootic disease that has had the single largest impact throughout Africa. A panzootic at the end of the nineteenth century swept across the entire continent killing millions of cattle and countless wild animals (Scott, 1981). Many anomalies in the distribution of wildlife in Africa are attributed to Rinderpest (Spinage, 1962). Rinderpest is a morbillivirus affecting cloven-hoofed artiodactyls and is now enzootic in domestic cattle and buffalo throughout northern equatorial Africa. Rinderpest is similar to Peste Des Petits Ruminants, which affect smaller ruminant animals (i.e., goats and sheep). The development of a vaccine has eradicated the disease from formerly affected areas like Europe, but periodic outbreaks continue to occur in Africa and southeast Asia, affecting both livestock and wildlife species.

The value of strategic cattle vaccination in areas where the disease remains endemic has been well demonstrated. Vaccination controlled cattle deaths and the spread of disease beyond the United Republic of Tanzania during a recent outbreak. Vaccination of cattle also protects wildlife. During the Tanzanian outbreak, vaccination prevented the spread of disease into the Serengeti system which was historically an enzootic area for Rinderpest and contains one of the most important migratory populations of wildlife on earth (Kock et al., 1999).

While vaccination has been successful in keeping disease at bay in domestic and wild species in most places, the only area where hirola is known to exist continues to be classified as an "infected zone" by the World Animal Health Organization (Office International des Epizooties, O.I.E., 2004). In a recent outbreak in Kenya, hirola were among the wildlife species included in a serosurveillance study to assess which species were affected (Kock et al., 1999). While hirola were not found to be infected in this particular study, some have suggested an outbreak in the animal's range during the early 1980s and may have been partially responsible for the decline of the species (Magin, 1996). Other rare animals potentially exposed during this outbreak include bongo (*Boocercus euryceus*), roan antelope (*Hippotragus equinus langheldi*), and Roosevelt sable (*H. niger rooseveltii*) (Kock et al., 1999).

This chapter investigates how *payments for environmental services* (PES) to private landowners and ranchers for the provision of various risk-reducing ecological investments can address the joint problem of wildlife habitat fragmentation and disease risks. PES have been applied to each of these problems separately. For instance, the Wildlife Conservation Lease (WCL) Program has paid landowners to provide corridors in the Kitengela plains to the south of Nairobi National Park (Gichohi, 2003). PES have been used to help construct the Mesoamerica Biological Corridor (MBC), an effort spanning eight countries from Mexico to Panama (Kaiser, 2001; Ewing, 2005). Other networks that use PES are under construction from the Yukon to Argentina, and in Brazil, Australia, Europe, and other locations (Kaiser, 2001). Payment programs have also been used to fund many livestock vaccination programs in the developing world, although with conservation being less of a concern than rural development and food security (Langa, 2001; Preslar, 1999; VSF, 2006). One example is the International Lookout for Infectious Animal Disease (ILIAD) program run by the Federation of American Scientists' Animal Health/Emerging Animal Diseases (AHEAD) project in sub-Saharan Africa. ILIAD focuses on communities located proximate to game reserves and national wildlife parks to confront diseases shared by wild and domestic animals, so as to aid rural communities and promote species preservation (Preslar, 1999).

Our analysis highlights the joint determination of economic and wildlife disease systems (see also Horan & Wolf, 2005), as disease risks – as with other ecological risks – are to some extent an endogenous function of human economic choices (Shogren & Crocker, 1991; Crocker & Tschirhart, 1992). We expand recent work on invasive species management and emphasize the benefits of allocating some PES for preventing new infections (i.e., livestock vaccination) in lieu of investing PES in *in situ* disease control or in habitat quality to increase species productivity (Leung et al., 2002; Finnoff et al., 2005). Earlier metapopulation models (Hess, 1996; McCallum & Dobson, 2002) indicated that investing solely in habitat connectivity could eventually increase disease risks. To counteract this risk, we find

Table 6.1 Examples of disease threats and recommended strategies

	Disease concern		
	C. tenuicollis	Canine parvovirus Canine adenovirus Rabies	Rinderpest
		Canine distemper	
Endangered species at risk	Huemul (*H. bisulcus*)	Ethiopian Wolf (*C. simensis*)	Hirola (*D. hunteri*)
			Bongo (*B. euryceus*)
			Roan antelope (*H. equinus langheldi*)
			Roosevelt sable (*H. niger rooseveltii*)
Domestic disease threat (reservoir)	Cattle and goats	Domestic dogs of livestock herders	Domestic cattle and buffalo
Protection strategies applied to:			
Wildlife	Create reserves	Establish buffer zones around wolf ranges	Conserve habitat and reduce fragmentation
	Establish habitat connectivity	Oral vaccination (controversial)	Anti-poaching activities
	Conservation management on private lands		Translocate animals outside of infected zone
Domestic animals	Vaccinate and treat infected livestock	Vaccinate domestic dogs	Vaccinate livestock
	Reduce livestock mobility	Buffer zones to prevent contact with wolves	Reduce infectious contacts (mobility)
	Habitat investments on grazing lands		

Sources: Huemul (Simonetti, 1995; Povilitis, 1998; Wikerhauser et al., 1971; Babiker & Eldin, 1987); Ethiopian Wolf (Sillero-Zubiri et al., 1996; Haydon et al., 2002); Hirola (Kock et al., 1999; O.I.E., 2004).

reallocating PES funds towards vaccination and reduced livestock movement becomes optimal after some degree of connectivity is achieved. No blanket subsidy toward habitat, vaccination, or livestock movement would work as well as the sequential portfolio approach of targeting habitat first, followed by vaccination and reduced livestock movement. The result is increased wildlife abundance, increased livestock health and abundance, and increased development opportunities. Bringing such an analysis to bear on endangered species facing livestock disease mortality risks in developing countries may provide useful insights to decision makers and funding agencies in a particular locale.

6.2 Prior Ecological Literature

The ecological literature on the joint problem of habitat fragmentation and disease is limited, whereas the separate literatures addressing each problem in isolation are substantial. This is perhaps surprising since the two problems share many analytical similarities. Consider first the fragmentation problem. The ecological literature (and the bioeconomic literature) addresses fragmentation through the use of metapopulation models. Researchers use these models to simulate population dynamics across a landscape consisting of multiple patches of land. Single-species models focus on productivity, mortality, intraspecific competition, and migration across different patches. Multispecies models add interspecific competition to this mix of ecological activities (and possibly economic, e.g., in the cases of poaching or land use). Given these activities, the essential questions are how much land is needed and in what configuration to support one or more wildlife populations. Ecological metapopulation models help determine alternative ways to accomplish this goal. Economic principles like cost effectiveness can be used to rank the "best" options given the many economic forces that also influence how the environment is managed and the economic and ecological feedbacks between these jointly determined systems (e.g., Parkhurst & Shogren, 2006; Sanchirico & Wilen, 2005).

Wildlife disease models deal with analogous issues. Even a single-host species consists of multiple, interacting subpopulations (susceptible, infected, and recovered) that can be modeled in a metapopulation framework. The basic question asked here is how to manage the various populations so the infected population is reduced or eliminated – the opposite of what is asked in traditional conservation models. Accordingly, management options differ but the ecological principles are similar. In contrast to the fragmentation literature, few economic analyses have addressed wildlife disease problems (e.g., Bicknell et al., 1999; Horan & Wolf, 2005), and none have addressed both fragmentation and disease.

Given the many similarities between habitat fragmentation models and disease models, we now describe a basic ecological model of disease transmission within a multi-host (e.g., wildlife and livestock) system. This sets the stage for understanding ecological and economic trade-offs that are important when allocating scarce resources, via subsidy payments, for wildlife conservation. For simplicity, we focus on the two-species case, although the model is generalizable to n species (Roberts & Heesterbeek, 2003; Dobson, 2004). We first focus on interactions on a single patch of land; we then expand the model to include multiple patches.

Suppose both a livestock and wildlife population inhabit a particular patch of land. Each population has three subpopulations: *susceptible* (but presently healthy), *infected*, and *recovered* (healthy and immune). Changes in these populations can be expressed mathematically or graphically, as in Fig. 6.1. Susceptible populations of each species expand due to births and due to lost immunity of animals within the recovered population, and they contract due to (natural or harvest-related) mortality and due to infectious contacts with infected animals of either species. Infected populations grow as new animals become infected, and these populations decline as

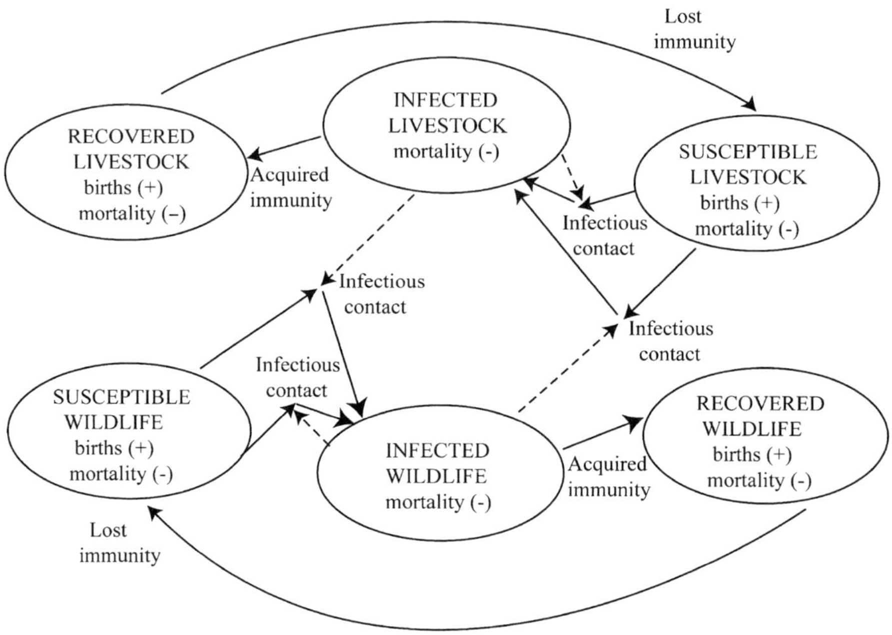

Fig. 6.1 Population dynamics for two hosts. Courtesy of Richard Horan, Jason Shogren, and Benjamin Gramig

animals recover or experience mortality. Finally, the recovered population expands due to acquired immunity (after infection) and contracts due to lost immunity or (natural or harvest-based) mortality.

Ecologists use a "next-generation" matrix to measure the capability of a pathogen to invade a healthy but susceptible multihost system (Diekmann et al., 1990). For our example of one livestock population (indexed by L) and one wildlife population (indexed by W), the next-generation matrix takes the form

$$K = \begin{bmatrix} K_{LL} & K_{LW} \\ K_{WL} & K_{WW} \end{bmatrix}. \tag{6.1}$$

Expression (6.1) takes into account that transmission that can occur both within and between species. For example, K_{LL} (K_{WW}) represents the expected number of secondary cases of infection in livestock (wildlife) that would arise from an initial infection within the susceptible livestock (wildlife) population; K_{LW} (K_{WL}) represents the expected number of secondary cases of infection in livestock (wildlife) that would arise from an initial infection within the susceptible wildlife (livestock) population (Diekmann et al., 1990). More specifically, K_{ij} represents the rate of transmission from the first infected member of species j to species i (i.e., number of infectious contacts per unit time) times the average duration of infection within host type j (Diekmann et al., 1990; Heesterbeek and Roberts, 1995; Dobson, 2004).

The average duration of infection is inversely related to mortality and disease recovery. A more easily transmitted disease increases K_{ij}; either greater mortality or faster recovery reduces K_{ij}. For the typical diseases with density-dependent transmission, each element K_{ij} is a function of the equilibrium (or steady-state) population densities that occur in the absence of disease, as this is the appropriate reference point for measuring the capability of a pathogen to invade a susceptible host system.

Even when focusing on a single wildlife species and a single livestock species, members of each species may be spatially distributed across a number of contiguous "patches" that make up the landscape. Accordingly, each species is treated as a set of spatially interacting populations, in which a particular population is defined as an individual species occupying a particular "patch" of land. With Z such populations (made up of potentially many livestock and wildlife species), the next-generation matrix takes the form

$$K = \begin{bmatrix} K_{11} & K_{12} & \dots & K_{1Z} \\ K_{21} & K_{21} & \dots & K_{2Z} \\ \vdots & \vdots & \ddots & \vdots \\ K_{Z1} & K_{Z1} & \dots & K_{ZZ} \end{bmatrix}, \tag{6.2}$$

where each element K_{ij} is defined as above.

Because the next-generation matrix K describes how a pathogen invades a non-infected host system, it provides clues on how to control or eliminate the pathogen from the host-pathogen system. The simplest case is when one of the *off-diagonal* elements is zero. For instance, if $K_{WL} > 0$ and $K_{LW} = 0$ in expression (6.1), the livestock transmits the disease to wildlife but not vice versa. Livestock is the disease reservoir in this case. Once wildlife become infected, however, further infection within the wildlife host is essentially driven by within-wildlife host-pathogen dynamics. The host-pathogen dynamics are almost independent across species (Dobson, 2004). The only way to permanently eliminate wildlife infections is to eliminate the disease reservoir within the livestock population. The opposite holds when $K_{LW}> 0$ and $K_{WL} = 0$ – disease spreads to livestock from wildlife.

If each element of K is non-zero, the host-pathogen dynamics are dependent across populations. Dobson (2004) defines the force of infection that population j exerts on population i as

$$\xi_{ij} = [K_{ij} + K_{jj}] \times 1/[\text{average duration of infection within host type } j]. \tag{6.3}$$

The force of infection is the sum of the rates of transmission from population j to all populations. One calculates this expression for each population to determine which populations make the largest contribution to disease outbreaks.

Roberts and Heesterbeek (2003) illustrate an alternate, slightly more complicated approach that can lead to additional insight. They use K to determine a host-specific disease reproduction ratio, R_{0i}, which is the expected number of all secondary cases resulting from a primary case in susceptible host i. They determine which population(s) is a reservoir for the disease. Focusing controls on this population(s), they use K to determine the constant time-invariant level of control (e.g., through vaccinations or culling) that can eradicate the disease in the reservoir host(s).

6.2.1 Economic-Ecological Trade-Offs

Recommendations based on the next-generation matrix K can be misplaced if they do not consider economic aspects of the problem. Incorporating economic principles into ecological models is important for two reasons. First, economic signals drive how people behave, such as farmers and poachers whose actions may influence disease dynamics. Second, scarce resources devoted to disease control are diverted away from other environmental or social problems. An explicit consideration of economic trade-offs helps society achieve its goals more efficiently, freeing up resources for other socially valued endeavors, e.g., education, nutritional programs, and health care.

The elements of the next-generation matrix K, as applied in the ecology literature, are exogenously determined, based on ecological relations and based on the assumption of a pre-disease steady state. Accordingly, if the K matrix is used to calculate the force of infection, the reservoir host, or the required level of control, these items are also exogenously determined, based on ecological relations and based on the pre-disease steady state. For instance, conventional models treat the required level of control as a time-invariant, exogenous function of ecological relations. In our view, this is an oversimplification, however, because it does not address the endogeneity of human choices that affect the host-pathogen system, as well as the associated ecological-economic feedbacks that drive the system.

Consideration of economic choices has several important impacts on the next-generation matrix K and its suitability for use in developing management recommendations. First, the matrix K is no longer exogenously determined by ecological relations because the non-infected steady state, on which K is based, is an endogenous function of economic choices. Harvesting and habitat management choices result in equilibrium values that differ from carrying capacity. Second, management choices may change over time during the post-invasion (control) phase in response to relative costs and benefits, which would reflect both market conditions and economic and ecological feedbacks between the two systems.

The overall level of control is also a choice variable. Management recommendations based on K generally advocate strategies to eradicate the pathogen in reservoir populations. But such an eradication approach might be too costly relative to the associated benefits (Horan & Wolf, 2005). Both economic and ecological factors determine the efficient allocation of controls across populations and the efficient overall amount of control.

Finally, the next-generation matrix K may be of limited guidance when disease control is only one element of a broader problem. The focus of this chapter – species conservation when a species is at risk due to both disease and habitat fragmentation – is one example in which this holds with particular force. Here the most efficient management choices might not target only the reservoir populations, as advocated by Roberts and Heesterbeek (2003); rather, efficient management might target both the reservoir population (e.g., vaccination) and the non-reservoir population (e.g., investing in habitat connectivity).

6.2.2 Management Options

The majority of prior disease management research has examined alternative situations in which wildlife (e.g., deer, badgers, possums) are not threatened but rather serve as the host reservoir that puts livestock (or human) populations at risk of infection (Daszak et al., 2000; Heesterbeek & Roberts, 1995; Roberts & Heesterbeek, 2003; Bicknell et al., 1999; Horan & Wolf, 2005; Fenichel & Horan, 2007; Barlow, 1991). When livestock populations are infected with diseases bearing sufficient production or mortality costs, there are both public and private incentives to implement on-farm control strategies. Vaccination is sometimes used, but not always due to vaccine unavailability or ineffectiveness, high cost – particularly in low-income countries (VSF, 2006; Preslar, 1999), or the risk of trade sanctions (e.g., foot-and-mouth disease; USDA-APHIS, 2002).[3] For many diseases, the primary livestock-focused strategies involve quarantines and "stamping out" infected livestock herds to stop additional spread to healthy domestic herds.

The situation differs for a livestock reservoir infecting endangered wildlife, the example we consider herein. Few (if any) private, market-based incentives exist to address problems in which livestock are primarily a reservoir that suffers little disease mortality or productivity loss – often the case for reservoir populations. The opposite story could also be told, in which PES are applied to problems in which wildlife affects livestock. Here conservation may be less of an issue, and so payments based solely on environmental benefits (as opposed to rural development benefits) might be comparatively small.

In the case of a livestock reservoir, vaccination of livestock (when available) may make more sense than culling. The strategy of culling reservoir livestock populations may impose excessive costs on ranchers in low-income countries that cannot afford to indemnify ranchers at market value, if at all. Even in developed countries, compensation for mandatory culling of livestock typically does not cover all of the private costs of depopulation, potentially creating an obstacle to early

[3] For instance, vaccinations for foot-and-mouth disease exist but are not widely used ex ante because: (1) they only protect against clinical signs of the disease and not the disease itself, making it harder to detect an actual outbreak; (2) they must be developed for particular strains of the disease, and may be ineffective against new outbreaks; (3) they are costly to administer, particularly when risks are low; and (4) countries that vaccinate are not considered FMD-free by trading partners, and so trade sanctions could ensue (USDA-APHIS, 2002).

reporting of infection (Gramig et al., 2006). Obstacles to early reporting may be exacerbated when vouchers issued for government compensation payments are perceived as incredible (McNeil, 2006). In contrast, subsidized vaccinations may now make sense because their use would not induce trade sanctions if (1) the market is otherwise unresponsive to the disease in question, or (2) sanctions on livestock-sector exports already exist due to other disease problems having greater market impacts – a common occurrence in many low-income countries (e.g., Perry et al., 2005). In the absence of significant disease-related productivity losses, the public sector may have to create or administer the needed incentives for a vaccination program.

Now consider wildlife interventions. Wildlife populations are seldom vaccinated because vaccines are unavailable or ineffective for many diseases (Smith & Cheeseman, 2002). Even when an effective vaccine does exist, it is controversial to vaccinate threatened and endangered species – in part due to the extinction of a study population of Serengeti wild dogs following interventions that used vaccination (Burrows, 1992; Burrows et al., 1994). The primary interventions for non-threatened species include culling the wildlife host population (e.g., Roberts & Heesterbeek, 2003; Smith & Cheeseman, 2002) and altering effective habitat availability to reduce density and infectious contacts (Hickling, 2002; Horan & Wolf, 2005; Fenichel & Horan, 2007). But reducing the density of an endangered population (via culling or translocation) to reduce infectious contacts may only put the population at greater risk. Culling is therefore not considered a viable intervention in the present case.

Changes in habitat availability, in contrast, may have multiple opposing effects. Greater habitat area or reduced habitat fragmentation can increase disease risks for endangered species, to the extent that there is a concomitant increase in infectious contacts between and within endangered species and a reservoir host (Hess, 1996; McCallum & Dobson, 2002).[4] Alternatively, McCallum and Dobson (2002) find that reducing habitat fragmentation (or increasing habitat connectedness) can increase the productivity of healthy animals. They show that some degree of increased habitat connectedness could simultaneously improve a species' chances for survival and increase the proportion of healthy animals within the population.

The parameters of McCallum and Dobson's (2002) model define the circumstances under which increased connectedness is beneficial. They investigate various combinations of fixed parameter values and illustrate why three parameters are critical: (1) the colonization rate of endangered species relative to reservoir hosts (livestock in our model), (2) the rate of infectious contacts between livestock and wildlife, and (3) the rate of recovery from infection (to an immune or recovered

[4] Similar trade-offs may impact farmers capable of habitat provision. Farmers appreciate income from conservation payments. This conservation, however, may put their own livestock at risk, as it is also possible that wildlife could be a reservoir for other diseases that could adversely affect livestock. The presence of or the potential for disease reservoirs in wildlife could make it harder to encourage private habitat investments. Habitat provision may also increase other wildlife conflicts such as livestock predation or crop damage.

state) within the livestock population. But these and other parameters are not fixed in reality – investments can be made to affect many parameter values, thereby altering the benefits of increased habitat connectedness. This holds when domestic animals are part of the ecological system. For instance, management of livestock movement affects the relative rate of recolonization, biosecurity efforts reduce infectious contacts, and livestock health management prevents infection and improves recovery rates. Farmers are unlikely to make any of these investments at socially efficient levels because doing so is costly, and they do not capture the full social benefits. Effective conservation may therefore require the public provision of incentives for these investments, such as PES.[5] Since provisioning payments to farmers requires reallocating public (and sometimes private for international conservation aid) funds away from other public good uses (e.g., public health, education), efficient conservation requires the PES to be allocated to align with the relevant economic and ecological trade-offs.

6.3 A Metapopulation Model of Disease and Habitat Fragmentation

We now develop a metapopulation model to investigate these issues in the context of a disease-habitat fragmentation problem. We follow the classic metapopulation modeling approach developed by Levins (1969), as this is the conventional approach to exploring disease-habitat fragmentation issues in ecology (e.g., Hess, 1996; Gog et al., 2002; McCallum & Dobson, 2002). Individual patches are initially assigned a status. In this framework, the relevant state variables are taken to be either the number or proportion of patches in each possible designation of health status and species occupancy.[6] Although the state variables are defined in terms of patches instead of populations, the next-generation matrix K can easily be adapted to this framework since this modeling framework captures basic ecological

[5] We assume disease impacts to reservoir populations are comparatively mild. But even if they were significant, a farmer would underinvest because he or she could not capture all the social benefits in the market price.

[6] An alternative approach is to model explicitly the dynamics of interacting susceptible, infected, and resistant populations on multiple patches of land, with migration between patches. Environmental and economic variables are taken to be homogeneous within a patch but heterogeneous between patches (e.g., Sanchirico & Wilen, 1999, 2005). This approach permits the most ecological and economic detail, particularly when the number of patches is large, but it comes at a price paid in more computation and less transparency. If n species live in one of three subpopulations (S, I, R) on each of N patches, the bioeconomic analysis must account for economic and ecological trade-offs arising among $3 \times n \times N$ interacting populations and possibly as many control variables. The solution could be so complex as to obscure any insight. The approach we adopt is more tractable, but it comes at a cost of reduced ecological and economic detail, as an underlying assumption is that the (ecological and economic) environment is homogeneous across the landscape.

processes. Our discussion and critique of the next-generation matrix remains relevant under this modeling framework.

We develop a slight variation of McCallum and Dobson's (2002) model, which provides opportunities for comparisons between bioeconomic and strictly ecological approaches. McCallum and Dobson's model is motivated by the problem of *C. tenuicollis* in Andean deer (Huemul; see Box 1), although it is a hypothetical simulation designed to provide a qualitative representation of the situation to produce insights into the problem.[7]

Assume there are N available patches, which are identical in species' carrying capacities and distances from each other. Each patch may be in one of seven states at time τ[8]

1. Empty. The proportion of patches in this state is E.
2. Occupied only by susceptible livestock; the proportion of patches is S_L.
3. Occupied only by susceptible wildlife; the proportion of patches is S_W.
4. Occupied only by susceptible livestock and wildlife; the proportion of patches is S_{LW}.
5. Occupied only by resistant or immune livestock; the proportion of patches is R.
6. Occupied only by resistant livestock and susceptible wildlife; the proportion of patches is R_W.
7. Occupied only by infected livestock (assume any infected wildlife within a patch immediately go extinct); the proportion of patches is I.

Wildlife never recovers from infection to become resistant. We also assume vaccination of wildlife is either unavailable or infeasible, which is common for many wildlife disease problems. These assumptions imply there are no patches containing resistant wildlife, and that livestock is the disease reservoir.

First consider factors influencing the status of livestock patches – susceptible (S), resistant (R), or infected (I). The rates of extinction of resistant and infected livestock patches are denoted X_R and X_I with $X_I > X_R > 0$. Following McCallum and Dobson (2002) and Hess (1996), these and all other rates in the model are expressed relative to the rate of extinction in susceptible livestock patches (denoted x_s), so that all rate variables are dimensionless. The term "extinction," when made in reference to a livestock patch, does not refer to a local biological extinction in the same sense as an endangered species; rather, it means the removal of livestock from one patch without relocation into another patch within the region. This would occur

[7] Their model extends Hess (1996) and Gog et al. (2002) to include reservoir species. The model is also relevant for the other problems presented in Box 1. Their use of hypothetical simulation is due to the lack of data, which is common among wildlife disease problems.

[8] Hess (1996) begins his analysis with a simple metapopulation model that models only between patch dynamics involving uniform patches (an island model), but then moves on to consider a more complicated model that considers dynamics both within and between patches, as well as different spatial configurations. The problem of Andean deer in Chile is more like the necklace configuration examined by Hess. But, at least when modeling a single host, Hess's results for the island and necklace models are qualitatively similar.

when a herd is either culled or sold. Migration of livestock across patches occurs at the rate M_L. If susceptible livestock enter an empty patch, that patch switches to susceptible livestock status. If infected livestock enter a patch of susceptible livestock, that patch has a probability δ of becoming infected. Resistant livestock entering a patch do not affect the patch status and neither do infected livestock entering a resistant patch. Infected patches recover to a resistant state at a rate Γ. Finally, susceptible livestock become resistant (e.g., due to vaccination) at a rate Ψ, and resistant patches revert back to susceptible status (as resistance is lost) at a decay rate Λ.

Now consider wildlife. The migration rate of endangered species is M_W. McCallum and Dobson (2002) refer to M_W as a measure of *wildlife habitat connectedness*, or an inverse measure of *wildlife habitat fragmentation*. For example, $M_W$is large for large contiguous reserves and small for fragmented patches. Endangered species become extinct on any patch at a rate X_W. They immediately become extinct on a patch if the livestock on their patch becomes infected or if an infected immigrant arrives. These assumptions imply that livestock is the disease reservoir species. The disease literature recommends targeting livestock to eradicate the disease. The habitat fragmentation literature, however, recommends investing in habitat connectivity. We show herein the bioeconomic solution is not as clear-cut. Livestock are not targeted at all under some situations, and the disease is not always eradicated.

The metapopulation dynamics of patches of susceptible livestock is given by the equation of motion

$$\begin{aligned}\frac{dS_L}{d\tau} &= M_L(S_L + S_{LW} + R + R_W)E - S_L - M_L\delta I S_L - \Psi S_L + \Lambda R \\ &\quad - M_W(S_W + S_{LW} + R_W)S_L + X_W S_{LW},\end{aligned} \tag{6.4}$$

where $E = 1 - (S_L + S_W + S_{LW} + R + R_W + I)$. The first right-hand side (RHS) term represents the colonization of empty patches by livestock migrants of other susceptible patches, i.e., it equals the migration rate of livestock, times the proportion of patches containing susceptible and resistant livestock, times the probability that these animals colonize empty patches. The second RHS term reflects extinction of susceptible livestock patches (e.g., due to sales). The third term is the proportion of susceptible livestock patches that become infected. The fourth term is the proportion of susceptible patches that become resistant due to vaccination, and the fifth term reflects the loss of resistance The sixth term is the proportion of susceptible livestock patches that become susceptible patches containing both species, i.e., it equals the migration rate of wildlife, times the proportion of patches containing wildlife, times the probability that these wildlife move onto patches of type L. Finally, the seventh term reflects switching of LW-type patches to L-type patches due to local extinctions of wildlife within those patches.

We define the equations of motion for the other state variables in a manner similar to (6.4):

$$\begin{aligned}\frac{dS_W}{d\tau} &= M_W(S_W + S_{LW} + R_W)E - X_W S_W - M_L\delta I S_W \\ &\quad - M_L(S_L + S_{LW} + R + R_W)\ S_W + S_{LW} + X_r R_W\end{aligned} \tag{6.5}$$

$$\frac{dS_{LW}}{d\tau} = M_L(S_L + S_{LW} + R + R_W)S_W + M_W(S_W + S_{LW} + R_W)S_L - M_L\delta IS_{LW} - \Psi S_{LW} + \Lambda R_W - (1 + X_W)S_{LW} \quad (6.6)$$

$$\frac{dR}{d\tau} = -R(X_r + \Lambda) + \Gamma I + X_W R_W - M_W(S_W + S_{LW} + R_W)R + \Psi S_L \quad (6.7)$$

$$\frac{dR_W}{d\tau} = M_W(S_W + S_{LW} + R_W)R - (X_r + X_W + \Lambda)R_W + \Psi S_{LW} \quad (6.8)$$

$$\frac{dI}{d\tau} = I(M_L E - X_i + \delta M_L(S_L + S_W + S_{LW}) - \Gamma). \quad (6.9)$$

The ability of wildlife to invade depends on the conditions present when wildlife are not initially present. We can calculate the equilibrium values S_L^*, R^*, I^*arising when $S_W = S_{LW} = R_W = 0$. The values depend on parameter values and also economic variables (vaccination and habitat connectivity, which we use as decision variables in the next section). These equilibrium values are plugged back into (6.5), (6.6), and (6.8) to determine the condition under which the number of wildlife patches increase from an initial value of zero, $d(S_W + S_{LW} + R_W)/d\tau|_{S_W=S_{LW}=R_W=0} > 0$, which is

$$M_W > X_W + \delta M_L I^*. \quad (6.10)$$

When the pathogen is absent, the endangered species is independent of the livestock species. Wildlife may invade now only if connectivity exceeds the extinction rate. When the pathogen is present, connectivity must be greater for wildlife to invade. The need for greater connectivity, however, can be tempered by investments in vaccination that reduce I^*.

Similarly, the ability of a pathogen to invade depends on the conditions present when no pathogen exists initially. We can calculate the equilibrium values of the state variables that arise when $I = 0$, and then plug these equilibrium values back into (6.9) to determine the condition under which $dI/d\tau|_{I=0} > 0$, which is

$$M_L > \frac{\Gamma + X_i}{1 + (1-\delta)(S_L^* + S_{LW}^* + S_W^*) + R^* + R_W^*}. \quad (6.11)$$

From (6.11), we find that equilibrium population levels are a function of economic choices. Less livestock connectivity is required for the pathogen to establish when there are more susceptible populations, while more livestock connectivity is required when there are more resistant populations. Wildlife habitat investment leads to more wildlife, making it easier for the disease to establish. Vaccination makes it harder for the disease to establish.

6.4 Incorporating Economic Choices

There are two differences between our model and McCallum and Dobson's (2002) model. First, we allow for preventative veterinary medicine (vaccination) to generate resistance within susceptible livestock.[9] Research suggests that vaccination is an effective direct preventive measure against *C. tenuicollis* in livestock (Wikerhauser et al., 1971; Babiker & Eldin, 1987); and an effective indirect tool to protect the huemul that comes into contact with livestock in grazing areas where the pathogen can be transmitted. Second, we allow wildlife and livestock habitat connectedness, M_W and M_L, to be endogenously affected by habitat management choices. For ease of notation and without loss, assume all landowners are ranchers who make investment decisions in habitat connectivity and livestock vaccination. Assume the existence of a representative rancher in each livestock-inhabited patch, and that ranchers are homogeneous across patches (at least, in the absence of any patch-specific subsidies they might receive).

The level of vaccination within a type k patch ($k = L, LW$) is Ψ_k. This is strictly a preventative measure. Vaccination does not directly affect the *in situ* productivity of the endangered species population, but it does provide indirect benefits by reducing disease propagule pressure. Net vaccination costs (i.e., actual vaccination costs less the private benefits of vaccination) in a type-k patch at time τ are given by $c(\Psi_k)$, which are increasing and convex in the level of vaccination, with $c(0) = 0$. Assume ranchers invest nothing in vaccination unless publicly provided incentives exist (i.e., $c'(0) = 0$) because they do not otherwise internalize any positive net benefits from vaccination. Many ranchers in poor areas cannot afford vaccination given other more pressing needs (Preslar, 1999; VSF, 2006; Langa, 2001).

Denote patch-specific investments in increased wildlife habitat connectedness by $Z_{Wk}(k = S_W, S_{LW}, R_W)$, so migration becomes patch-specific: $M_{W,k} = M_{W0} + Z_{Wk}$, where M_{W0} is the pre-investment level of migration. Wildlife habitat connectivity investments establish corridors that connect patches of ideal habitat for the target species. In practice, this may involve agreeing not to fence, quarry, subdivide, or use slash-and-burn agriculture, and to manage the land for wildlife and sustainable livestock grazing (Gichohi, 2003; Kaiser, 2001). We allow habitat connectivity investments to be targeted to facilitate certain types of desired wildlife migrations, as will be determined by the bioeconomic model.

[9] Veterinary medicine can also increase the rate of livestock recovery from infection, Λ, but prevention of disease occurrence is the only way to avoid costs associated with the loss of endangered species. Biosecurity, under some situations, is also a preventative measure ranchers could invest in to reduce the rate of infectious contact between wild and domestic species. This usually involves separating wildlife from livestock by a physical barrier (fences) or some other means. Since livestock in close proximity to Chilean parks tend to be free ranging (Povilitis, 1998), physical separation is not straightforward unless wholesale cultural and production system changes are made – changes that would probably be untenable at least in the short run. We take these systems as given and do not consider biosecurity as a choice variable.

We also consider investments in livestock connectivity. Specifically, we consider investments that *reduce* migrations out of infected patches, as this is the only source of infectious contacts in the model. Denote the patch-specific decrease in habitat connectedness for infected livestock by Z_{LI}, so that $M_{L,I} = M_{L0} - Z_{LI}$, where M_{L0} is the pre-investment level of migration. This decreased connectedness is costly. For instance, livestock in close proximity to Chilean parks tend to be free ranging (Povilitis, 1998). Actions to reduce the mobility of infected animals would be costly – particularly if infections have little impact on livestock productivity.

Habitat connectivity costs are denoted by g_i, and these are increasing in the level of changes in habitat connectivity, with $g_i(0) = 0$. These are net costs, as constructing habitat corridors can yield some benefits such as reduced human-wildlife conflicts (Gichohi, 2003), and restricting movement of infected livestock can reduce infectious contacts with susceptible herds.[10] Assume ranchers invest nothing in additional habitat connectivity (or disconnectivity, in the case of livestock) unless publicly provided incentives exist (i.e., $c'(0) = g_i'(0) = 0$). Given these specifications for habitat and vaccination costs, total rancher costs in a type-k patch at time τ are given by the separable cost function, $C_k = c(\Psi_k) + \sum_{i=M,L} g_i(Z_{ik})$.

6.5 A Bioeconomic Model

We develop a bioeconomic model of livestock health management and wildlife mobility to examine resource allocation when the presence of endangered wildlife has social benefits and conservation funds are in short supply. Suppose an agency – governmental or non-governmental organization (NGO) – values the wildlife population in any given period by $U(S_W + S_{LW})$ (with $U' > 0, U'' < 0$), which can be thought of as a combination of existence and ecotourism values. To promote this value, the agency subsidizes private investments in habitat connectivity and livestock vaccination. The subsidy rate for the jth investment ($j = \Psi, Z_i$) in patch k is denoted σ_{jik}, so that post-subsidy costs in patch k are[11]

$$[c(\Psi_k) - \sigma_{\Psi k}\Psi_k] + \sum_{i-W,L} [g_i(z_{ik}) - \sigma_{Z,ik}Z_{ik}]. \tag{6.12}$$

[10] The benefits of reduced human-wildlife conflicts may be largely external to an individual farmer with limited landholdings. Rather these benefits emanate from the joint habitat investment decisions by many farmers in an area.

[11] We assume vaccination subsidies go to ranchers, though it would make no difference to assume veterinarians were paid. Except for distributional differences, the same outcome should arise regardless of who is paid, provided that animal health providers certify herds are vaccinated in order for payments to be made. But we do note many existing payment programs actually fund local veterinarians and community members enlisted to provide animal health services (Preslar, 1999; VSF, 2006; Langa, 2001).

Minimization of (6.12) leads to the first-order conditions and response functions:

$$c'(\Psi_k) - \sigma_{\Psi,k} = 0 \Rightarrow \Psi_k(\sigma_{\Psi,k}),\ k = S_L, S_{LW} \tag{6.13}$$

$$g_i'(Z_{ik}) - \sigma_{Z,ik} = 0 \Rightarrow Z_{ik}(\sigma_{Z,ik}),\ k = \begin{cases} S_W, S_{LW}, R_W & \text{for i = W} \\ I & \text{for } i = L \end{cases}. \tag{6.14}$$

Assume the agency is concerned with intertemporal efficiency in managing the wildlife resource and disease risks. Given a discount rate of φ, the agency chooses subsidy rates to solve

$$\begin{aligned} \underset{\sigma_{\Psi,k},\sigma_{Z,ik}}{Max} \int_0^\infty & \left[U(S_W + S_{LW}) - \sum_k k \left[c(\Psi_k(\sigma_{\Psi,k})) + \sum_{i=W,L} g_i(Z_{ik}(\sigma_{Z,ik})) \right] \right. \\ & \left. - \beta \sum_k k \left[\sigma_{\Psi,k} \Psi_k(\sigma_{\Psi,k}) + \sum_{i=W,L} \sigma_{Z,ik} Z_{ik}(\sigma_{Z,ik}) \right] \right] e^{-\varphi\tau} d\tau \end{aligned} \tag{6.15}$$

for $k = S_W, S_{LW}, R_W, I,$ subject to the equations of motion (6.4)–(6.9) and the rancher's response functions defined by (6.13) and (6.14). The term $\beta \sum_k k$ $[\sigma_{\psi,k}\psi_k(\sigma_{\psi,k}) + \sigma_{Z,k}Z_k(\sigma_{Z,k})]$ represents the social welfare impacts of the subsidy payments – this captures the opportunity cost notion that allocating more subsidies to this conservation problem implies less money is available for other conservation or public goods investments. The parameter β represents the (constant) marginal cost of diverting funds to this conservation problem, which may include transactions costs (Alston & Hurd, 1990).[12]

The Hamiltonian associated with this problem is

$$\begin{aligned} H = U(S_W + S_{LW}) & - \sum_k k \left[c(\Psi_k(\sigma_{\Psi,k})) + \sum_{i=W,L} g_i(Z_{ik}(\sigma_{Z,ik})) \right] \\ & - \beta \sum_k k \left[\sigma_{\Psi,k} \Psi_k(\sigma_{\Psi,k}) + \sum_{i=W,L} \sigma_{Z,ik} Z_{ik}(\sigma_{Z,ik}) \right] + \mu_{S_L} S_L \\ & + \mu_{S_W} \dot{S}_W + \mu_{S_{LW}} \dot{S}_{LW} + \mu_R \dot{R} + \mu_{R_W} \dot{R}_W + \mu_I \dot{I} \end{aligned} \tag{6.16}$$

[12] These transactions costs could include rancher education on the benefits of vaccination. We do not explicitly model monitoring and enforcement problems, although they would exist in any payment program (and also in command-and-control programs). Existing programs that pay for habitat connectivity and vaccinations employ personnel for program monitoring (Gichochi, 2003; Preslar, 1999; VSF, 2006). If the associated expenses are fixed regardless of the level of payments (e.g., one worker per participating community), these costs would not affect the optimal plan. If these costs are proportional to the level of payments, they could be captured by *b*.

where μ_j represents the costate variable associated with the jth state variable. The first-order conditions associated with an interior solution are

$$\frac{\partial H}{\partial \sigma_{\Psi,k}} = 0,\ k = S_L, S_{LW} \tag{6.17}$$

$$\frac{\partial H}{\partial \sigma_{Z,ik}} = 0,\ k = S_W, S_{LW}, R_W, I. \tag{6.18}$$

For instance, in the case of $k = S_{LW}$, conditions (6.13) and (6.14) can be used to rewrite conditions (6.17) and (6.18) as

$$c'(\Psi_{LW}) = \frac{\mu_{R_W} - \mu_{S_{LW}}}{[1 + \beta(1 + 1/\varepsilon_\Psi)]} \tag{6.17a}$$

$$g'_{S_W}(Z_{S_W}) = \frac{\mu_{S_W}E + [\mu_{S_W} - \mu_{S_L}]S_L + [\mu_{R_W} - \mu_R]R}{1 + \beta(1 + 1/\varepsilon_Z),} \tag{6.18a}$$

where ε_Ψ is the elasticity of supply of vaccination and ε_Z is the elasticity of supply of habitat connectivity. Condition (6.17a) equates the marginal cost of vaccination (which equals the subsidy rate) with a net price defined by the ratio of marginal external benefits of vaccination relative to the marginal costs of subsidization. The numerator is the net marginal value of an increase in vaccination in type-*LW* patches, $[\mu_{R_W} - \mu_{S_{LW}}] > 0$, which converts these susceptible patches into resistant patches. The denominator represents the marginal costs of vaccination in type-*LW* patches, as a larger subsidy reduces the funds available for other conservation activities. The denominator captures how the agency acts like a monopsonist in the market for vaccination (with $1/\varepsilon_\Psi$ representing the degree of monopsony power). The optimal subsidy for vaccinations in type-*L* patches is derived in a similar manner, but will generally take on a different value due to a different value in the numerator.

Condition (6.18a) has a similar interpretation. The RHS is the net marginal external benefit of increased connectivity divided by the marginal costs of subsidization. The numerator represents the marginal net benefits of adding wildlife to non-infected patches (a productivity effect). For each type of possible conversion of a non-wildlife patch to one that includes wildlife, the marginal benefits are calculated as the net price of conversion, times the proportion of the associated non-wildlife patches. Though we have allowed the subsidies $\sigma_{Z,ik}$ to be patch-specific, it can be verified that these should be applied uniformly. Uniform application arises because connectivity is defined by outward migration, and once wildlife leaves one patch they are free to move to any other patch.

Finally, the following adjoint conditions are also necessary for an optimum

$$\dot{\mu}_j = \varphi\mu_j - \frac{\partial H}{\partial j},\ j = S_L, S_W, S_{LW}, R, R_W, I. \tag{6.19}$$

These conditions in (6.19) ensure no intertemporal arbitrage opportunities arise (see Clark, 1990), and they link the costate variables to the marginal utility of healthy populations.

6.6 Numerical Example

The ecological component of our numerical example is based on the example in McCallum and Dobson (2002), which allows us to compare directly our bioeconomic model with the ecological-only model. Their numerical example is based on a series of equilibrium equations identical to steady-state versions of our equations (6.4)–(6.9), except that (1) $\Psi_k = Z_{ik} = 0$ in their model, and (2) they specify the following proportional relation between the connectivity parameters M_L and M_W: $M_W = M_\varsigma M_L$, where M_ς is a scaling parameter. We also use a scaling parameter for connectivity, but only for the initial value of wildlife connectivity: $M_{W0} = M_\varsigma M_L$. Investments in Z decouple M_L and M_W from the proportionality constant. This means wildlife take advantage of greater connectivity and ranchers do not, even though opportunities may exist for ranchers to do so. Subsidies for increased habitat connectivity can therefore be viewed as having a secondary impact of effectively reducing livestock connectivity, at least to some extent.

We focus on McCallum and Dobson's (2002) scenario a, as that scenario best illustrates a case in which an endangered species can survive with only moderate connectivity.[13] Figure 6.2 presents our derivation of their results (using their parameter values, with $\Psi_k = Z_{ik} = 0$). The figure illustrates equilibrium outcomes for different values of $M_W = M_{W0}$. No disease outbreak occurs under conditions of low connectivity because there are too few infectious contacts with the disease reservoir to support spread of the disease. The endangered species goes extinct, however, because habitat fragmentation makes it problematic for the species to re-colonize extinct patches.

The endangered species also goes extinct under conditions of high connectivity, as this leads to more infectious contacts with the reservoir species, creating significant disease pressures on the endangered species. Only under intermediate/moderate levels of connectivity, when the proportion of infected patches is small, can the endangered species survive.

Now consider our bioeconomic model. The economic components of the model are specified to conduct a numerical analysis. All economic relations take on constant elasticity forms:

$$U = \alpha_u (S_W + S_{LW})^{\eta_u}, \; c(\psi_k) = \alpha_c \psi_k^{\eta_c}, \; g_i(Z_{ik}) = \alpha_g Z_{ik}^{\eta_g},$$

[13] McCallum and Dobson's (2002) graph is qualitatively the same as ours, but some values appear to differ slightly. This could be a result of the accuracy of the numerical methods being used (we have solved the model using Mathematica 5.0; Wolfram Research Inc, 2003).

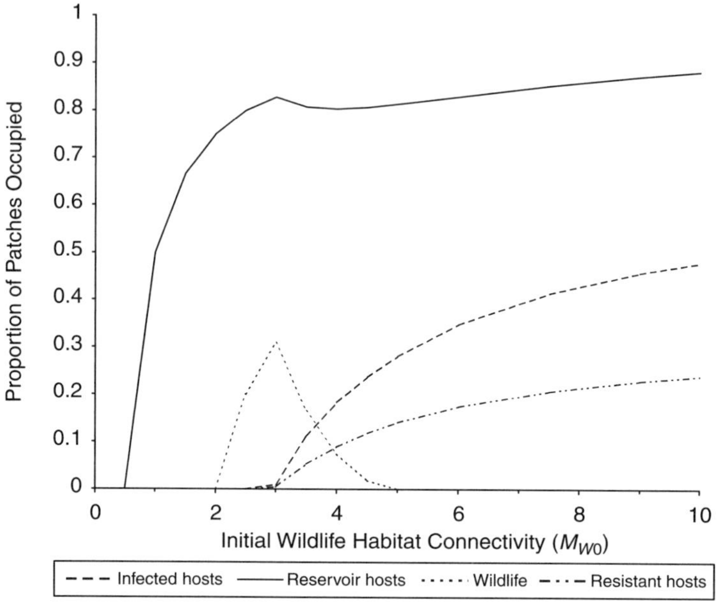

Fig 6.2 McCallum and Dobson's ecological scenario *a*. Parameter values: $\delta = 0.8$, $X_i = 4.0$, $X_r = 1.0$, $X_W = 2.0$, $M_\varsigma = 0.5$, $\Gamma = 1.0$, $\Lambda = 2.0$. Reservoir hosts are defined as all patches including livestock. Wildlife is defined as all patches including wildlife. Resistant hosts include all patches with resistant livestock. Courtesy of Richard Horan, Jason Shogren, and Benjamin Gramig, as derived from McCallum and Dobson's (2002) model

where $\alpha_i = N^{\eta_i}$ are parameters and η_i are elasticities. Unlike the ecological model, the economic model is not dimensionless – the values of N and $x_s$matter. Using the values presented in Fig. 6.3, we now explore steady-state solutions for the necessary conditions (6.4)–(6.9) and (6.17)–(6.19). Figures 6.3 and 6.4 and Table 6.2 present the equilibrium results.

Comparison of the bioeconomic results (Fig. 6.3) with the ecological-only results (Fig. 6.2) illustrates the impacts of economic investments on ecological out-comes.[14] The most obvious results are: (a) the endangered species is now prevalent for all values of initial connectivity, M_{W0}, and (b) the disease emerges at significantly larger values of connectivity. Habitat connectivity investments enable the endangered species to survive at lower values of M_{W0}. At low values of M_{W0}, Table 6.2 and Fig. 6.4 illustrate that subsidies are optimally targeted at increasing wildlife habitat connectivity, which increase wildlife productivity. The

[14] While not presented, comparisons of bioeconomic results against McCallum and Dobson's other scenarios *b*–*d* are qualitatively similar to scenario *a*. Quantitative differences do arise, however, there are fewer incentives to subsidize conservation activities when endangered species are not really in danger (scenario *b* and a range of scenario *c*), and there are greater incentives to subsidize conservation when the endangered species is in even more danger (scenario *d* and part of scenario *c*).

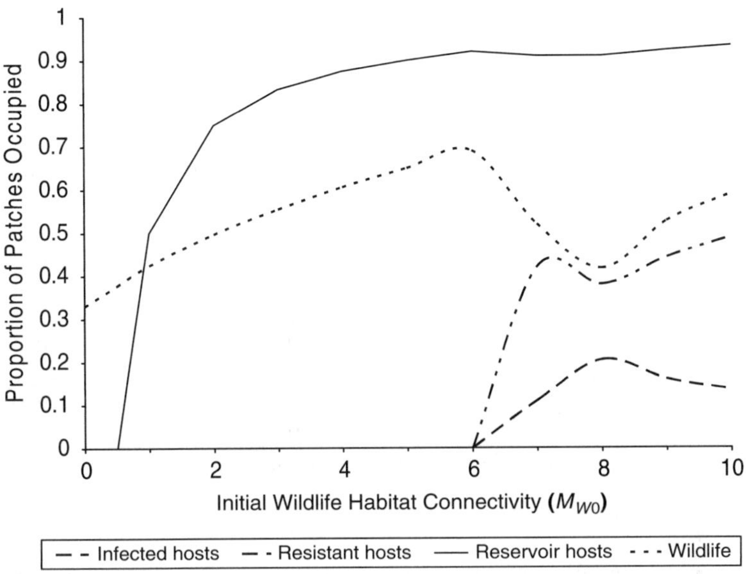

Fig. 6.3 Bioeconomic results. Parameter values: $\eta_c = 2$, $\eta_g = 2$, $\eta_u = 0.5$, $\beta = 0.25$, $x_s = 0.1$, $N = 10.0$, $\varphi = 0.05$, all other parameters as in Fig. 6.2. Courtesy of Richard Horan, Jason Shogren, and Benjamin Gramig

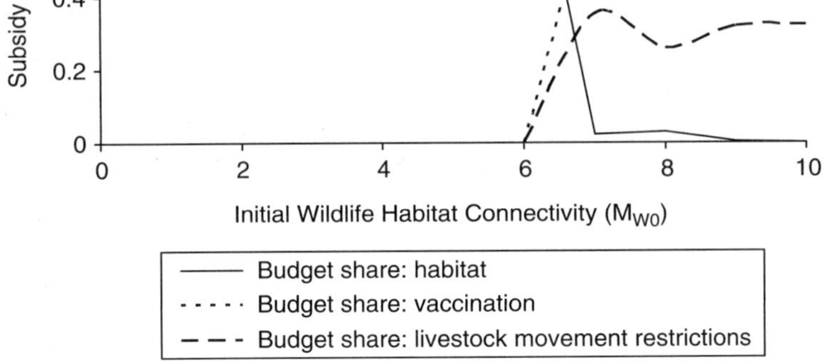

Fig. 6.4 Optimal shares of subsidy payments. Parameters are the same as in Figs. 6.2 and 6.3. Courtesy of Richard Horan, Jason Shogren, and Benjamin Gramig

Table 6.2 Equilibrium outcomes for welfare measures and rancher investment choices

	Initial connectivity (M_{W0})					
	0	2	4	6	8	10
Net social welfare	1.37	1.93	2.37	2.60	1.08	0.58
Recreation/tourism benefits	1.82	2.23	2.37	2.63	1.48	1.6
Vaccination and habitat management costs	0.3	0.2	0.07	0.02	0.27	0.68
Subsidy payments	0.59	0.39	0.14	0.04	0.53	1.35
Aggregate connectivity investments						
$S_W Z_W$	0.99	0.25	0.08	0.03	0.02	0.01
$S_{LW} Z_{LW}$	0	0.74	0.56	0.33	0.08	0.02
$S_{R_W} Z_{RW}$	0	0	0	0	0.09	0.03
IZ_I	0	0	0	0	0.38	0.54
Aggregate vaccination investments						
$S_L \Psi_L$	0	0	0	0	0.41	0.41
$S_{LW} \Psi_{LW}$	0	0	0	0	0.35	0.77

Source: Richard Horan, Jason Shogren, and Benjamin Gramig

largest investments (in aggregate) initially occur on *W*-type patches, and then become larger on *LW*-type patches (and on *RW*-type patches, once a disease outbreak occurs), as these become the most prevalent type of patch that the endangered species occupies. These investments decline, however, for larger values of M_{W0} as the marginal benefits of additional connectivity decline.

Disease outbreaks in the bioeconomic model are delayed until $M_{W0} > 6$, compared to the threshold of $M_{W0} \approx 2.5$ in the ecological model (Fig. 6.2). This result arises from disinvestments in mobility of infected livestock, although this is not immediately apparent from the equilibrium results in Table 6.2. If an infection were to emerge, a small reduction in livestock mobility would prevent the disease from becoming established in the livestock population. At the margin, the subsidy for reduced livestock mobility is available, but it is not used in equilibrium since the disease is unsustainable.

Once the disease establishes, investments are optimally made in both livestock movement restrictions and vaccination. Payments for these activities increase as M_{W0} increases, with a corresponding decrease in payments for wildlife habitat connectivity. Figure 6.4 illustrates this substitution of payment types. Vaccination and livestock immobility both increase the safety of endangered species, even at high levels of connectivity, by reducing the proportion of infected patches and increasing the proportion of resistant patches (compare Figs. 6.2 and 6.3), thereby significantly reducing infectious contacts relative to the case of no vaccination and greater livestock mobility.

Over the range of M_{W0} for which the disease becomes established, the connectivity in both infected and resistant patches initially increase, whereas wildlife patches decrease. Infected patches peak at $M_{W0} = 8$, beyond which infected patches decrease and resistant and wildlife patches increase. This is similar to McCallum and Dobson's (2002) scenario *c* – infected livestock recovered quickly produce a large resistant subpopulation relative to the infected population. The difference here is that resistance is created through vaccination instead of recovery from infection, and spread of infection is reduced through mobility restrictions.

Our model illustrates the role of habitat connectivity and vaccination as prevention and control strategies against disease. Traditionally, disease prevention involves lowering the chance of an outbreak, whereas disease control involves minimizing post-outbreak damages (see, for example, Leung et al., 2002). This distinction fits into Ehrlich and Becker's (1972) self-protection and self-insurance framework. Prevention is self-protection; control is self-insurance. But Ehrlich and Becker point out this distinction can be blurry, and this is true in our model. Here prevention should imply keeping the disease out of any given patch; control should imply minimizing damages if the disease makes it to any given patch. But habitat connectivity has both effects: the disease does not spread without connectivity (prevention), yet some degree of connectivity permits spread while simultaneously improving conservation (control). Similarly, livestock vaccination implies both prevention of its spread and control, in that the disease has no impact and dies away. The methods of prevention and control are intertwined, and here one cannot make general claims about "preventive ounces" versus "pounds of cure."

6.7 Distributional and Development Impacts

Table 6.2 presents the equilibrium welfare results. Net social welfare is concave in M_{W0}: It initially increases in M_{W0} as fewer connectivity investments are necessary, then it decreases in M_{W0} after the disease emerges and funds are needed for livestock movement restrictions and vaccination. Conservation costs and subsidy payments are inversely related to these results.

The conservation subsidies result in several development benefits to low-income ranchers. First, the subsidies generate rents for ranchers so they will bear the conservation costs. These rents are double the costs incurred due to the linear, increasing marginal cost functions. The rents would be more than double if marginal costs were increasing and convex. The rents provide ranchers with an additional source of income to help alleviate other pressing needs. For instance, the WCL program in Kitengela makes payments at the beginning of school terms to encourage education, which has worked to increase school enrollment, particularly among girls (Gichohi, 2003). The rents also help poor ranchers keep their land rather than selling it off to be subdivided (Gichohi, 2003; Ewing, 2005).

Second, the subsidies lead to a slight increase in the total number of patches containing livestock, in combination with a smaller number of infected patches – particularly at higher levels of M_{W0} (compare Figs. 6.2 and 6.3). While we do not explicitly model any benefits associated with more livestock patches, it is reasonable to assume they exist. Ranchers also benefit from reduced numbers of infected livestock, with these benefits embedded in the net cost functions for vaccination. Gichohi (2003) suggests these benefits arise from the WCL Program and provide additional incentives to discourage cultivation that would otherwise degrade wildlife habitats. Moreover, the combination of reduced cultivation and larger corridors can reduce human-wildlife conflicts.

Finally, the costs of vaccines are small relative to total vaccination costs, which include labor, travel, and other operating costs (VSF, 2006; Preslar, 1999). This raises the possibility of marginal costs to provide additional vaccines, and perhaps other basic veterinary services might also be small. A vaccination program for conservation purposes may induce cross-subsidizing of other vaccination or veterinary activities needed by low-income ranchers.

6.8 Concluding Remarks

We build a stylized model to gain insight into PES designed to protect endangered species given wildlife-livestock disease risks and habitat fragmentation. We consider subsidies to (a) increase wildlife habitat connectivity, (b) vaccinate livestock, and (c) reduce movement of infected livestock. Our results suggest the cost-effective policy is to subsidize habitat connectivity first rather than vaccinations (assuming little initial connectivity); this increases the contiguousness of habitat, which eventually also increases disease risks. Once habitat is sufficiently connected, disease risks increase so as to make a vaccination subsidy and subsidies to reduce infected livestock movement worthwhile. Highly connected habitat requires nearly all the government budget be devoted to these disease-related subsidies. No blanket subsidy toward habitat, vaccination, or livestock movement alone would work as well as the sequential portfolio approach of targeting habitat first, followed by vaccination and reduced livestock movement. The conservation payments result in significantly increased wildlife abundance and increased livestock health and abundance, resulting in spillovers to broader development objectives.

How should such a program be implemented? First, the question of who should pay whom revolves around who has the best access to farmers to implement the strategy. Local governments have the most inherent knowledge, but could also be the most susceptible to the misallocation of funds that were targeted for conservation. If corruption is an issue, NGOs might be a more appropriate administrator of funds. The question here is whether NGOs would have the ability to monitor and enforce the PES when it was to switch from habitat to vaccination and reduced livestock movement. This switching property suggests it might take a combined effort by local governments to monitor and enforce the switch and NGOs to allocate the funds.

Second, implementation of our scheme in the real world will be restricted (as are all policies) by the ability to monitor the level of risk to wildlife. Switching from habitat to vaccination and livestock movement controls requires an objective assessment of the increased risks to wildlife, such that the switching point can be roughly identified. This requires resources to construct and parameterize an integrated bioeconomic risk assessment model that works for the specific site in question. Such integrated risk assessment will be landscape specific in many cases, which requires more resources to get the assessment up to task. Nowadays, however, the field of rapid risk assessment has advanced significantly and can provide new tools for more timely assessment of wildlife risks.

Our model has three main caveats, which do not easily allow for insight into specific issues. First, the "island" metapopulation model we use is a restriction both for ecological relations and for the ability to target management activities. For instance, in reality, one could develop corridors between specific patch types but not between others due to economic and physical constraints. This would change the basic model from an island model to something else (maybe a "necklace"). Second, metapopulation modeling based on the proportion of patches in various states is elegant for ecological modeling and for bioeconomic maximization because it is parsimonious in state variables, but it creates complications when trying to incorporate management because we cannot keep track of individual patches. We do not know where these patches are located in space, and we do not know which patch is in which category at any particular point in time. Finally, ranchers have many other choices that could be included, at some loss in model parsimony. Future work should look to create powerful and tractable ways to relax these current restrictions on modeling disease risk between wildlife and livestock.

References

Alston JM, Hurd BH (1990) Some neglected social costs of government spending on farm programs. Am J Agric Econ 72:149–156.

Babiker HAS, Eldin EI SAZ (1987) Preliminary observations on vaccination against bovine cysticercosis in the Sudan. Vet Parasitol 24:297–300.

Barlow ND (1991) Control of endemic bovine Tb in New Zealand possum populations: Results from a simple model. J Appl Ecol 23:794–809.

Bicknell KB, Wilen JE, Howitt RE (1999) Public policy and private incentives for livestock disease control. Aust J Agric Resour Econ 43:501–521.

Burrows R (1992) Rabies in wild dogs. Nature 359:277.

Burrows R, Hofer H, East ML (1994) Demography, extinction and intervention in a small population: The case of the Serengeti wild dogs. Proceedings of the Royal Society, London. Biol Sci 256:281–292.

Ceballos G, Ehrlich PR, Soberon J, Salazar I, Fay JP (2005) Global mammal conservation: What must we manage? Science 309:603–607.

Clark CW (1990) Mathematical bioeconomics. Wiley, New York.

Crocker T, Tschirhart J (1992) Ecosystems, externalities, and economies. Environ Resour Econ 2:551–567.

Daszak P, Cunningham AA, Hyatt AD (2000) Emerging infectious diseases of wildlife – Threats to biodiversity and human health. Science 287:443–448.

Diekmann O, Heesterbeek JAP, Metz JAJ (1990) On the definition and computation of the basic reproduction ratio R_0 in models of infectious disease in heterogeneous populations. J Math Biol 28:365–382.

Dobson A (2004) Population dynamics of pathogens with multiple host species. Am Nat 164:S64–S78.

Ehrlich I, Becker G (1972) Market insurance, self-insurance, and self-protection. J Politi Econ 80:623–648.

Ewing J (2005) The Mesoamerican biological corridor: A bridge across the Americas. EcoWorld Mag, 19 December. Available at http://www.ecoworld.com/home/articles2.cfm?tid=377. Accessed November 3, 2006.

Fenichel EP, Horan RD (2007) Jointly-determined thresholds and economic tradeoffs in wildlife disease management. Nat Resour Model 20:511–547.

Finnoff D, Shogren JF, Leung B, Lodge D (2005) Risk and nonindigenous species management. Rev Agric Econ 27:475–482.

Gaydos JK, Gildardi KVK (2004 March 2-6) Addressing disease risks when recovering species at risk. In: Hooper, TD (ed.) Proceedings of the species at risk, 2004. Pathways to recovery conference, Victoria, BC, Species at Risk 2004, pp. 1–10.

Gichohi HW (2003) Direct payments as a mechanism for conserving important wildlife corridor links between Nairobi National Park and its wider ecosystem: The Wildlife Conservation Lease Program, Vth World Parks Congress. Sustainable Finance Stream, Durban, South Africa.

Gog J, Woodroffe R, Swinton J (2002) Disease in endangered metapopulations: The importance of alternative hosts. Proc R Soc Lond B 269:671–676.

Gramig BM, Barnett BJ, Skees JR, Black JR (2006) Incentive compatibility in livestock disease risk management. In: Koontz SR, Hoag DL, Thilmany DD (eds.) The economics of livestock disease insurance: Concepts, issues and international case studies. CABI Publishing, Oxfordshire, UK.

Haydon DT, Laurenson MK, Sillero-Zubiri C (2002) Integrating epidemiology into population viability analysis: Managing the risk posed by rabies and canine distemper to the Ethiopian wolf. Conserv Biol 16(5):1372–1385.

Hess G (1996) Disease in metapopulation models: Implications for conservation. Ecology 77:1617–1632.

Heesterbeek JAP, Roberts MG (1995) Mathematical models for microparasites of wildlife. In: Grenfell BT, Dobson AP (eds.) Ecology of infectious diseases in natural populations. Cambridge University Press, New York.

Hickling G (2002) Dynamics of bovine tuberculosis in wild white-tailed deer in Michigan. Michigan Department of Natural Resources, Wildlife Division.

Horan RD, Wolf CA (2005) The economics of managing infectious wildlife disease. Am J Agric Econ 87:537–551.

Horan RD, Shogren JF, Gramig BM (2008) Wildlife conservation payments to address habitat fragmentation and disease risks. Environ Dev Econ 13:415-439.

Kaiser J (2001) Conservation biology: Bold corridor project confronts reality. Science 21:2196–2199.

Kock RA, Wambua JM, Mwanzia J, Wamwayi H, Ndungu EK, Barrett T, Kock ND, Rossiter PB (1999) Rinderpest epidemic in wild ruminants in Kenya 1993-97. Vet Rec 145(10):275–283.

Langa JSR (2001) Mozambican field experience - Gaza Province. In: Alders RG, Spradbrow PB (eds.) SADC Planning Workshop on Newcastle Disease Control in Village Chickens. Australian Center for International Agricultural Research, Proceedings PR103.

Laurenson K, Sillero-Zubiri C, Thompson H, Shiferaw F, Thirgood S, Malcolm J (1998) Disease as a threat to endangered species: Ethiopian wolves, domestic dogs and canine pathogens. Anim Conserv 1(4):273–280.

Laurenson MK, Shiferaw F, Sillero-Zubiri C (1997) Disease in domestic dogs and the Ethiopian wolf: The current situation. In: Sillero-Zubiri C, Macdonald DW (eds.) The Ethiopian wolf: Status survey and conservation action plan. IUCN/SSC Canid Specialist Group, Gland, Switzerland and Cambridge, UK.

Leung B, Lodge DM, Finnoff D, Shogren JF, Lewis MA, Lamberti G (2002) An ounce of prevention or a pound of cure: Bioeconomic risk analysis of invasive species. Proc R Soc Lond 269:2407–2413.

Levins R (1969) Some demographic and genetic consequences of environmental heterogeneity for biological control. Bull Entomolog Soc Am 15:237–240.

Magin C (1996) Hirola recovery plan. Species Survival Commission, The World Conservation Union (IUCN).

Marino J, Sillero-Zubiri C, Macdonald DW (2006) Trends, dynamics and resilience of an Ethiopian wolf population. Anim Conserv 9(1): 49–58.

McCallum H, Dobson A (1995) Detecting disease and parasite threats to endangered species and ecosystems. TREE 10(5):190–194.

McCallum H, Dobson A (2002) Disease, habitat fragmentation and conservation. Proc R Soc Lond 269:2041–2049.

McNeil DG (2006) In war on bird flu, U.N. looks to recruit killer army. New York Times, Late Ed. (1):4(January 29).

Michigan Department of Natural Resources (2002) Taenia hydatigena. Available at http://www.michigan gov/dnr/0,1607,7-153-10370_12150_12220-27283–,00.html. Accessed January 5, 2006.

Millennium Ecosystem Assessment (MEA) (2005) Ecosystems and human well-being: Biodiversity synthesis. World Resources Institute, Washington, DC.

O.I.E. (2004) Rinderpest in Kenya. World Anim Health Organ Dis Inf 17:282–284.

Parkhurst G, Shogren J (2006) Smart subsidies for conservation. Working paper, University of Wyoming.

Perry B, Pratt AN, Stones K, Stevens C (2005) An appropriate level of risk: Balancing the need for safe livestock products with fair market access. Narobi, Kenya: International Livestock Research Institute and UN-FAO, Pro-Poor Livestock Policy Initiative, Working Paper No. 23.

Peterson MJ, Grant WE, Davis DS (1991) Bison-brucellosis management: Simulation of alternative strategies. J Wildl Manag 55:205.

Povilitis A (1983) The huemul in Chile: National symbol in jeopardy? Oryx 17:34–40.

Povilitis A (1998) Characteristics and conservation of a fragmented population of huemul Hippocamelus bisulcus in central Chile. Biol Conserv 86:97–104.

Preslar DB (1999) Lessening the impact of animal disease on developing country agriculture: A proposed program using developed country technologies' in sustainable agriculture solutions. Action Report 1999. Novello Press, London.

Pybus MJ (1990) Survey of hepatic and pulmonary helminths of wild cervids in Alberta, Canada. J Wildl Dis 26:453–459.

Research Network for Environment and Development (ReNED) (2005) Ecosystem services and biodiversity in developing countries: Proceedings and conclusions of the ReNED Conference, Copenhagen.

Roberts MG, Heesterbeek JAP (2003) A new method for estimating the effort required to control an infectious disease. Proc R Soc Lond B 270:1359–1364.

Sanchirico J, Wilen JE (1999) Bioeconomics of spatial exploitation in a patchy environment. J Environ Econ Manag 37(2):129–150.

Sanchirico JN, Wilen JE (2005) Optimal spatial management of renewable resources: Matching policy scope to ecosystem scale. J Environ Econ Manag 50:23–46.

Scott GR (1981) Rinderpest. In: Davis JW, Karstad LH, Trainer DO (eds.) Infectious diseases of wild animals, 2nd ed. The Iowa State University Press, Ames, IA.

Shogren JF, Crocker T (1991) Risk, self-protection, and ex ante economic value. J Environ Econ Manag 20:1–15.

Sillero-Zubiri C, King AA, Macdonald DW (1996) Rabies and mortality in Ethiopian wolves (Canis simensis). J Wildl Dis 32:80–86.

Simonetti JA (1995) Wildlife conservation outside parks is a disease-mediated task. Conserv Biol 9:454–456.

Smith GC, Cheeseman CL (2002) A mathematical model for the control of disease in wildlife populations: Culling, vaccination and fertility control. Ecol Model 150:45–53.

Smith-Fleuck JM, Fleuck WT (1995) Threats to the huemul in the southern Andean Nothofagus forest. In: Bissonette JA, Krausman PR (eds.) Integrating people and wildlife for a sustainable future. The Wildlife Society, Bethesda, MD.

Spinage CA (1962) Rinderpest and faunal distribution patterns. Afr Wildl 16:55.

Texera WA (1974) Algunos aspectos de la biología del huemel (Hippocamelus bisulcus) an cautividad. Ann Inst Patagon, Chile, 155–188.

United States Department of Agriculture, Animal and Plant Health Inspection Service (USDA-APHIS) (2002) Foot-and-mouth disease vaccine. Factsheet.

Vétérinaires sans Frontiers (VSF) (2006) Community based animal health program. Available at http://www.vsf-belgium.org/dzf/view/en/395. Accessed November 3, 2006.

Wikerhauser T, Zukovic M, Dzakula N (1971) *Taenia saginata* and *T. hydatigena*: Intramuscular vaccination of calves with oncospheres. Exp Parasit 30:36–40.

Wolfram Research Inc. (2003) Mathematica. Version 5.0, Champaign, IL.

Chapter 7
Payments for Ecosystem Services, Poverty and Sustainability: The Case of Agricultural Soil Carbon Sequestration

John M. Antle and Jetse J. Stoorvogel

Abstract This chapter explores the potential impacts of payments for ecosystem services on poverty and sustainability of farm households, using the example of agricultural soil carbon sequestration. Economic analysis shows that there is a variety of technical and economic factors affecting adoption of practices that increase soil carbon and their impacts on poverty, hence, the net effect of these factors is an empirical question. The evidence suggests that carbon payments could have a positive impact on the sustainability of production systems while also raising incomes and reducing poverty. However, carbon contracts are found to have only modest impacts on poverty, even at relatively high carbon prices. Moreover, the participation of poor farmers in carbon contracts is likely to be constrained by the same economic and institutional factors that have inhibited their use of more productive, more sustainable practices in the first place. Thus, payments for ecosystem services are most likely to have a positive impact on poverty and sustainability when they are implemented in an enabling economic and institutional environment.

7.1 Introduction

Throughout the world the focus of agricultural policy is shifting from traditional subsidy and trade policies to conservation and environmental aspects of agriculture. This shift in policy focus has been encouraged by a growing public demand for the ecosystem services provided by agricultural lands. Ecology tells us that ecosystems perform various functions that contribute to human well-being, water availability and quality, nutrient cycling, habitat for plants and animals, flood control, and regulation of the global atmosphere (National Research Council, 2004). Agriculture can be viewed as a managed ecosystem, and the ecosystem services provided by agriculture are known to depend on agricultural land use and associated management practices. Consequently, it can be argued that the appropriate role for agricultural policy is to maximize the social value of both conventional agricultural

L. Lipper et al. (eds.), *Payment for Environmental Services in Agricultural Landscapes*, Natural Resource Management and Policy.
doi: 10.1007/978-0-387-72971-8_7;

products and the ecosystem services provided by agriculture (Antle & Capalbo, 2001). The movement towards an environmental focus for agricultural policy has also been encouraged by the incorporation of agriculture into the General Agreement on Tariffs and Trade in the mid-1990s and the recent Doha Round of multilateral trade negotiations led by the World Trade Organization. These international policy negotiations have encouraged countries to reduce or eliminate subsidies that are trade distorting.

Payments for ecosystem services (PES) is one mechanism being used to provide farmers with incentives to increase the supply of ecosystem services from agriculture. In developing countries, an additional possible co-benefit of providing farmers with PES would be to contribute to broader economic development objectives such as poverty alleviation, food security, and sustainability. As yet, there is insufficient experience with ecosystem service payments to know what their effects are likely to be on these development objectives.

The objective of this chapter is to explore the potential impacts of PES on poverty of farm households and on the sustainability of agricultural systems. We do this using the case of agricultural soil carbon sequestration. We focus specifically on agricultural soil carbon sequestration for several reasons. First, soil degradation – in many cases the result of or resulting in declining soil carbon contents – is widely regarded as a major factor contributing to the persistent problems of poverty and food insecurity, particularly in the most agriculturally marginal areas of the developing world (Lynam et al., 1998). Second, soil carbon sequestration has been proposed as a way to meet the joint goals of mitigating greenhouse gas emissions while enhancing the productivity and sustainability of agricultural lands, both in the industrialized and developing countries (Lal et al., 1998; Soil Management Collaborative Research Support Program, CRSP, 2002). Moreover, due to the likely positive correlation between soil degradation and rural poverty, soil carbon sequestration might be a way to target farmers in the poorest, most environmentally vulnerable areas. Third, as yet, soil carbon sequestration has not been widely implemented in the context of international agreements such as the Kyoto Protocol, or in national policies, so there is little information available from actual projects about the likely impacts.

This chapter begins with a discussion of the demand for greenhouse gas mitigation services, and how farmers can contribute to the supply of those services. Next we focus more specifically on factors affecting farmers' willingness to supply ecosystem services, and how PES could impact farmers' incomes and poverty. Finally, we present results from three case studies of agricultural carbon sequestration in Kenya, Peru, and Senegal. The chapter concludes with a discussion of policy implications.

7.2 Demand and Supply of Greenhouse Gas Mitigation Services

Most ecosystem services are public goods – they provide benefits to a large group of people, and this enjoyment is non-rival (one person's use does not prevent another person from also benefiting from them). In some cases, ecosystem service benefits are local, as in the case of regulating water supply and water quality. But in other cases, such as maintenance of biodiversity and regulating the concentration of greenhouse gases in the atmosphere, the benefits are global. The scope of the public good is important because it defines the group of people who should help pay for their provision.

A key problem in ensuring an appropriate provision of public goods is determining how much people value them. Because the benefits of public goods are enjoyed by a large number of people, and this enjoyment is non-rival, individuals do not have an incentive to reveal their willingness to pay for them. Individuals have an incentive to be "free riders" and let others pay to provide the public good since everyone benefits from them whether they help pay for them or not. The free-rider problem is one reason why governments often use taxes or user fees to pay for the provision of public goods. In the context of ecosystem services, governments can use tax revenues to pay landowners to manage their land in ways that protect or enhance the provision of those services.

Another way that public goods can be provided is through a combination of government regulation and markets. For example, the government can create a law or regulation that sets limits on the amount of a pollutant that can be emitted into air or water. But instead of government using so-called "command-and-control" regulations that define how much emissions each individual polluter (say, an electric generating plant) can release, the government simply establishes the total amount of emissions to be allowed, and allocates to each individual an initial share of emissions permits so that the total amount of permits equals the total amount of emissions allowed. These permits can be bought and sold. Therefore, plants that are efficient at reducing emissions will have an incentive to do so, and sell some of their permits to plants where it is costly to reduce emissions. The net result is that the cost of meeting a given emissions goal is much cheaper than when every polluter has to meet the same emissions reduction standard.

So how do emissions permits and the trading of these permits relate to the provision of ecosystem services? Think of pollution as reducing the amount of ecosystem services being provided. One way to increase the supply of ecosystem services is to simply reduce the amount of pollution by reducing the activity that causes the pollution. But another way to increase the supply of ecosystem services is to increase some other activities that offset the effects of the pollution. This idea of offsetting activities is how agriculture and forestry could perform the service of reducing greenhouse gas emissions. Government regulations would establish a total amount of greenhouse gas emissions allowed. Emitters would either have to reduce their emissions directly, or could purchase "emissions reduction credits" from other entities that take actions that offset their emissions. Landowners could sequester carbon, the most abundant greenhouse gas, by planting trees. Farmers could sequester carbon by changing crops or crop management in ways that increase the amount of carbon in the soil (Lal et al., 1998; Paustian et al., 2006). In addition, governments could provide incentives to landowners and farmers to further reduce greenhouse gas concentrations beyond current emissions levels to offset past emissions.

There are many issues that would have to be addressed in creating regional or global greenhouse gas emissions trading systems. It is beyond the scope of this chapter to discuss those issues in detail, but the reader can find a wealth of information about emissions trading on the World Wide Web. In the remainder of this chapter, our goal is to discuss how we can assess the potential for farmers to participate in greenhouse gas emissions trading, and what the impact might be on poverty and sustainability of small-scale agriculture in the developing world.

7.3 Three Case Studies

To put the remainder of our discussions in context, we outline three case studies that we use to assess the potential poverty and sustainability impacts of agricultural soil carbon sequestration.

7.3.1 Machakos, Kenya

The Machakos study area includes the Machakos, Makueni, and Mwingi districts; is located southeast of Nairobi; and ranges in altitude between 400 and 2,100 m above sea level. The area is ~20,000 km^2 in size and is located between 0°70′′ and 3°00′′ southern latitude and between 36°87′′ and 38°51′′ eastern longitude. The semi-arid climate in the study area has low, highly variable rainfall, distributed in two rainy seasons. The annual rainfall average ranges from 500 mm to 1,300 mm and mean annual temperature varies from 15°C to 25°C. Soils in the region are rather shallow, generally deficient in nitrogen and phosphorus, and low in organic matter. Moreover, low infiltration rates and susceptibility to sealing makes them prone to erosion, especially since most of the rains occur at the beginning of the growing season when the land is still bare. The region suffered extensive soil degradation in the early to mid-twentieth century, at which time government programs caused large areas to be terraced. The success of these programs has been documented by Tiffen et al. (1994). Though the region is highly dependant on agriculture, its population obtains significant income from non-farming activities inside and outside the district's boundary. The farms can be characterized as subsistence-oriented mixed farming systems that include both crop and livestock production. Maize is the most important staple crop that is sold for cash, and a wide variety of subsistence crops are grown, such as vegetables, fruits, and tubers.

Farm survey data were obtained from studies conducted in the 1997–2001 period. The data covered 120 households in six villages with detailed input and output data for nearly 2,700 fields. Further description of the data can be found in de Jager et al. (2001) and Gachimbi et al. (2005). Two of the villages in the study produce vegetables with irrigation and market them to urban areas. Maize yields are generally low and crop failure is widespread. Livestock was traditionally managed by letting it graze freely, but intensive zero-grazing units are proliferating in the region in the last years, and their importance in nutrient recycling is considerable.

Details of the economic models are provided in Antle et al. (2003b). The carbon contracts modeled require farmers to utilize minimum amounts of organic fertilizer (600 kg/ha/season)[1] and mineral fertilizer (60 kg/ha/year).

7.3.2 *Cajamarca, Peru*

The study focuses on the La Encañada watershed in the Cajamarca region in northern Peru. The 10 km^2 watershed ranges between 2,950 m and 4,000 m above sea level and is located between 7°00′′ and 7°07′′ southern latitude and between 78°15′′ and 78°22′′ western longitude. Average annual rainfall is low ranging between 430 mm/year in the valleys and up to 550 mm/year in the higher parts of the watershed (Romero & Stroosnijder, 2001). Soils are shallow and calcaric clay matured on limestone parent material, and deeper low-calcium clay soils matured on claystone parent material. This region is characterized by three agro-ecozones – the valley floors, the lower hillsides, and the upper hillsides. Milk production dominates in the valley floors where access to irrigation allows for cultivation of permanent pastures. In the lower hillsides where little irrigation is available, field crops dominate the production system, including Andean tubers, legumes, cereals, and pasture. Cultivation in this zone occurs in two seasons, December to May and June to September/November. In the upper hills where risk of frost is high, natural pastures dominate the landscape.

The data used in this analysis were collected through farm surveys conducted in 1997–1999 for a random stratified sample of 40 farm households in five communities in the watershed (see Valdivia, 1999, 2002; Valdivia & Antle, 2002, for further details). The data show that crop yields are low and parcel size is small, as is typical of this type of semi-subsistence agriculture. Size distributions of the parcels and farm sizes are highly skewed, with a large number of very small parcels and farms and a small number of much larger parcels and farms. The analysis reported here is based on the lower-hillside region where cropland is the principal land use. Valdivia (2002) and Antle et al. (2005b) provide details on the economic models used in the simulations. Antle et al. (2005a) provide details of the carbon sequestration analysis, which is based on the adoption of terraces and terraces with agroforestry (trees planted on the tops of terrace walls).

7.3.3 *Southern Peanut Basin, Senegal*

The Nioro region of Senegal is in the southern part of Senegal's "peanut basin" occupying the central part of the country. Nioro contains about 103,000 ha of

[1] Kilogram per hectare per season.

cropped area, or about 5% of Senegal's agricultural area, and lies in the Sudano–Sahelian zone of the peanut basin, situated between 13°35′′ and 13°50′′ northern latitude and 16°00′′ and 16°30′′ western longitude with an average elevation of 40 m above sea level. The rainy season lasts from June to October, and the total annual rainfall is about 750 mm. Annual temperatures average 27.5°C and the mean maximum and minimum temperatures are, respectively, 38°C and 15°C. Most soils in the Nioro area have been formed in materials that originate from ironstone or the underlying sandstones. On the ironstone plateaus, soils are stony and shallow. On the glacis, terraces, and bas-fonds, soils are deep. In general, the clay content increases with soil depth. Millet and peanut grown in annual rotation are the two main crops. These two crops represent almost 90% of Senegal's cropped area in most years.

The data used in this study are cross-sectional and come from farm surveys organized and conducted by the Ecole Nationale d'Economie Appliquée in 2001. More than a hundred households in 13 villages in the Nioro area were surveyed to collect detailed socioeconomic and agricultural production data including household demographic characteristics, labor availability, annual food grain production and consumption, annual income and expenses, and agricultural inputs and outputs. Diagana et al. (2005) provide a detailed description of the economic models used and the specification of the carbon contracts based on incorporation of crop residues and application of mineral fertilizer.

7.4 Economic Analysis of Soil Carbon Sequestration Contracts

In this section we present the economic analysis used to assess farmers' willingness to participate in contracts that provide payments for agricultural soil carbon sequestration. We begin with a characterization of the initial conditions before farmers have the option to participate in carbon contracts. Following Antle et al. (2003a), in large-scale commercial agricultural settings, such as in the United States, it can be assumed that when carbon contracts are not available, farmers adopt those land-use and management practices that maximize economic returns (adjusted for risk if farmers are risk averse), under the assumption of well-functioning factor and capital markets and well-informed farmers. These assumptions imply that, from the farmer's perspective, the initial conditions represent an efficient allocation of resources, absent payments for carbon sequestration. Importantly, this does not mean that farmers are managing soil carbon stocks efficiently from a social perspective if a reduction of greenhouse gas concentrations has a positive social value.

In the context of developing countries, there is much evidence that productivity is constrained by low levels of soil organic matter and consequently soil fertility (Kherallah et al.,2002; Koning et al., 2001; Sanchez, 2002; Scherr, 1999). The literature identifies many factors contributing to this situation, including: policies that discriminate against agriculture; high transportation costs, coupled with imperfect factor and capital markets; high population densities and rapidly growing populations; lack of accurate information about the long-term consequences of management decisions, particularly when it involves factors such as soil fertility that are difficult to observe. In the analysis presented here, we assume that farmers are rational and make management decisions to maximize economic returns, but we recognize that those decisions may be the result of various factors that lead to a loss of soil productivity over time. Indeed, in the case studies considered here, field measurements show that productivity is constrained by low levels of soil organic matter and low rates of nutrient use. The goal of the analysis is to simulate the effects of introducing soil carbon contracts that require farmers to increase incorporation of organic matter into the soil and to increase the use of mineral fertilizer, and to adopt soil conservation investments such as terraces and agroforestry. However, it is important to note that in the baseline conditions that are observed without carbon contracts, some farmers already apply relatively high rates of organic and mineral fertilizers, or have constructed terraces, but in most cases adoption rates are low. Table 7.1 shows that in the three case studies 20–76% of farms did not use any mineral fertilizer on their cash crops. The data also show that on subsistence crops most farmers used lower rates of organic fertilizer and almost no mineral fertilizer. The data from the Peru case study show that about 18% of the fields in the region are terraced, even though the average field slope in the region is over 20%.

7.4.1 Designing Contracts for Carbon Sequestration

In the case studies, model simulations were conducted in which it is assumed that carbon contracts provide payments to farmers and require them to adopt land-use or management practices that increase the amount carbon is stored in the soil. In the cases of Senegal and Kenya, the contracts are based on incorporation of crop residues and application of organic and mineral fertilizers at specified rates, whereas the Peru study considers adoption of terracing and agroforestry practices. When fertilizer use is required, a key assumption made is that farmers participating in carbon contracts have access to fertilizer at the market price when they are planting their crops, and have the cash available to purchase the fertilizer when it is needed. Thus, if farmers' access to fertilizer is being constrained by imperfections in fertilizer markets, we assume that the organization (either governmental or nongovernmental) acting as an intermediary to facilitate carbon contracts takes whatever actions are needed to make the quantities of fertilizer required under the contract available to farmers. In the Peru study, we consider the case wherein farmers must pay the full price of the soil conservation investments.

The economic simulations for the three case studies show that farmers using low levels of fertilizer inputs would generally benefit economically from using at least as much as required in the carbon contracts. This finding supports the general view that factors such as credit and fertilizer availability at planting time constrain profitable use of fertilizer. One way that fertilizer use could be financed is by

Table 7.1 Fertilizer use and farm characteristics in three case studies

	Farms (%)	Fertilizer (kg/ha)	Farm size (ha)	Family labor (md/sea)	Hired labor (md/sea)	Cash crop share (%)	Crop revenue ($/ha)	Off-farm income ($/ha)
Kenya								
Rain fed	59 (41)	0 (79)	2.9 (3.1)	127 (166)	8 (10)	40 (40)	30 (54)	46 (78)
Irrigated	33 (77)	0 (97)	1.6 (1.9)	179 (154)	8 (30)	71 (89)	463 (612)	52 (62)
Peru								
Unterraced	63 (37)	0 (121)	4.9 (7.6)	85 (224)	[a]	20 (97)	50 (127)	NA
Terraced	70 (30)	0 (87)	7.0 (7.8)	122 (192)	[a]	25 (98)	27 (81)	NA
Senegal	19 (81)	0 (57)	6.5 (8.4)	34 (50)	[a]	50 (50)	157 (125)	NA

Notes: First number is value for farms not using mineral fertilizer; second number in parentheses is for farms using mineral fertilizer
Fertilizer = average total kg active ingredient (N, P, K) applied on cash crop fields
Off-farm income = off-farm income divided by farm size, converted to USD
Cash crop share = share of cash crop revenue in total revenue
Source: Antle and Stoorvogel (2008)
[a] = Hired labor included in family labor
NA = Not available

providing the carbon payments in the form of fertilizer (Antle & Diagana, 2003). However, calculations show that carbon payments at the beginning of the season would not be sufficient to provide all of the fertilizer needed for the contracts. For example, the simulations carried out for the Kenya study assume farmers utilize at least 60 kg N per hectare per season and at least 600 kg of organic fertilizer per hectare per season. Under these assumptions, simulations show that farmers who do not use any mineral fertilizer and low rates of organic amendments would obtain an increase of about 0.6 MgC/ha/year, or about 0.3 MgC/ha per season (throughout we use MgC to denote mega-gram or metric ton of carbon). If the price of carbon were $50/MgC, then the payment would provide a payment of $15/ha per season. With a fertilizer price of about $0.40/kg, this would provide the farmer with about 38 kg of fertilizer if the payments were made in kind, thus falling short of the 60 kg required under the contract. In addition, most farmers would also need to increase their use of organic fertilizer.

In the case study of terracing in Peru, the issue of financing adoption of the conservation practices may be even more critical. The cost of constructing terraces on 1 ha is estimated to be about $300/ha, and the cost of maintaining them is about $65/ha/year (Valdivia, 2002). With an average carbon rate of less than 1 MgC/ha/year, at a carbon price of $50/MgC farmers would receive less than $50/ha/year, thus the carbon payments would cover part of the maintenance costs but not the initial investment. Thus, we have to consider the following analysis in light of these possible constraints on adoption.

The carbon payments each season could be based either on the number of hectares on which these practices are adopted (a *per hectare* payment), or on the expected amount of carbon sequestered. In the latter case, the contract is based on a *per-ton* payment mechanism. As Antle et al. (2003a) show, the per-ton payment mechanism is economically more efficient because it pays farmers per unit of environmental service provided rather than per hectare of land under contract regardless of the amount of carbon sequestered. Accordingly, the case studies presented below simulate per-ton contracts based on carbon rates estimated by agro-ecozone. In other words, the carbon contract specifies a payment based on the price of carbon and the carbon rate estimated for the zone in which each field is located. We assume that many individual farm fields are aggregated to make up a standard marketable contract (e.g., 1,000 metric tons of carbon). Carbon rates are verified for each contract using periodic randomly sampled soil measurements. Analysis by Mooney et al. (2004) indicates that these measurement and monitoring costs are likely to be small relative to the value of the carbon sequestered. In the case of contract default, several possible mechanisms could be used. One option would be for the entity aggregating contracts to discount carbon rates for risk of default (in effect, maintaining an insurance pool of sequestered carbon to offset defaults). Another option would be to require repayment by defaulting farmers, although that may not be feasible for small, poor farms.

7.4.2 Transaction Costs

Setting up and verifying carbon contracts and insuring against default will involve transaction costs that also must be estimated and factored into the analysis. Few data are available to estimate transaction costs for agricultural soil carbon sequestration (see Mooney et al., 2004; International Energy Agency, 2005; Paustian et al., 2006). Some analysts argue that these transaction costs could be high for organizing small-scale farmers to adopt practices on enough hectares to constitute marketable quantities of carbon, but as yet no actual pilot projects have been implemented in which such costs could be quantified. In the case studies presented here, because reliable data on transaction costs are not available, transaction costs are included in a sensitivity analysis.

Important informational issues arise in defining and verifying compliance with carbon contracts. Soil carbon accumulation is a function of past land-use and management practices. Whereas it is relatively low-cost to verify adoption of soil conservation investments such as terracing and agroforestry, basing carbon payments on use of variable inputs such as organic and mineral fertilizer raises the problem of knowing past practices as well as monitoring compliance with the contract. Essentially, there is an asymmetric information problem because farmers know their past and current practices, but the entity responsible for verifying compliance with contract does not. Efficient solutions to the asymmetric information problem depend on designing incentive mechanisms that lower the cost of verifying compliance. For example, successful micro-credit programs have utilized self-enforcement mechanisms. However, if these information problems cannot be addressed at low cost, then there will be incentives for many farmers to default on carbon contracts, similar to the problems encountered in credit markets (e.g., see Blackman, 2001).

7.4.3 Risk and Adoption of Carbon-Sequestering Practices

Much research has addressed the impact of risk and risk aversion on farmers' adoption of technology, particularly in developing countries (Sunding & Zilberman, 2001). In the case studies risk is not formally incorporated, and it is important to note that risk could impact farmers' willingness to participate in carbon contracts both positively and negatively. On the negative side, the use of inputs such as mineral fertilizer is often said to increase production risk. However, increased use of organic fertilizers and incorporation of crop residues and other organic matter is typically assumed to stabilize production (e.g., by improving water-holding capacity of the soil). Similarly, the use of terracing and other soil conservation practices is generally believed to improve water availability and thus both stabilize and increase productivity. Thus, the net risk effect of the set of practices being adopted is not clear. In the case studies from Kenya and Senegal, econometric tests did not support the hypothesis that either organic or mineral fertilizers were risk-increasing inputs. Also on the positive side, carbon payments would appear to represent a stable source of income as compared to income from risky crops, although there could be some risk of default on the contract as well as possible policy risk if the payments were being made by an unreliable governmental or non-governmental entity. Finally, because of concerns about permanence of soil carbon, some have argued that carbon contracts would require farmers to adopt and maintain appropriate land-use and management practices for long periods of time, say, 20 years or longer. Such long-term contracts would impose costs on farmers in the form of forgone option value due to uncertainty about the long-term productivity benefits of the practices, price uncertainty, and political risk. However, it is not correct that carbon contracts would have to require such long-term commitments by farmers. Instead, farmers can be offered relatively short-term contracts with the option to renew, with the price appropriately adjusted to reflect the implied non-permanence of the carbon (e.g., see Lewandrowski et al., 2004). Thus, while the net effect of carbon contracts on farmers' perceptions of production and income risk are not entirely clear, both logic and available evidence do not suggest that farmers would perceive them as substantially increasing the risk they face, and may well decrease risk.

7.4.4 Modeling Farmer Participation in Carbon Contracts

The Appendix presents the way that carbon contract participation is modeled in the simulation studies discussed below. These models assume that to increase the stock of soil carbon on a land unit, a farmer must make a change from a conventional production system (say, one with low levels of organic and mineral fertilizer applications, or without soil conservation practices such as terraces) that had been followed over some previous period (the historical land-use baseline) to some alternative system (say, one with higher fertilizer application rates or with soil conservation practices). We assume that utilization of the conventional management practice up to time 0 results in an initial soil carbon level, and adoption of the alternative practice causes the level to increase over time to a new, higher equilibrium level that is maintained until further changes in management occur. In defining ex ante carbon contracts, the expected change in carbon accumulation is the relevant variable; the actual rate of carbon accumulation will typically only be verified for the land units aggregated into a contract, as discussed by Antle et al. (2003a). This expected change in carbon is assumed to be estimated by agro-ecozone and past land-use practices, with all farmers in the contract in that zone receiving credit for the same rate, as explained further below.

With carbon contract that pays the farmer for each metric ton of carbon sequestered, the farmer receives a payment of $\$P$ per ton of carbon sequestered each time period. If the farmer changes from the conventional practice to the alternative practice and Δc tons per hectare are expected to be stored in the soil, the farmer receives a total return of $\mathrm{NR_s} + P\Delta c$, where $\mathrm{NR_s}$ is the return to the alternative practice. If the farmer does not participate in the contract and continues producing with the conventional practice, he receives $\mathrm{NR_i}$. Thus, assuming for the moment that there are no other costs to changing from the conventional practice to the alternative practice, the farmer will switch to the alternative practice if

$$\mathrm{NR_s} + P\Delta c > \mathrm{NR_i}. \tag{7.1}$$

This expression has several implications for analysis of adoption of soil carbon sequestration practices.

In the initial equilibrium in which there are no payments available for carbon sequestration, $P = 0$, and the farmer adopts the alternative practice only if it provides higher net returns than the conventional practice. When a carbon contract is offered for adoption of practices that sequester carbon, $P > 0$ and we can rewrite Eq. (7.1) as

$$P > (\mathrm{NR_i} - \mathrm{NR_s})/\Delta \mathrm{c}. \tag{7.2}$$

The expression in the numerator on the right-hand side is the opportunity cost for switching from the conventional system to the alternative system. Thus, expression (7.2) states that the farmer will be willing to enter a carbon contract when the price per ton of carbon is greater than the opportunity cost per ton. An important fact to keep in mind is that both the numerator and the denominator of Eq. (7.2) vary across the landscape, from one field to another. Net returns to the conventional and alternative practices are clearly site-specific due to spatial variation in both prices and productivity. Similarly, the expected rate of carbon accumulation is specific to the agro-ecozone where the land unit is located. Thus, the participation by farmers in carbon contracts depends on the spatial distribution of the opportunity cost of changing practices. Those land units with opportunity cost less than P will participate in the contract, and those land units with a higher opportunity cost will not participation. In the simulation studies presented below, the regional carbon supply curve is derived by summing the quantities of carbon across participating land units at each carbon price.

7.4.5 Carbon Sequestration, Poverty, and Food Insecurity

Once the analysis of farmer participation in carbon contracts is carried out, we can investigate the question of whether farmers are better off, in terms of income and food security, by participating in a carbon contract. When there is a net benefit, there is the question of how those benefits are distributed.

Carbon contracts that provide cash payments or payments in kind contribute to household income. However, the impact on farm production and income is less clear. We assume that rational farmers who participate do perceive a net economic gain, but for farmers facing a positive opportunity cost to adoption of the carbon-sequestering practice, the net impact on income is less than the payment for all except the marginal land unit. The impact on food security will depend on the production impacts of the alternative practice. In most cases, practices that increase soil carbon are expected to improve both average productivity and stabilize production, thus enhancing food security of semi-subsistence households that depend on their own production for food security.

The distributional effects of PES depend on a number of factors as well. From the regional or national perspective, it is a well-established fact that rural households in developing countries typically have lower incomes and are less food-secure than urban households. Data from recent poverty-mapping research (Government of Kenya, 2003) show this fact clearly for Kenya, where rural poverty rates exceed 50% in most areas and 90% in some areas. Data from Peru show similar patterns, where poverty among rural households occurs at much higher rates than among urban households (Interinstitutional Commission, 2005; Zeller et al., 2005), and the survey data utilized in the case study show that poverty rates in rural Senegal are also extremely high. Therefore, ecosystem service payment schemes should contribute to poverty reduction and food insecurity in rural areas. However, because the PES primarily benefit the owners of land, the impact will also depend on the pattern of landownership and the prevalence of landless poor in rural areas. On the one hand, payments for afforestation or improved forestry management may largely go to landowners with relatively high incomes when landownership is highly skewed. On the other hand, in areas where the principal land use is small-scale agriculture, and payments are based on adoption of agricultural practices, PES will go primarily to rural households, and most of these households will have low incomes.

There is also the question of how PES will be distributed within rural farm household populations. As explained in the Appendix, the rate of carbon sequestration credited to a farmer in a carbon contract depends on prior adoption of the practice. To the extent that adoption of more sustainable practices is constrained by factors associated with poverty and food insecurity, carbon contracts based on adoption should tend to target farm households that are poor and food insecure. The data in Table 7.1 indicate that in the three case studies there is a tendency for farm households, which do not use mineral fertilizer on cash crops and that do not adopt terraces to be smaller, to have lower farm and off-farm income, and to be less specialized in cash crop production (although the direction of causality in these relationships is not clear).

Equation (7.2) shows that the opportunity cost of adopting the carbon-sequestering practice s depends on two factors. In the numerator are the forgone returns from changing from the conventional practice to the alternative practice. Typically, the conventional practice (e.g., not using soil conservation practices or using low rates of organic matter incorporation) has the highest productivity on the land with inherently favorable soil and climatic properties, and the value of the conservation practice may be relatively low in these favorable conditions. Therefore, the forgone returns to adopting the carbon-sequestering practices are likely to be high on relatively good land and low on marginal lands. The opportunity cost also depends on how much carbon is sequestered per hectare, the denominator of Eq. (7.2). Land with favorable properties may have the highest potential for carbon sequestration, even if the land is not highly degraded, as compared to marginally productive land. An interesting side effect is that those lands produce the largest quantities of crop residues and consequently farmers will have larger amounts of organic amendments available for incorporation. Therefore, marginal lands are not necessarily more economically efficient at sequestering carbon, and indeed the opposite may be true.

In some cases, the opportunity cost of adopting carbon-sequestering practices may actually be negative when factor market distortions or imperfect information cause farmers not to adopt profitable conservation practices. For example, the farmer may perceive that the opportunity cost of adoption is positive due to uncertainty about the future productivity of the conserving practice. Payments for carbon sequestration may induce such a farmer to adopt, and then learn that the practice is profitable even without an incentive payment. To the extent that these uncertainties are correlated with poverty and food insecurity, then carbon payments would indeed target benefits to the most poor and food-insecure farmers.

Finally, transaction costs could also impact the participation of farms differentially in terms of land quality and size. Larger, wealthier farms are more likely to be located on more favorable land where carbon rates are higher. The Appendix shows that the effect of a fixed transaction cost will be smaller per unit of carbon sequestered where carbon rates are higher. Similarly, if there is a component of transaction costs that is fixed per farm (e.g., associated with learning about carbon contracts), then the average transaction cost per hectare will be lower for larger farms (Antle, 2002).

7.4.6 Hypotheses: Poverty, Food Security, and Sustainability

The preceding discussion shows that there are a variety of factors affecting adoption of practices that increase soil carbon and the sustainability of production systems. The impact of these practices and carbon payments on poverty also depends on a number of factors. We can conclude that the net effect of these various factors is an empirical question. In summary, we have the following hypotheses about the impacts of carbon sequestration on poverty, food security, and sustainability:

Hypothesis 1: Carbon contracts increase adoption of sustainable practices.

Hypothesis 2: Carbon contracts transform unsustainable agricultural systems into sustainable systems.

Hypothesis 3: Carbon contracts increase aggregate rural income.

Hypothesis 4: Carbon contracts reduce poverty and food insecurity in the rural farm population.

Hypothesis 5: The impacts of carbon contracts on poverty and food insecurity are greatest in the poorest regions and households.

Hypothesis 6: Transaction costs substantially reduce participation in carbon contracts.

7.5 Evidence on Carbon Sequestration, Poverty, Food Security, and Sustainability

All three of the case studies are based on the simulation methods described in Stoorvogel et al. (2004) and Antle and Stoorvogel (2006) using the Tradeoff Analysis simulation software. The Tradeoff Analysis software can be used to analyze the potential for soil carbon sequestration contracts as shown in Fig. 7.1. The first step is to assemble the needed data, including the data for implementation of crop growth and soil carbon models and for estimation of site-specific economic simulation models for the region. In addition, any relevant scenarios regarding alternative production technologies that could be used to sequester soil carbon and the types of contracts that would be used would need to be assembled. The crop and carbon simulation models are executed for the set of fields that was being used in the analysis (this could be a set of fields randomly sampled from the region being analyzed using a map of the region, or a set of fields randomly sampled in a production survey). Crop yields and soil carbon values are saved in a file that becomes an input into the econometric-process simulation model. This economic model simulates farmer's land-use and management decisions for the baseline case

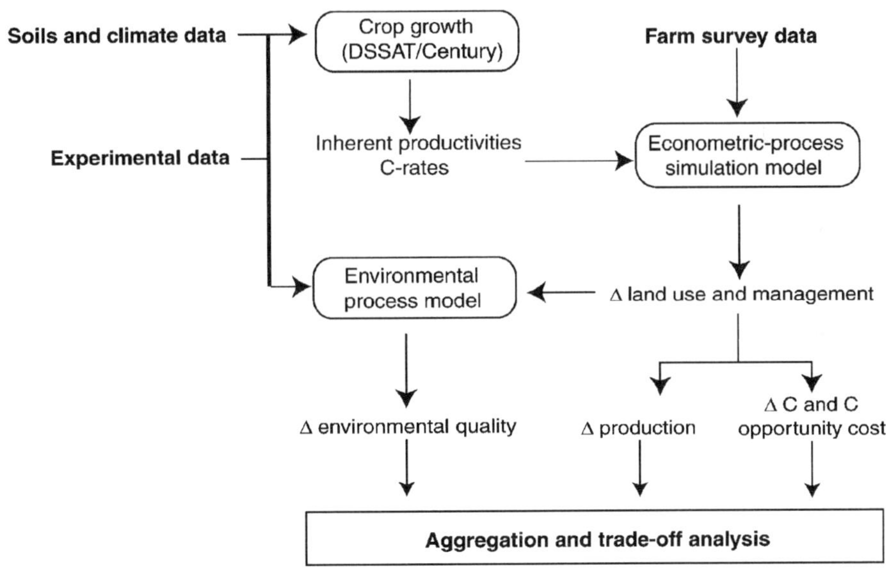

Fig. 7.1 Integrated assessment of soil carbon sequestration (Antle, 2002)

of no carbon contracts, and for the types of contracts that farmers could be offered. The economic model creates an output file containing the farmer's land-use and management decisions and the changes in soil carbon associated with those decisions. This information is passed to other environmental process models to analyze other environmental impacts such as nutrient depletion, soil erosion, or fate of pesticides. Finally, the results of the various models are combined into an output file that can be aggregated to represent the region and used for various types of analysis. For the analysis of soil carbon sequestration, a principal use of this output is to construct a supply curve for soil carbon corresponding to each type of contract that was simulated. If other environmental process models were included in the analysis, it is also possible to assess tradeoffs with other environmental impacts, such as nutrient depletion, to represent the sustainability of the system.

7.5.1 Hypothesis 1: Carbon Contracts Increase Adoption of Sustainable Practices

Figures 7.2–7.4 show simulated contract participation rates for the three case studies. All three studies support the hypothesis that carbon contracts would substantially increase adoption of carbon-sequestering practices, although the degree of participation would depend importantly on the price of carbon and other factors such as transaction costs, and the rate of participation would vary spatially according to local biophysical and economic conditions.

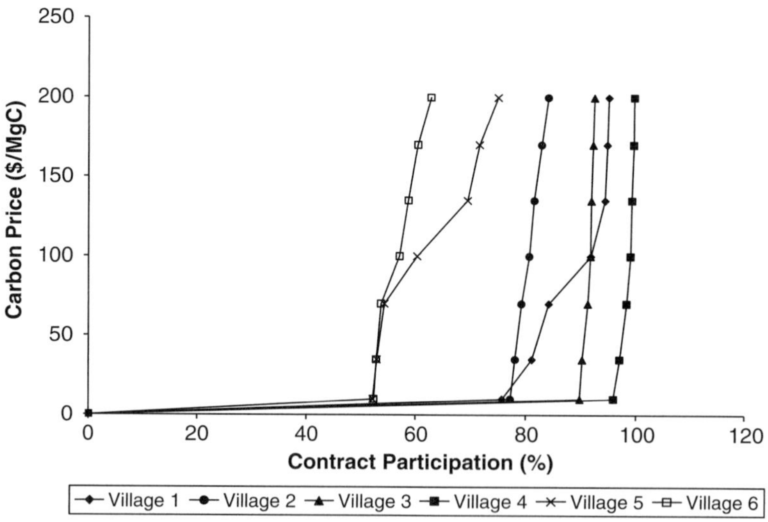

Fig. 7.2 Simulated participation in carbon contracts, Machakos, Kenya (Antle & Stoorvogel, 2008)

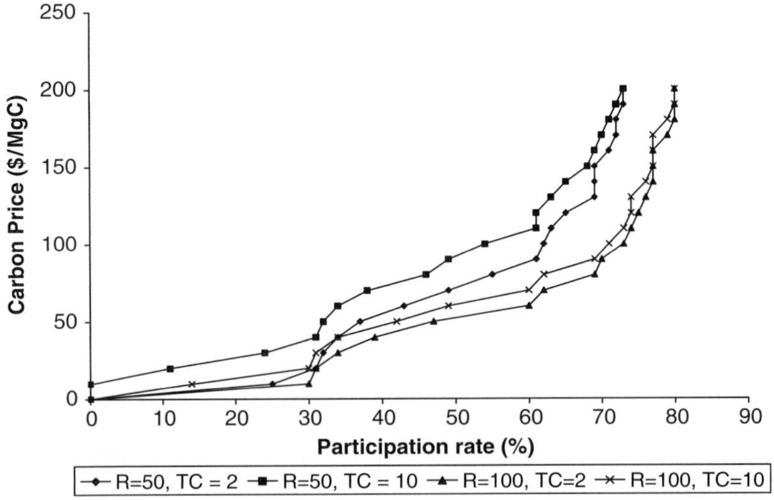

Fig. 7.3 Simulated participation in carbon contracts, Senegal peanut basin (*R* denotes percent of crop residue incorporation, *TC* denotes transaction cost in dollars per hectare per season) (Antle & Stoorvogel, 2008)

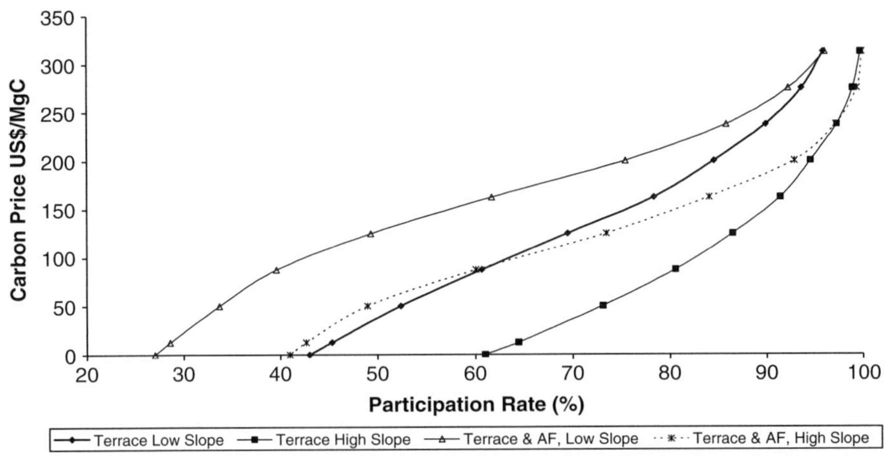

Fig. 7.4 Simulated participation in carbon contracts, Cajamarca, Peru, for adoption of terraces and agroforestry on fields with low and high slopes (Antle & Stoorvogel, 2008)

The results on contract participation in Kenya are stratified by village (Fig. 7.2). Villages 1–4 are characterized by rain-fed agriculture, whereas villages 5 and 6 are predominantly irrigated vegetable production. As Table 7.1 shows, fertilizer use is relatively high in irrigated agriculture, but very low in rain-fed crops. This fact explains the pattern shown in Fig. 7.2, with very high participation rates in villages 1–4 and lower rates in villages 5 and 6. Recall that the simulations are based on the assumption that the fertilizer required by the contract is available to farmers at the prevailing market price, and that they have the resources available to buy it, possibly by making the carbon payments in the form of fertilizer. The economic simulations show that most farms that are utilizing zero or low rates of fertilizer would earn higher returns by using more fertilizer, even if they pay the market price. Thus, these findings are consistent with the hypothesis that farmers in this region generally are under-utilizing fertilizer because of constraints on fertilizer availability or financing, not because the fertilizer price makes fertilizer unprofitable.

Figure 7.3 shows participation in carbon contracts simulated in the peanut basin of Senegal. The analysis is not stratified by region due to the relatively small amount of spatial variation in conditions in the study area. The figure shows results for simulations assuming farmers increase use of mineral fertilizer and also increase incorporation of crop residues into the soil, with two assumptions about transaction costs (discussed below). The simulations show a pattern similar to Kenya, but with generally lower participation rates, presumably because a much higher percentage of farms use fertilizer without carbon contracts (81% in Senegal, compared to 41% of rain-fed farms in Kenya, Table 7.1).

Figure 7.4 shows carbon contract participation in Peru for terracing investments alone, and for terracing combined with agroforestry. The simulations were conducted for terraces on fields with low slopes and high slopes, to represent the effects on land with more and less favorable productivity characteristics. The points on the horizontal axis with a zero carbon price represent the rates of adoption without carbon payments. Terracing alone is profitable for a larger proportion of fields at zero or low carbon prices, and profitable for a substantially higher proportion of steeply sloped fields.

The various case studies show that carbon payments do increase the adoption of more sustainable practices. However, it should be noted that, depending on the agro-ecological conditions, carbon contracts do not necessarily result in positive carbon gains but rather result in a decrease in carbon losses over time.

7.5.2 *Hypothesis 2: Carbon Contracts Transform Unsustainable Agricultural Systems into Sustainable Systems*

The results from the three studies suggest that in some cases, the combination of appropriate practices and sufficiently high carbon payments could move production systems to a much higher degree of sustainability, but in some areas that are experiencing high rates of degradation this could not be attained at plausible carbon prices.

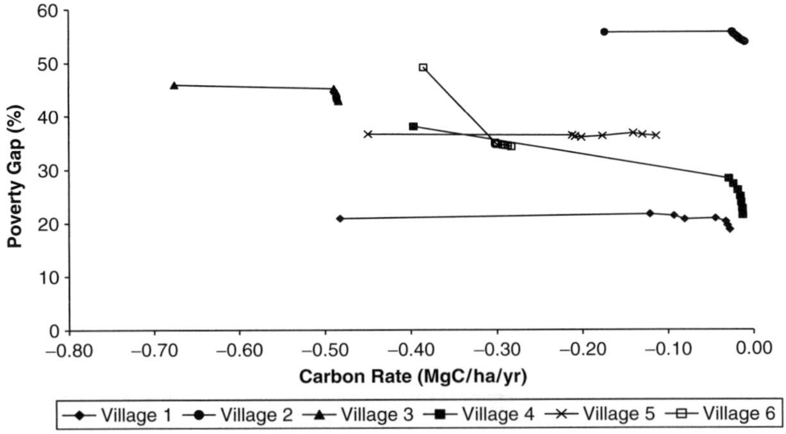

Fig. 7.5 Rate of change in soil carbon versus poverty gap with carbon contracts, Machakos, Kenya (left-most point corresponds to a zero carbon price, the price increases to $200/MgC at the right-most point) (Antle & Stoorvogel, 2008)

Figure 7.5 shows the impact of carbon contracts on the average rate of change in soil carbon simulated for farms in the Machakos, Kenya villages, in the base case (the value on the *x*-axis at a zero carbon price) and with farmers participating in carbon contracts. The data show that without carbon contracts, the rate of change in soil carbon ranges from −0.17 MgC/ha/year to −0.68 MgC/ha/year across the six villages. With introduction of carbon contracts, this rate approaches zero for villages 1, 2, and 4, and is reduced from about −0.46 MgC/ha/year to about −0.20 MgC/ha/year for village 4. Villages 3 and 6 see their rates of carbon loss reduced but remain relatively high. In this case the carbon contract results in a reduction in the rate of soil carbon loss, but the system remains unsustainable because there is an ongoing net loss of soil organic carbon. The implication is that the system will eventually approach a low-level equilibrium for both soil carbon stocks and crop productivity. In this case the carbon contract results in lower rate of soil carbon loss, but the system remains unsustainable.

Data from the Senegal study showed a baseline rate of change in soil carbon of about −0.60 MgC/ha/year (Fig. 7.6), similar to the high rates of loss found in some of the Kenyan sites. When 50% of crop residues are incorporated, the rate of carbon loss declines by about half but is still near −0.30 MgC/ha/year. Under the scenario of 100% residue incorporation, however, the average rate of change in soil carbon is greater than 0.10 MgC/ha/year. The main effect on the carbon rate comes from the increased residue incorporation by farmers who are induced to enter contracts at a low carbon price in order to gain access to fertilizer, as shown by the fact that the average carbon rate is little affected by a higher price of carbon.

Taken together, the Kenya and Senegal studies tell a similar story about the impacts of contracts based on increased use of organic material and mineral fertilizer. In areas where rates of carbon loss are relatively low, carbon contracts

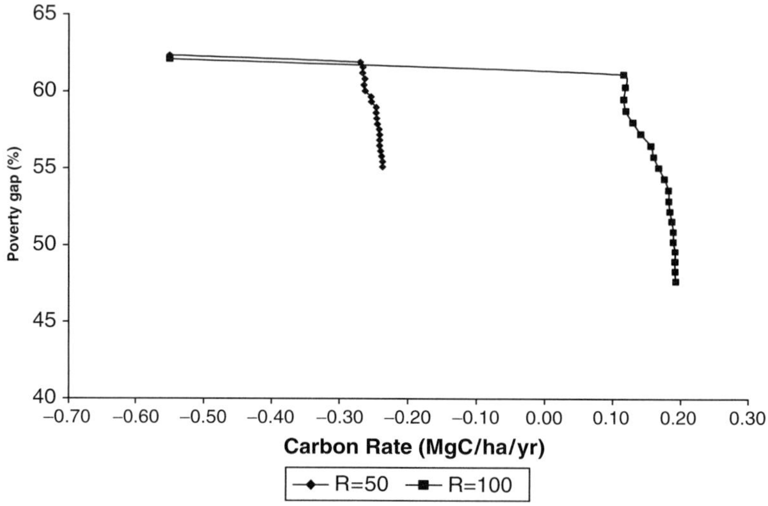

Fig. 7.6 Rate of change in soil carbon versus poverty gap with carbon contracts, Senegal peanut basin (left-most point corresponds to a zero carbon price, the price increases to $200/MgC at the right-most point; *R* denotes percent of crop residue incorporation required in the carbon contract) (Antle & Stoorvogel, 2008)

appear to have the potential to stabilize soil carbon stocks, but in areas where the rates of loss are relatively high, carbon contracts are unlikely to transform unsustainable systems to sustainable ones unless carbon prices are extremely high and farmers radically increase the amount of crop residue being incorporated into the soil (i.e., in the range of $200/MgC or higher).

Field research in Peru showed that terracing would increase soil carbon by about 6 MgC/ha/year over 10 years, and then stabilize soil C at that level or continue to increase gradually until a somewhat higher equilibrium soil C stock was attained. The terracing study showed that carbon contracts would increase adoption of terracing from 43% to 61% on low-slope fields at a carbon price of $100/MgC, and from 61% to 81% on high-slope fields at $100/MgC, but would not approach 100% adoption until the carbon price were as high as $300/MgC (Fig. 7.4). The Peru study also showed that, due to the costs of agroforestry investments, the adoption rate of terraces with agroforestry would be lower without carbon contracts, but due to the higher carbon rates associated with the combination of terracing and agroforestry, the increase in adoption would be greater, so that at sufficiently high carbon prices the overall rate of adoption could be higher. Thus, we can conclude that in the case of terracing and agroforesty in Peru, carbon contracts would increase the sustainability of the system, but the degree of improvement would be sensitive to the price of carbon and the vulnerability of the field to degradation.

7.5.3 Hypothesis 3: Carbon Contracts Increase Aggregate Rural Income

Figures 7.7, 7.8, and 7.9 show net returns per hectare in the three study areas, with the point at a zero carbon price indicating the returns without carbon contracts. These figures show that returns respond somewhat differently in each case. In Kenya, the main effect comes from farmers entering into contacts and using more fertilizer, hence the carbon price has a relatively small effect on revenue. In Senegal and Peru, participation increases more gradually with the carbon price, and consequently the revenue effect is greater, particularly in the scenarios in which carbon rates are higher.

7.5.4 Hypothesis 4: Carbon Contracts Reduce Poverty and Food Insecurity in the Rural Farm Population

Data were available for household income for the Kenya and Senegal studies. Figures 7.5 and 7.6 show the poverty gap for the Kenya and Senegal study areas, using the poverty gap defined by Foster et al. (1984) as the average percentage that poor households fall below the poverty line. The data from Kenya show that carbon contracts have a relatively small impact on poverty in most of the villages, even as the carbon price increases towards the upper limit of $200/MgC in the simulation. Two villages show a more substantial effect of both the initial entry into contracts

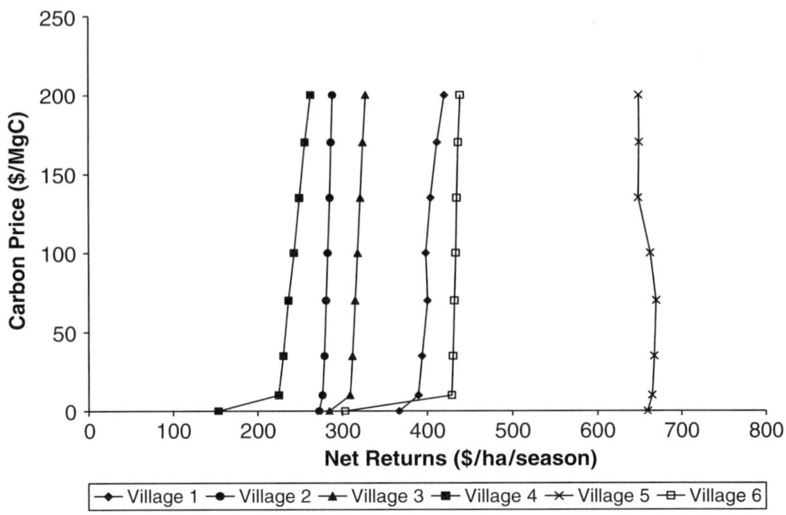

Fig. 7.7 Net returns with carbon contracts in Machakos, Kenya (Antle & Stoorvogel, 2008)

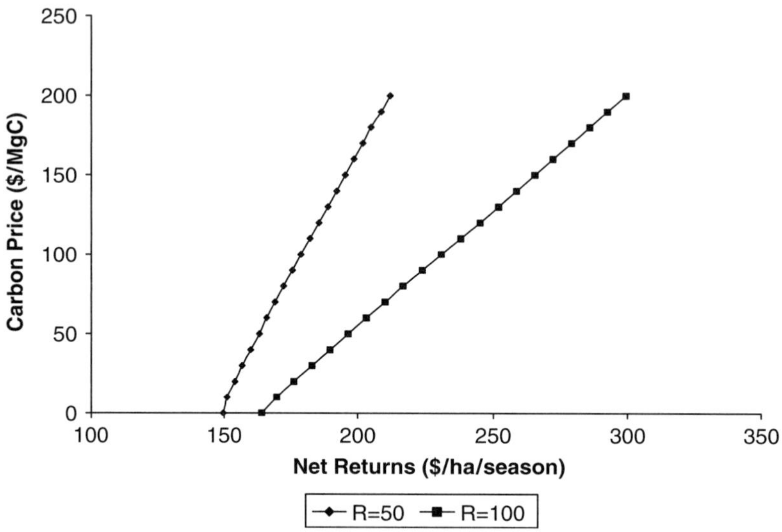

Fig. 7.8 Net returns per hectare with carbon contracts in the Senegal peanut basin (Antle & Stoorvogel, 2008)

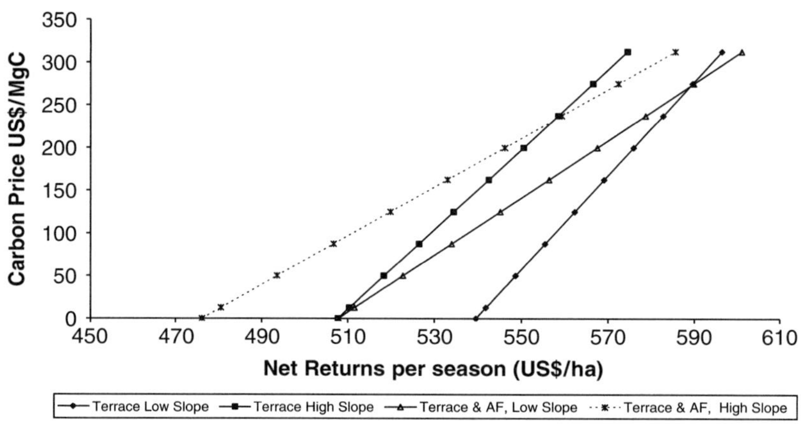

Fig. 7.9 Net returns with carbon contracts in Cajamarca, Peru, for adoption of terraces and agroforesty on fields with low and high slopes (Antle & Stoorvogel, 2008)

and a higher price. The simulations for Senegal show little effect of the initial entry into contracts at a low price. However, a higher carbon price has some impact on poverty, particularly for the scenario of 100% residue incorporation, lowering the poverty gap index from over 60% to less than 50%.

7.5.5 Hypothesis 5: The Impacts of Carbon Contracts on Poverty and Food Insecurity Are Greatest in the Poorest Regions and Households

Figure 7.5 provides little evidence to support this hypothesis, as it shows that carbon contracts reduce poverty the most in villages 4 and 6, yet village 2 has the highest initial poverty gap and carbon contracts appear to have little effect on poverty there. This fact is confirmed by Fig. 7.10, which shows the relationship between income per person and the poverty gap as the carbon price varies from \$0/MgC to \$200/MgC. In Senegal, where the poverty level is similar to the poorer villages in Kenya, the impact on poverty is also small unless the carbon price is above \$100/MgC. Although household data were not available for the Peru study, Fig. 7.9 shows little difference in the effect of carbon payments in the low-slope and high-slope fields, indicating that the effect on poverty would not be different between farms with predominantly lower or higher slopes.

7.5.6 Hypothesis 6: Transaction Costs Substantially Reduce Participation in Carbon Contracts

Figure 7.3 shows results from Senegal with transaction costs at a relatively low value (\$2/ha) and a relatively high value (\$10/ha). The simulations show that the effect of the higher transaction cost is small for the case in which all crop residues

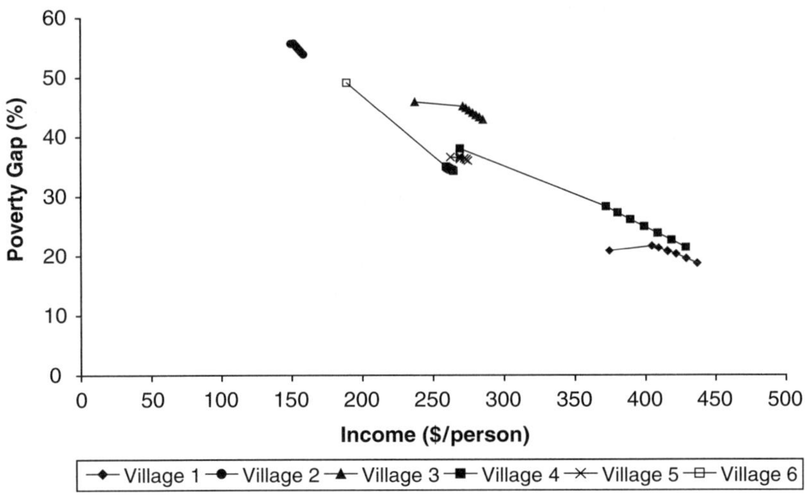

Fig. 7.10 Average income per person versus poverty gap with carbon contracts in Machakos, Kenya (left-most point corresponds to a zero carbon price, the price increases to \$200/MgC at the right-most point) (Antle & Stoorvogel, 2008)

are incorporated, because the carbon rate is sufficiently high to offset the effect of the transaction cost on the opportunity cost per ton of carbon. However, in the scenario with 50% residue incorporation, the higher transaction cost does have a substantial effect on the participation rate, reducing it from 25% to 0% when the carbon price is $10/MgC, but having a smaller impact as the carbon price increases. Similar results were obtained in the Kenya simulations.

7.6 Conclusions

This chapter explores the potential impacts of payments for agricultural soil carbon sequestration on poverty of farm households and on the sustainability of agricultural systems, using economic theory combined with evidence from three case studies in Kenya, Peru, and Senegal. The first section discusses the concept of carbon sequestration as a type of environmental service that could be provided by farmers in exchange for certain economic incentives. The second section of the chapter introduces three case studies, and then uses economic analysis to show that there are a variety of technical and economic factors that could affect adoption of practices that increase soil carbon and the sustainability of the production systems in the three case studies. Likewise many of these factors will impact how PES such as carbon sequestration could affect poverty in the farm population of developing countries. Therefore, the net effect of these various factors on participation in carbon contracts and the impact on poverty and sustainability is an empirical question. Five hypotheses were identified which were then tested using simulations from the three case studies.

All three studies support the first hypothesis that carbon contracts would substantially increase adoption of carbon-sequestering practices, although the degree of participation would depend importantly on the price of carbon and other factors such as transaction costs, and the rate of participation would vary spatially according to local biophysical and economic conditions.

The second hypothesis is that carbon contracts transform unsustainable agricultural systems into sustainable systems. The results from the three studies suggest that in some cases, the combination of appropriate practices and sufficiently high carbon payments could move production systems to a much higher degree of sustainability and stabilize carbon stocks at higher levels than would have otherwise been the case. However, in areas that are experiencing high rates of degradation, this transition to a more sustainable system is not likely to be attained at plausible carbon prices.

The case studies support the hypothesis that carbon contracts would increase aggregate income in rural areas, but the impacts on poverty were found to be relatively small. Moreover, neither the economic analysis presented, nor the results of the case studies, support the hypothesis that the impacts of carbon contracts on poverty and food insecurity are necessarily greatest in the poorest regions and households.

Finally, transaction costs were found to have a substantial effect on participation in carbon contracts in areas where expected rates of carbon accumulation are low and when carbon prices are low. This finding means that the impacts of transaction costs on participation are likely to be greatest in marginal areas – such as semi-arid areas with sandy soils – where soil carbon accumulation rates are typically lower than in areas with better soils and more rainfall or access to irrigation.

In conclusion, the economic analysis presented in this chapter, and the empirical results of the three case studies, all suggest that the likely impact of carbon contracts will be to raise rural incomes and reduce the rate of soil carbon loss. In some cases, for example when it is feasible to substantially increase the incorporation of organic matter at relatively low cost, carbon contracts may be able to stabilize soil carbon stocks at a higher level than would otherwise be economically feasible. Given that rural areas dominated by small farms are typically the poorest parts of most developing countries, these findings suggest that carbon payments could have a positive impact on the sustainability of these systems while also reducing poverty. However, these conclusions must be tempered by the finding that the impacts on poverty are likely to be relatively small, and in areas where degradation is highest and people are often poorest, carbon payments do not appear to be capable of transforming unsustainable systems into sustainable ones. Additionally, as noted in the economic analysis presented in this chapter, the participation of poor farmers in carbon contracts is likely to be constrained by the same factors that have inhibited their use of more productive, more sustainable practices in the first place. Thus, PES are not a panacea and are most likely to have a positive impact when they are implemented in an enabling economic and institutional environment.

Appendix

In this appendix we provide a more formal discussion of how farmer participation in carbon contracts is modeled in the case studies. Following Antle and Diagana (2003), the analysis is formalized by assuming that to increase the stock of SOC on a land unit, a farmer must make a change from production system i (conventional) that had been followed over some previous period (the historical land-use baseline) to some alternative (conservation) system s. We assume that utilization of management practice i up to time 0 results in a SOC level of $C(i)$, and adoption of practice s at time 0 causes the level to increase to an equilibrium $C(s)$ at time T. At time T, the soil reaches a new level at which the level of soil C stabilizes until further changes in management occur. In defining ex ante carbon contracts, we emphasize that the expected change in carbon accumulation is the relevant variable; the actual rate of carbon accumulation will typically only be verified for the land units aggregated into a contract, as discussed by Antle et al. (2003a). This expected change in carbon is assumed to be estimated by agro-ecozone and past land-use practices, with all farmers in the contract in that zone receiving credit for the same rate, as explained further below.

With a per-ton carbon contract, the farmer receives a payment of $\$P_t$ per ton of C sequestered each time period, so if the farmer changes from practice i to practice s and soil C is expected to increase by $\Delta c_t(i,s)$ tons/ha per period, the farmer receives a payment of $P_t\Delta c_t(i,s)$ per hectare per period. The net present value (NPV) of changing from system i to system s for T periods is given by:

$$NPV(i,s) = \sum_{t=1}^{T} D_t[NR(p_t, \mathbf{w}_t, z_t, s) + g_t(i,s) - M_t(i,s)] - I(i,s) \qquad (7.A1)$$

where $D_t = (1/(1+r))^t$ and r is the interest rate per time period, $NR(p_t, \mathbf{w}_t, z_t, s)$ is expected net returns per hectare for system s in period t, given product price p_t, input prices $\mathbf{w}_t$ and capital services z_t; $g_t(i,s) = g_t$ if a per-hectare contract, or $g_t(i,s) = P_t\Delta c_t(i,s)$ if a per-ton contract; $M_t(i,s)$ is the variable cost per period for changing from system i to s; and $I(i,s)$ is the fixed cost for changing from system i to system s (both variable and fixed costs of adoption may include transaction costs). If the farmer does not participate in the contract and continues producing with system i, then $g_t(i,s) = M_t(i,s) = I(i,s) = 0$ and the farmer earns $NPV(i)$. The farmer enters the contract if and only if $NPV(i,s) > NPV(i)$, and does not enter the contract otherwise.

To simplify this discussion, it is useful to consider the special case where $NR(p, w, z, s)$, P, $\Delta c(i,s)$, and $M(i,s)$ are constant over time. If we also let the fixed investment be converted into an equivalent annuity of $fc(i,s)$ dollars per period, then the expression $NPV(i,s) > NPV(i)$ can be simplified to

$$NR(p,w,z,s) + g(i,s) - M(i,s) - fc(i,s) > NR(p,w,z,i). \qquad (7.A2)$$

Note that under these assumptions, if it is profitable to enter the contract in one period, it is profitable in all periods regardless of the discount rate. More generally, the discount rate will play an important role, as in the analysis of terracing in Peru. This expression has several implications for analysis of adoption of soil carbon sequestration practices.

In the initial equilibrium in which there are no payments available for carbon sequestration, $g = 0$, and the farmer adopts the conservation practice s only if it provides higher net returns than the conventional practice i. When a carbon contract is offered for adoption of practices that sequester carbon, $g > 0$ and we can rewrite Eq. (7.A2) as:

$$g(i,s) > NR(p,w,z,i) - NR(p,w,z,s) + M(i,s) + fc(i,s). \qquad (7A3)$$

The expression on the right-hand side is the opportunity cost for switching to system s from system i. The farmer will switch practices when the opportunity cost is less than the payment per period. In the case of a per-ton contract, $g(i,s) = P\Delta c(i,s)$ and the condition for participation in the contract can be expressed as:

$$P > \{NR(p,w,z,i) - NR(p,w,z,s) + M(i,s) + fc(i,s)\}/\Delta c(i,s) \qquad (7.A4)$$

showing that the farmer will be willing to enter a carbon contract when the price per ton of carbon is greater than the opportunity cost per ton.

A critical feature of Eq. (7.A4) is the spatial variation in the opportunity cost. Net returns to the conventional and alternative practices are site-specific. Some components of the variable and fixed costs of changing practices may be site-specific (e.g., the cost of constructing a terrace), whereas transaction costs may be spatially invariant. The denominator of Eq. (7.A4), the expected rate of carbon accumulation, is specific to the agro-ecozone where the land unit is located, as noted above. Thus, the participation by farmers in carbon contracts depends on the spatial distribution of the opportunity cost of changing practices. Those land units with opportunity cost less than P will participate in the contract, and those land units with a higher opportunity cost will not participate. Summing the quantities of carbon across participating land units at each price gives the carbon supply curve for the region.

In the discussion thus far, we have assumed that the practices i and s involve a binary choice, such as the use of terracing on a field. In the case of incorporation of organic matter and use of fertilizer, however, while it is true that many farmers use no fertilizer, many farmers may use positive amounts but less than the quantities required by the carbon contract. In that case, the carbon rate used to calculate the payment is adjusted to reflect the fact that a smaller amount of carbon will be added to the soil before the new equilibrium stock of carbon is attained. The simulation studies discussed below assume that for a required input rate x_c specified in the contract, farmers who have been using a baseline rate x_b less than x_c receive credit for a carbon rate in proportion to the difference between the base rate and the contract rate, and receive zero credit otherwise:

$$\Delta c(i,s,x_c,x_b) = \Delta c(i,s)(x_c - x_b)/x_c, x_c - x_b > 0 \tag{7.A5}$$

$$= 0 \text{ otherwise.}$$

The baseline rate of input use is defined as the average rate used by the farmer on a field, over a specified period of time, before the field was entered into a carbon contract.

References

Antle JM (2002) Economic analysis of carbon sequestration in agricultural soils: An integrated assessment of carbon sequestration in agricultural soils. In: A soil carbon accounting and management system for emissions trading, Special Publication SM CRSP 2002–2004. Soil Management Collaborative Research Support Program, University of Hawaii, Honolulu, Hawaii. Available at http://www.tradeoffs.montana.edu.

Antle JM, Capalbo SM (2001) Econometric-process models for integrated assessment of agricultural production systems. Am J Agric Econ 83(2):389–401.

Antle JM, Capalbo SM, Mooney S, Elliott ET, Paustian KH (2003a) Spatial heterogeneity, contract design, and the efficiency of carbon sequestration policies for agriculture. J Environ Econ Manag 46(2):231–250.

Antle JM, Diagana B (2003) Creating incentives for the adoption of sustainable agricultural practices in developing countries: The role of soil carbon sequestration. Am J Agric Econ 85(5): 1178–1184.

Antle JM, Meijerink GW, de Jager A, Stoorvogel JJ, Vallejo AM (2005a, June) Econometric-process models of semi-subsistence agricultural systems: An application of the nutrient monitoring data for Machakos, Kenya. Paper presented at the Ecoregional Fund Workshop, Nairobi. Available at http://www.tradeoffs.montana.edu.

Antle JM, Stoorvogel JJ, Valdivia RO, Yanggen D (2003b) Assessing the economic potential for soil carbon sequestration: Terraces and agroforestry in the Peruvian Andes. Working paper, Department of Agricultural Economics and Economics, Montana State University, Bozeman, MT. Available at http://www.tradeoffs.montana.edu.

Antle JM, Valdivia RO, Crissman CC, Stoorvogel JJ, Yanggen D (2005b) Spatial heterogeneity and adoption of soil conservation investments: Integrated assessment of slow formation terraces in the Andes. J Int Agric Trade Dev 1(1):29–53.

Antle JM, Stoorvogel JJ (2006) Agricultural soil carbon sequestration, poverty, and sustainability. Tradeoff Analysis Project, Soil Management Collaborative Research Support Project, Montana State University. Available on-line at http://www.tradeoffs.montana.edu.

Antle JM, Stoorvogel JJ (2008) Agricultural carbon sequestration, poverty, and sustainability. Environ Dev Econ 13(3):327–352.

Blackman A (2001) Why don't lenders finance high-return technological change in developing-country agriculture? Am J Agric Econ 83(4):1024–1035.

de Jager A, Onduru D, van Wijk MS, Vlaming J, Gachini GN (2001) Assessing sustainability of low-external-input farm management systems with the nutrient monitoring approach: A case study in Kenya. Agric Syst 69(1–2):99–118.

Diagana B, Antle J, Stoorvogel JJ, Gray K (2005) Economic potential for soil carbon sequestration in the Nioro Region of Senegal's Peanut Basin. Agric Syst 94:26–37.

Foster J, Greer J, Thorbecke E (1984) A class of decomposable poverty measures. Econom 52: 761–765.

Gachimbi LN, van Keulen H, Thuranira EG, Karuku AM, Jager A, Nguluu S, Ikombo BM, Kinama JM, Itabari JK, Nandwa SM (2005) Nutrient balances at farm level in Machakos (Kenya), using a participatory nutrient monitoring (NUTMON) approach. Land Use Policy 22(1):13–22.

Government of Kenya (2003) Geographic dimensions of well-being in Kenya: Where are the poor? From districts to locations, Vol. 1. Central Bureau of Statistics, Ministry of Planning and National Development, Nairobi.

Interinstitutional Commission (2005) Estudio de linea de base de la Provincia de Cajamarca. Published by Los Andes de Cajamarca, Cajamarca, Peru.

International Energy Agency Greenhouse Gas R&D Program (2005) Assessment of the costs and enhanced potential for carbon sequestration in soils. Technical Report No. 4, IEA Greenhouse Gas R&D Programme, Glos, UK.

Kherallah M, Delgado C, Gabre-Madhin E, Minot N, Johnson M (2002) Reforming agricultural markets in Africa. Johns Hopkins and IFPRI, Baltimore, MD.

Koning N, Heering N, Kauffman S (2001) Food insecurity, soil degradation and agricultural markets in West Africa: Why current policy approaches fail. Oxford Dev Stud 29(2):189–207.

Lal R, Kimble LM, Follett RF, Cole CV (1998) The potential of U.S. cropland to sequester carbon and mitigate the greenhouse effect, Ann Arbor Press, Chelsea, MI.

Lewandrowski J, Peters M, Jones C, House R, Sperow M, Eve M, Paustian K (2004) Economics of sequestering carbon in the U.S. agricultural sector. Technical Bulletin No. (TB1909), Economic Research Service, USDA, Washington, DC.

Lynam JK, Nandwa SM, Smaling EMA (1998) Nutrient balances as indicators of productivity and sustainability in Sub-Saharan African agriculture: Introduction. Agric Ecosyst Environ 71:1–4.

Mooney S, Antle JM, Capalbo SM, Paustian K (2004) Design and costs of a measurement protocol for trades in soil carbon credits. Can J Agric Econ 52(3):257–287.

National Research Council (2004) Valuing ecosystem services: Toward better environmental decision making. National Research Council, National Academy Press, Washington, DC.

Paustian K, Antle J, Sheehan J, Paul E (2006) Agriculture's role in greenhouse gas mitigation, Pew Center on Global Climate Change, Arlington, VA.

Romero C, Stroosnijder L (2001) A multi-scale approach for erosion impact assessment in the Andes. In: Methodologies for interdisciplinary multiple scale perspectives. Proceedings of the SAAD-III Third International Symposium on Systems Approaches for Agricultural Development, Lima, Peru.

Sanchez P (2002) Soil fertility and hunger in Africa. Science 295:2019–2020.

Scherr SJ (1999) Soil degradation: A threat to developing-country food security by 2020? Food, Agriculture, and the Environment Discussion Paper No. 27, International Food Policy Research Institute, Washington, DC.

Soil Management CRSP (2002) A soil carbon accounting and management system for emissions trading, Special Publication SM CRSP 2002-04, Soil Management Collaborative Research Support Program, University of Hawaii, Honolulu. Available at http://www.tradeoffs montana. edu.

Stoorvogel JJ, Antle JM, Crissman CC, Bowen W (2004) The tradeoff analysis model: Integrated bio-physical and economic modeling of agricultural production systems. Agric Syst 80(1): 43–66.

Sunding D, Zilberman D (2001) The agricultural innovation process: Research and technology adoption in a changing agricultural sector. In: Gardner BL, Rausser GC (eds.), Handbook of agricultural economics, Vol. 1A: Agricultural Production, Elsevier, Amsterdam, the Netherlands.

Tiffen M, Mortimore M, Gichuki F (1994) More people, less erosion: Environmental recovery in Kenya, John Wiley & Sons, Sussex.

Valdivia RO (1999) Farm data collection in la Encañada, Cajamarca, Peru, to support analysis of tradeoffs in sustainable agriculture. A Report to the Project, Trade-offs in Sustainable Agriculture and the Environment in the Andes: A Decision Support System for Policy Makers, Montana State University, Bozeman, MT. Available at http://www.tradeoffs.montana.edu.

Valdivia RO (2002) The economics of terraces in the Peruvian Andes: An application of sensitivity analysis in an integrated assessment model. MS Thesis, Department of Agricultural Economics and Economics, Montana State University, Bozeman, MT.

Valdivia RO, Antle JM (2002) La Encañada: Description of the data preparation process for use with the tradeoff analysis model. CD detailing the documentation, programs, and databases used in the tradeoff projects. Montana State University, Bozeman, MT. Available at http:// www tradeoffs.montana.edu.

Zeller M, Johannsen J, Alcarez GV (2005) Developing and testing poverty assessment tools: Results from accuracy tests in Peru. MicroReport, Accelerated Microenterprise Advancement Project, The IRIS Center at the University of Maryland, College Park, MD. Available at http:// www.microfinancegateway.org.

Chapter 8
Lessons Learned from Mexico's Payment for Environmental Services Program

Jennifer Alix-Garcia, Alain de Janvry, Elisabeth Sadoulet, and Juan Manuel Torres with the assistance of Josefina Braña Varela and Maria Zorilla Ramos

Abstract This chapter outlines the evolution of Mexico's payments for hydrological services program from its inception through the first 2 years of the program's implementation. Background information on forests, deforestation, and potential environmental services provide context for a political economy analysis of the path the program traveled through Mexico's legislative and administrative structures. We also analyze the characteristics of the recipients during the first 2 years, including results from a survey of participants and community case studies. A final section extracts lessons from the Mexican experience, including possible alternative program designs to address some of the problems encountered in its implementation.

8.1 Introduction

Programs of payments for environmental services (PES) are becoming a popular way of creating, conserving, and restoring natural resources that provide public benefits. These programs encompass a variety of strategies, including payments for the continued existence of a forest, for the planting of native species on fallowed land, or for working-lands projects. Though the term "payments for environmental services" is relatively new, such programs have been in existence for quite some time. The Nature Conservancy pioneered one type of PES strategy, having purchased 116 million acres around the world since 1951 (The Nature Conservancy, 2003). In the United States, the water supply of New York City is partially guaranteed by the subsidized conservation efforts of working farmers in the watershed that feeds the metropolis, an effort which began in the 1980s.

In recent years, such programs have increasingly been introduced by developing countries, with one of the earliest efforts occurring in Costa Rica in 1997, and pilot programs mushrooming throughout Latin America and Asia (World Bank, 2005). In 2002, more than 300 such schemes were inventoried (Mayrand & Paquin, 2004). Despite the increasing number of such projects, there is a scarcity of rigorous

L. Lipper et al. (eds.), *Payment for Environmental Services in Agricultural Landscapes*, Natural Resource Management and Policy.
doi: 10.1007/978-0-387-72971-8_8;

studies analyzing their effectiveness in providing environmental services (ES) and their impacts on the people and communities receiving the payments. This chapter intends to partially address this gap by presenting an analysis of the first 2 years of the Mexican PES program for hydrological services (PSA-H), which began in 2003, where payments are made to individuals and communities as incentives to preserve existing forests. Although the program has not been in place long enough to assess results in terms of forest conserved, sufficient time has passed to extract various lessons from both the political process that led to the program as well as the impact of the payments on recipient communities and, to some extent, on their forest management behavior.

The following pages will outline the evolution of the Mexican PSA-H from the original proposal through the first 2 years of the program's implementation.[1] The second section provides background information on forests, deforestation, and potential ES in Mexico. Section 8.3 presents a political economy analysis of the path the program traveled through Mexico's legislative and administrative structures. The fourth section focuses on the recipients of the first 2 years of the program, including a summary of results from a survey of participants and community case studies. Section 8.5 extracts lessons from the Mexican experience, including possible alternative program designs to address some of the problems encountered in its implementation.

8.2 Deforestation and Environmental Services in Mexico

According to the National Forest Commission (CONAFOR), forests and areas with natural vegetation (including arid and semi-arid environments) cover 72% of the Mexican territory (CONAFOR, 2001). Mexico is among the most biologically diverse countries in the world, with first place in reptilian diversity, third in bird, and fourth in mammal diversity. Its plant diversity exceeds that of the United States and Canada combined. The area in temperate and tropical forests (covering over 50% of the country in 2000) is shown in Fig. 8.1.

These biological riches and the hydrological services associated with forests are threatened by deforestation, which has reduced the extension of forests by 50% over the past 5 decades. Velázquez et al. (2003) estimate the overall deforestation rate at 1.2% per year, a rate that, if it continues, would eliminate all forests in the country within a century. This deforestation is not uniformly distributed across forest types. Table 8.1 shows the distribution of deforestation across forest types between 1994 and 2000. Clearly, deforestation in tropical forests is progressing at a much faster rate, 2.4% per year, than in temperate forests, 1.2% per year, and in scrub forests, 0.6% per year.

[1] Note that this chapter reviews the program through 2005. The Instituto Nacional de Ecología (INE) is currently undertaking an updated review of the program. Also note that since the implementation of the PSA-H, several other federal and local programs to conserve environmental services have begun in Mexico, and the administration has changed.

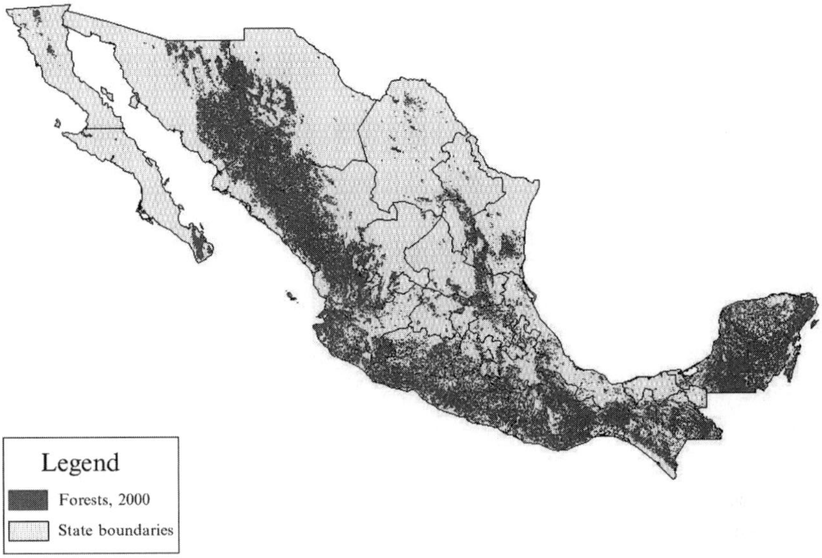

Fig. 8.1 Mexican forest cover, 2000. Courtesy of SEMARNAT (2000)

Table 8.1 Change in forest cover by forest type from 1993 to 2000

Forest type	Km^2 in 1994	Km^2 in 2000	Average annual rate of change
Temperate forests (pine, oak, and cloud)	352,969	328,471	−1.2
Tropical forests (rainforest and dry tropical)	356,228	308,001	−2.4
Scrub forests	578,841	558,077	−0.6
All forests	1,288,038	1,194,549	−1.2

Source: Veláquez et al. (2003)

About 5% of Mexico's remaining forest is located in the National System of Protected Areas (SINAP), while private owners control around 15–20% of the forest. The remainder of the forested land (75–80%) is found in the *ejidos* and *comunidades*, rural communities resulting from a drawn-out land reform that extended from the end of the 1910 Revolution until the constitutional reform of 1994. In general, these types of communities hold their forests in common and have private parcels for farming.

Where and what are the ES provided by Mexican forests? The PSA-H focuses on a service that the forests provide strictly within its national boundaries – the growing scarcity of water. Although the relationship between forest cover and water flows is highly debated, there is clearly a positive effect of forests on water quality, if not always on quantity. For this reason, the original PES program proposal focused on the watersheds defined as overexploited, as well as on cloud

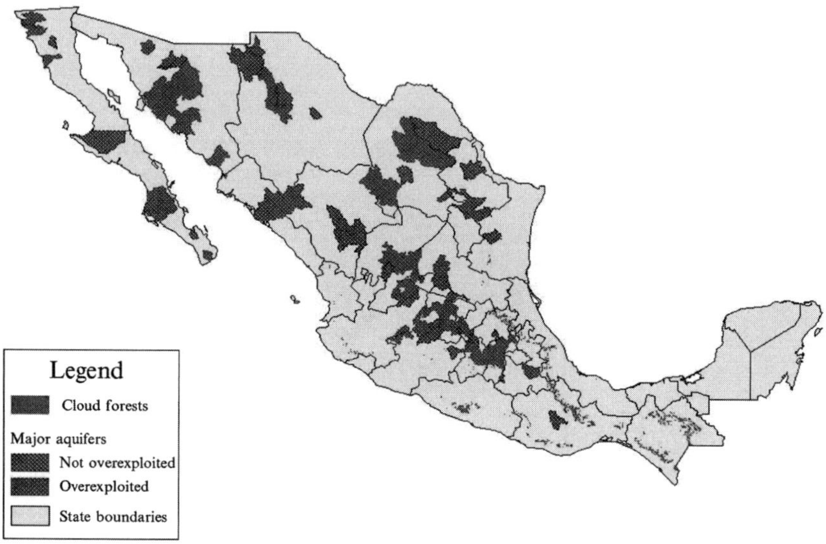

Fig. 8.2 Major aquifers and cloud forests in Mexico. Courtesy of National Institute of Geography, Statistics and Information Systems, Mexico (INEGI, 2005)

forests, which are thought to have a particularly strong relationship with water quantity (García Coll, 2002). According to the National Water Commission, 66% of the most important aquifers in Mexico are overexploited, with an average extraction 190% above the replacement rate. It is estimated that 28.7% of the country's population located in the aquifer area defined as very high or extremely high overexploitation (Muñoz et al., 2005). Around 17,000 hectares of cloud forest, or about 3% of the total forest, are found in Mexico, all of them in the central and southern zones of the country. As Fig. 8.2 shows, the distribution of these areas is highly regionalized, with major concentrations of overexploited watersheds in the central and northern areas of the country and the bulk of cloud forests in the states of Oaxaca and Chiapas.

With the intent of comparing the total forested area with the area prioritized by the national scheme, Fig. 8.3 shows the distribution of all forests, both tropical and temperate, overlaid with the overexploited aquifers. This figure highlights several important issues. First, it shows that there is little overlap of the forests with the overexploited aquifers, although a forest may help with aquifer recharge without being located within the aquifer itself. It is very important that the forests located in the recharge zone for these aquifers be identified in order to establish which provide potential water services. The cloud forest is clearly located in areas where it is highly unlikely that they are recharging the aquifers of concern.

The forested area is very large, which implies that the potential to provide alternative services with potential international markets, like carbon sequestration and biodiversity, may also be. These types of services may be particularly important

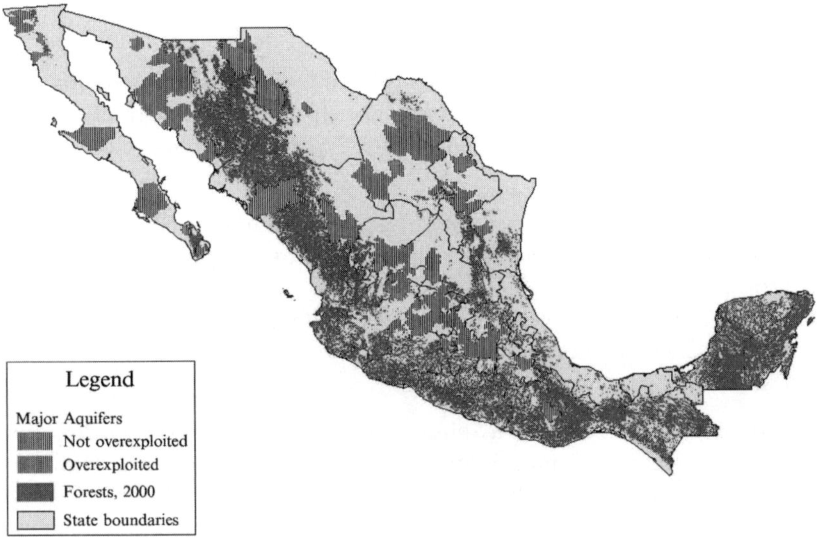

Fig. 8.3 Forested areas and overexploited aquifers. Courtesy of National Institute of Geography, Statistics and Information Systems, Mexico (INEGI, 2005)

for the tropical forests of Southern Mexico, given their lack of overlap with critical watersheds. There are large areas of the country – Baja California, Nuevo Leon, San Luis Potosí, and Zacatecas – which possess very little forest (though they have overexploited aquifers) and that would not benefit from an ES program targeted at forest conservation. The water-focus of the PSA-H in Mexico can only justify payments to very specific tracts of forest. However, the large tracts of remaining forest may still house important benefits, including reducing soil erosion, maintaining biodiversity, and improving air quality.

8.3 The Evolution of Mexico's PES Program for Hydrological Services[2]

This section details the evolution of the PES program from the beginning of 2000 to the close of the second year of payments in 2005. The initial idea, proposed by the Instituto Nacional de Ecología (INE) together with academics from the Universidad

[2]This section is a summary of the analysis conducted by Josefina Braña Varela and María Zorilla Ramos. All monetary references are expressed in US dollars unless otherwise noted.

Iberoamericana (UIA), was to target payments towards areas of the country defined as "high" or "very high marginality" according to a municipal marginality indicator based on information from the population census (CONAPO, 2000). The INE hoped to begin with a pilot program administered by an outside institution before launching into a larger, nationwide payment scheme. In October of 2002, the proposed pilot project was intended to last for 2 years beginning in the spring of 2002 with the following features:

- The pilot would be the responsibility of a Subsecretariat of SEMARNAT, and would include 100 *ejidos* and an annual payment of $20 per hectare.
- The project would be focused on water services.
- The beneficiaries would be *ejidos* and *comunidades* with forests in "priority watersheds," meaning those that are both overexploited and serving as the main water source for large population centers.

This proposal was presented to a Subsecretariat of SEMARNAT, the directors of which initially allocated $2 million over 2 years beginning in 2002. However, a budget cut to the Secretariat left the program without funding for the following year. Given this lack of support, the Secretary of the Environment presented the project to Felipe Cárdenas, the president of the Comisión Nacional Forestal (CONAFOR), who agreed to take responsibility for it. INE, with technical support from UIA, Centro de Investigación y Docencia Económica (CIDE), and UC Berkeley, proposed to link the financing of the program directly to the services obtained through an additional payment on water use, ideally calculated by watershed. The final objective of the program was to calculate a budget by watershed in order to link the benefits more closely to the costs. Unfortunately, this was impossible to do with the existing data. The lack of data on water services led to the proposal of designating 2.5% of annual water fees, which for 2002 were around $20 million, to finance the project. Water fees in Mexico are collected at the municipality level, but because water is officially state property, the fees are sent to the federal government, which then returns them to the municipalities to invest in infrastructure.

The proposal was presented to the Secretariat of Hacienda and Public Credit (SHCP) (similar to the US Treasury Department), which opposed the idea of using water fees to pay for the program, arguing that SHCP had an informal agreement with the municipalities to devolve 100% of their water fees to be invested in infrastructure projects. In the face of resistance in SHCP, Cárdenas presented the proposal directly to the National Congress, where it was accepted. SHCP, through the Finance Commission, succeeded in converting the 2.5% levy on water fees into a fixed amount equivalent to $20 million per year taken from the water fees collected. This eliminated the possibility for the program to benefit from future increases in water fee revenues. It has been estimated that, had the 2.5% levy remained in place, the program would have tripled its budget by 2005.

At this point, several changes occurred in the original proposal. First, the idea of targeting marginalized communities was removed from the discussion. The second important change was that the program would no longer be targeted toward overexploited watersheds, but instead implemented nationwide. Finally, the pilot project was cancelled as a result of the progress of the political calendar, which made it risky for the agency to run the pilot and then advocate for a national level program.

Soon after, CONAFOR initiated a national tour to promote the program, though at this point it was unclear exactly what the shape of the program would be. This premature promotion was undertaken because the responsible parties were worried that the program would fail due to lack of demand for the budgetary resources, given that the target audience might never have heard of ES. Unfortunately, this strategy created a variety of problems. The promotion failed to adequately convey the concept of ES, but was very successful in generating false expectations. Because the policy had yet to be well defined, many of the concepts described by the CONAFOR representatives were not incorporated into the final program.

Meanwhile, SHCP classified the new program as a subsidy, which required the submission of "rules of operation" which would have to be debated in a public forum. In April–May 2003, internal negotiations over the rules of operation within CONAFOR resulted in the following changes. Wanting to avoid the technical problem of measuring forest density, the payment schedule changed from three payments ($40 per hectare for high-density cloud forests, $30 for medium-density cloud forests and other forests of high density, and $20 for forests of medium density) to two ($40 per hectare for cloud forest and $30 for others). The liberalization of most of the agricultural products under the North American Free Trade Agreement (NAFTA), set for 2003, also affected the program, as an organization composed of various rural opposition groups gained strength. After several weeks of negotiations, President Fox signed an agreement through which he gave the right to a commission of representatives to review and discuss all government programs having to do with the rural sector. The biggest impact on the program was the inclusion of lands under management for timber harvest, which had previously been excluded from consideration. At the end of the negotiations, the rules were sent to the Federal Commission of Regulation, and published in the Federal Registry on October 3, 2003.

At this date, implementation of the program began under the responsibility of CONAFOR. The fact that the rules of operation were published in October posed a substantial problem for CONAFOR, which, due to the rigidity of the governmental fiscal year, had to spend 4 million pesos in less than 3 months. Normally, funds allocated to federal programs must be spent within the fiscal year, but the managers of the PES program wanted to use the 2003 budget allocation to guarantee payments to participant communities for five consecutive years. Fortunately, the Mexican Forest Fund (FFM) facilitated this process by allowing the set-aside of the remaining $16 million from the program's annual budget to cover the next 4 years. Although the existence of the FFM was a great advantage, allocating even the relatively smaller budget in such a short period of time is a difficult task, and it was complicated by a lack of personnel for program implementation – in October, only three staff members of CONAFOR had been assigned to promote the program and review requests for payments.

Application for the program was very simple – all it required was a two-page form and proof of legal ownership. For *ejidos*, a document verifying that a general assembly had been called in the participating community and that a vote had taken place was also required. The program contracts gave payments for a specified area of forest within each community's boundaries according to the dual price system of $40 per hectare for cloud forest and $30 per hectare for other types. In most cases the contract specified that removal of trees from the community's entire forested area (even outside of the area for which payments were being made) constituted a contract violation and subsequent non-payments. Contracts were assessed and renewed on a yearly basis based upon contract compliance the previous year. Monitoring was to be conducted on a random sample of participants using satellite imagery. The criteria for selecting properties were three: (1) Properties with forests with more than 80% density (i.e., hectares with more than 80% tree cover), (2) located in overexploited aquifers, and (3) with nearby population centers greater than 5,000 inhabitants.

CONAFOR hired supplemental workers to assist in the promotion of applications and the selection of recipients. Unfortunately, by the time the hiring and training process was over, there was only 1 month left for these activities. As a result, the promotion of the program was only done to CONAFOR's traditional constituency – *ejidos* and private landowners with wood extraction projects supported by its other programs.

CONAFOR received many more demands than it could finance. With only three employees to review, catalogue, and evaluate 900 proposals, several changes were made in order to expedite the process. First, a combination of misinterpretation of the rules and the fact that there was only one geographical technician to analyze the satellite images resulted in the elimination of the criterion of forest density in favor of forest coverage, meaning that only properties that were more than 80% covered with forest were selected. This resulted in the selection of much larger properties, and with lower population density and probably a lower probability of deforestation than if the 80% forest density criterion had been used.

CONAFOR had considered monitoring the program through high-resolution satellite images. However, insufficient time and staff meant that satellite images of potential properties were not purchased, with the result that properties located in regions where images had not been purchased were not allowed in the program. In addition, if the properties were not already georeferenced, they could not receive payments since placing them on a satellite image would be impossible. Finally, in the communities with forest extraction activities, it was often impossible to determine if the area chosen for environmental payments overlapped with area earmarked for tree harvests.

At the beginning of 2004, two important selection criteria were added as a result of an internal shift of responsibilities within CONAFOR. A piece of land could be in a National Protected Area or in a "Priority Mountain" and receive the same priority as a property in an overexploited watershed. The Priority Mountain program was also administered by CONAFOR, and focuses on protection of water production, carbon capture, and biodiversity in 60 mountains throughout

Table 8.2 Changes in the targeting strategy

Original targeting rules (SEMARNAT/INE)	Final targeting rules (SEMARNAT/CONAFOR)
• Pilot program with an experimental design	• Nationwide program – Rules of operation – Establishment of a trust fund
• Beneficiaries *ejidos* and *comunidades* located in priority watersheds – Overexploited – Serving large populations	• Beneficiaries augmented to include private owners
• Other selection criteria – Forest cover – Clear property rights – Ecosystem type – Marginalization	• Added selection criteria – Priority mountains – Availability of satellite image – Protected areas
• Priority given to forest with high deforestation risk	• Subtracted selection criteria – Marginalization – Deforestation risk

Source: Jennifer Alix-Garcia, Alain de Janvry, Elisabeth Sadoulet, and Juan Manuel Torres

the country. Table 8.2 summarizes the changes in the targeting criteria from the original proposal to the program's 2003 implementation.

In 2004 CONAFOR again received applications far in excess of what it could finance. By this time, however, a shift of management within CONAFOR had resulted in a point system approach: payments were allocated by giving a point for each of the criteria listed in the rules of operation and contracts awarded to those properties with the highest point values.

8.4 Results of Implementation, 2003–2004

8.4.1 Summary Statistics for Participating Communities

This section describes the recipients of the initial payments made by the program. The data used to characterize the participants come from an evaluation of the program conducted by the Colegio de Posgraduados (COLPOS, 2004) and a survey conducted by INE (INE, 2004). The COLSPOS survey was comprised of over 300 randomly selected participants (common property and private owners), while the INE survey covered 27 participant *ejidos* selected to reflect the mean characteristics of the *ejidos* participating in the PES program in 2003. Except where otherwise noted, the statistics presented come from the INE survey.

Applications for the program were received from 25 states, but only 15 actually received PES contracts, with nearly 127,000 hectares enrolled. Table 8.3 shows the

Table 8.3 Distribution of PES contracts by state, 2003

State	Number of contracts[a]	Hectares enrolled[a]	Hectares forested[b]	Percentage enrolled	Payments[a] (US dollars)
Baja California Sur	2	2,231	442,874	0.50	63,749
Coahuila	29	7,188	514,771	1.40	205,368
Chihuahua	8	11,279	7,702,586	0.15	322,269
Distrito Federal	4	5,058	38,301	13.21	144,507
Durango	16	15,224	5,870,668	0.26	434,959
Estado de México	2	709	740,205	0.10	20,271
Jalisco	24	11,801	4,407,937	0.27	337,175
Michoacán	10	8,633	3,510,806	0.25	254,317
Nayarit	9	3,222	1,731,879	0.19	96,721
Nuevo León	1	1,450	571,327	0.25	41,424
Oaxaca	20	28,469	6,392,049	0.45	813,396
Puebla	19	5,655	1,599,605	0.35	168,641
Querétaro	45	4,664	419,098	1.11	143,792
San Luis Potosí	7	9,874	857,912	1.15	282,121
Veracruz	75	11,361	1,135,089	1.00	328,434
TOTAL	271	126,818	35,935,107	0.35	3,657,143

[a]CONAFOR (2004)
[b]Estimate for year 2000 (SEMARNAT, 2004)

Table 8.4 Physical characteristics of participating *Ejidos*

Characteristics	Estimate
Average size of forested area, in hectares	3,961
Average hectares enrolled in the program	466
Total hectares of cloud forest in the sample	1,830
Total hectares of temperate forest in the sample	55,280
Total hectares enrolled in sample	12,680
Percentage of participants with cloud forest (from total)	2.9
Average annual forest loss in hectares, 1994–2000	38
Percentage of participants harvesting wood for sale	63

Source: Own estimates with data from INE (INE, 2004)

distribution of payments by state. A few states – Oaxaca, Durango, and Veracruz – got a large share of the budget (43%). The states with the smallest number of hectares enrolled were the Distrito Federal, Nuevo León, Baja California Sur, and Nayarit.

For the first year of operation, *ejidos* and *comunidades* accounted for 47% of the contracts and for 93% of the area contracted. Table 8.4 shows the main characteristics of participating communities. The average size of participant *ejidos* was 3,961 hectares. The mean number of hectares enrolled in the program is 466, with 2.8% of the total hectares in the sample being cloud forests. On average, 75% of the land in participating *ejidos* is considered common property. Out of the sample of 23 common properties receiving payments, 15 (65%) had experienced deforestation

Table 8.5 Distribution of payment recipients by aquifer type, 2003 and 2004

Aquifer type	Total area (%)	Population living in area (%)	Hectares in PES, 2003 (%)	Hectares in PES, 2004 (%)
Extremely overexploited (+100% to +800%)	0.05	9.2	0.02	0.00
Strongly overexploited (+50% to +100%)	0.04	19.5	0.00	0.00
Moderately overexploited (+5% to +50%)	18.6	14.5	13.3	9.6
In equilibrium (−5% to +5%)	2.9	11.3	0.01	0.00
Not overexploited (<−5%)	65.1	45.4	78.7	85.0
No data	13.4	0.1	8.0	5.3
TOTAL	100	100	100	100

Source: Muñoz et al. (2005)

over the 1994–2000 period. The average yearly rate of forest loss amongst those with positive deforestation was 1.5%. Sixty-three percent of the participants harvest wood for sale, and within these *ejidos*, 74% have reported illegal logging in their properties. In some of these *ejidos*, the legal harvest volumes exceed 32,000 m^3, far beyond the national average of 4,546 m^3 a year.

Table 8.5 details the distribution of PES hectares according to watershed, where the population is all the participants in the program. According to this information, the payments have not been going to areas where the aquifers are overexploited. Essentially no hectares under PES are forests in aquifers qualifying as extremely or strongly overexploited. Seventy-nine percent and 85% of the PES hectares, in 2003 and 2004, respectively, are in aquifers that are not *over*exploited, with the remainder of the hectares in aquifers that qualify as moderately overexploited. Just because a property is not directly on top of an aquifer, however, does not mean that it is not in the recharge zone of that aquifer.

Table 8.6 shows the distribution of PES hectares according to forest type. Recall that cloud forests are given a slightly higher payment per hectare under the current scheme, with the hope that a proportionately higher number of cloud forest hectares be enrolled. Again, these results are based on a census of the payment recipients. The effort to enroll a larger proportion of cloud forests was successful; in 2003, 6.8% of the enrolled hectares were cloud forests and, in 2004, 16.3%, relative to the overall percentage of 3.4 and the eligible area of 6.6%. The temperate forest categories of pine, oak, and fir are over-represented as a group, both relative to the eligible areas and the national distribution. This may be because there are more commercial forests in these ecosystems, and the owners of these forests are likely to have a closer relationship with CONAFOR through other programs administered by the Commission. It is impossible to tell whether this bias results from greater promotional efforts by CONAFOR with these types of forest holders or is simply the result of self-selection.

Table 8.6 Comparison of forest types enrolled in PES, 2003 and 2004

Forest type	Distribution at the national level (%)	Hectares enrolled in PES, 2003 (%)	Hectares enrolled in PES, 2004 (%)	"Eligible" area[a] CONAFOR, 2004 (%)
Pine and oak-pine forests	37.8	60.1	43.9	46.4
Oak–fir forests	23.0	17.2	24.9	18.0
Cloud forests	3.4	6.8	16.3	6.6
Low tropical forests	25.0	3.0	4.9	2.4
Medium and high tropical forests	10.8	12.9	10.4	26.6
TOTAL	100	100	100	100

Source: Muñoz et al., 2005

[a]In 2004, CONAFOR used three criteria to define eligibility: overexploited aquifers, priority mountains, and protected areas

Table 8.7 Distribution of deforestation risk in participant communities

Deforestation risk index	2003 recipients		2004 recipients		National
	%	Hectares	%	Hectares	%
Very high	3.6	5,922	10.9	18,550	20
High	6.7	11,034	16.8	28,529	20
Medium	17.3	28,446	20.5	34,953	20
Low	30.4	50,046	29.9	50,940	20
Very low	41.9	68,815	21.8	37,133	20
Total	100	164,263	100	170,105	100

Source: Muñoz et al. (2005)

In order to predict how effective the payments might be in reducing deforestation, it is interesting to consider the distribution of the payments according to predicted deforestation risk. Table 8.7 shows the distribution of forest area among participants according to deforestation risk estimates, where risk is determined by exogenous community characteristics. It can be observed that most of the participant forests have low and very low deforestation risk indices, suggesting that they would have been conserved even in the absence of the program.

Table 8.8 describes the distribution of PES hectares according to the level of marginalization of the participating communities. The definition of marginal is given by Mexico's National Population Council (CONAPO, 2000) and is based upon a combination of nine indicators encompassing literacy, education, employment, and quality of dwelling. Interestingly, even though marginality was removed from the program as a selection criterion, the majority of the enrolled hectares – 71.9% in 2003 and 82.9% in 2004 – are located in areas with high or very high marginality. It is important to emphasize that the correspondence between payments and poverty is *purely coincidental*, reflecting the fact that 80% of the forest in Mexico is held by *ejidos* and *comunidades*, and that within this group, 86.3% of the forest is located in communities with high or very high marginality. Some bias does appear to exist towards including areas of high, rather than very high, marginality. One explanation for this is that the former communities are less likely to have

Table 8.8 Marginalization and PES

Level of marginalization	PES 2003		PES 2004		Proportion in *Ejidos* with >100 hectares of forests (%)	National distribution across forests (%)
	Hectares	%	Hectares	%		
Very high	41,282	25.0	36,567	21.5	69.1	31.2
High	77,339	46.9	104,362	61.4	17.2	16.3
Medium	29,924	18.1	13,521	7.9	8.6	22.2
Low	13,018	7.9	9,741	5.7	3.3	10.1
Very low	3,386	2.1	5,839	3.4	1.8	20.3
Total	164,948	100	170,030	100	100	100

Source: Muñoz et al. (2005) and own estimates with data from CONAPO (2000)

commercial forests (and hence contact with CONAFOR), and are probably more remote and therefore difficult to reach.

It is also interesting to consider how payments received were distributed within communities. The use of the 2003 payments varied from distributing 100% equally between all members to investing all the money into public goods for the community, with many intermediate cases where the allocation included a combination of direct distribution of payments, payment for guarding the forest and fire prevention, and investment in local public goods. The survey shows that 18% of the *ejidos* decided to distribute all payments directly among *ejido* members, 22% invested the entirety in forest activities related to conservation, 18% allocated the full amount to public goods not related to forestry, while the remaining 42% adopted a combination of the three strategies.

In 87% of the communities surveyed, participants declared that they had respected the contract, while 26% stated that they had deforested over the past 2 years. Note that deforestation is not necessarily a breach of contract, given that some contracts are not specified to be inclusive of all the forested area. In most cases, the activities implemented as a result of the program included increasing the surveillance of forestlands and revision of rules regarding the extraction of forest resources. In no cases were new activities introduced as a result of the program. Payments had not been withheld from any of the survey participants, suggesting that either compliance is very good or the monitoring system is not very effective. Monitoring of the contract after the first year of operation was performed randomly in 28 *ejidos* (22%) in November 2004. All monitored *ejidos* met contract requirements. The annual cost of operation and monitoring for the first year of operation was estimated at $714,285, yielding an average cost of $5.6 per hectare absorbed by CONAFOR. Compared to payments of $30 per hectare, this indicates that administrative costs represent 19% of the PES budget. In addition, there is an annual evaluation of program objectives, processes, and expenses made by an external institution. For the first year, this evaluation amounted to $98,214.

8.4.2 Case Studies

8.4.2.1 Basic Findings

In the winter of 2004–2005, case studies of 11 communities receiving the pilot payments were undertaken in the states of Michoacan, Puebla, Veracruz, Durango, Chihuahua, and Coahuila. Given that the majority of the forest in the program is from *ejidos* and *comunidades*, all case studies were conducted in these types of properties. This section summarizes the overall findings.[3] The intention of these studies was to detail the experience of the recipient communities in order to understand how they were managing their forests before the PES program, if this behavior had changed with the payments, and if the payments had affected the internal dynamics of the communities.

The studies cover a variety of communities with varying membership and size in different institutional situations. The membership size ranges from 40 to 225, while the area ranges from 493 hectares to over 10,000 hectares. The forest area enrolled in the PES program in each community varies from 73 hectares to 1,400 hectares. Four *ejidos* included areas of forest that are organized for wood extraction under the permit system. Three communities were located in the Biosphere Reserve las Tuxtlas, which constitutes a unique institutional context within Mexico, as there are rules specifying limitations on certain extractive uses.

With regards to our first question, we found that 5 of the 11 *ejidos* had deforested in the period prior to receiving payments. As was suggested by the statistics of the previous section, the case studies imply that a significant proportion of the budget may be being paid to people who were not planning to cut down the forest in the first place. All of the *ejidos* interviewed were already engaged in some form of conservation activity before implementation of the program. This suggests a selection bias in the program design – it is highly likely that communities with some experience in conservation would volunteer to participate in a program requiring conservation activities. The three forestry *ejidos* of the Northern states all participate in conservation activities, which are actually part of their forest management plan.

In the three Northern forestry *ejidos,* the main use of the forest is extractive forestry under a management plan designed by a forest technician from outside the community. These communities exhibited a high percentage of forest loss (12.4%, 12.1%, and 6.6%) over the period from 1994 to 2000. However, their extractive activities began after 1994, and the *ejidos* practice a rotation style of forestry that involves harvesting a parcel and then allowing it to rest for 10 years. It is unclear whether the large initial forest loss came from the first phase of the forestry process or whether it is the result of an unsustainable deforestation path. In addition to the extraction itself, there is some pressure on these forests originating in the expansion

[3]The studies were conducted by Adán Martínez Cruz, Josefina Braña Varela, and Jaime Sainz Santamaría.

of the urban area of these communities, from subsistence agriculture, forest fires, and pest infestation. In one particular case, the forest loss is mainly the result of a forest fire that occurred in 1998. Much of the area that is recovering from the fire is currently being included in the program. The other activity taking place within the forest is the grazing of a small cattle herd whose owners reside within the *ejido*. This community used the PES as a way to induce the cattle owners to move the small herd away from the recuperating land.

In most cases, communities stated that they had intensified their conservation efforts by increasing their frequency as a result of the program. These facts were not corroborated by outside sources, and in some instances the case study teams perceived that community members had trouble locating firebreaks and forest roads that they claimed to be maintaining. The forestry communities seem to see the program as a way of subsidizing their forestry project – the hectares of land integrated in the program are in fact hectares which are part of a 10-year rotation and happen to be in fallow at the moment. In one case, however, the PES land is located in what is considered a sensitive area for water conservation, and it has therefore been fenced in and is monitored to ensure that no one enters into the area. This *ejido* does not intend to put this particular piece of land back into production. The results of this section suggest that significant behavior change induced by forest conservation payments is unlikely since many communities were already preserving the forest.

Another situation where payments were used to provide a sustained incentive for mandated conservation activities was found in the Biosphere Reserve, where communities are forbidden to continue extractive projects in their forests. The provision of payments to these communities could be seen as replacing the command-and-control approach, which is difficult to enforce and seen as unfair by forest owners.

It is also important to note that, with the exception of the two cases in Northern Mexico, the communities received no technical assistance in the design of their PES implementation schemes, and in fact were not even aware of the contractual requirements of such a scheme. It is unreasonable to expect communities without technical assistance or experience in forest management to be able to create an effective management plan.

It is possible that in many cases, the amount of money received by the communities was not sufficient to induce any sort of behavioral change. The total annual payments by community vary widely, from $2,200 up to nearly $45,000, as do the ways in which communities decided to divide up this money. In over half of the cases, the majority of the allocation was divided up and given to individual *ejido* members. Per capita payments, under the assumption that the allocation was equally divided between all *ejido* members, vary from $60 per member to $1,100. Given that GDP per capita in Mexico in 2003 was around $6,000, these amounts very from totally insignificant to substantial, with the majority falling on the low end (CIA, 2005).

In only one case was 100% of the money distributed equally among all *ejido* members. In all of the other communities, a percentage ranging from 3 to 100 was invested in some kind of public good, where public goods in this case include equipment used to monitor the forest commons (radios, trucks), infrastructure like school classrooms, and road maintenance. There are several reasons why *ejidos* might choose to invest the majority of the payments in public goods. First, it is possible that there are returns to scale in these investments. That is, giving a transfer of $75 per year may not be as valuable to a family as using the same money to build new classrooms for the school where the family sends its children to study. Second, these goods can be enjoyed by non-members of the community who would not normally have rights to cash transfers from *ejido* funds. Finally, there is evidence of sharing norms present in the *ejidos* and *comunidades* of Mexico, and it may simply be that it is preferable for them to distribute this money in a more egalitarian fashion through investment in public goods. One very interesting development in two of the communities interviewed was the proposal to form local microbanks using the PES money as seed capital.

Although in most cases there was no obvious change in the social dynamics within the *ejido*, in two cases there was a shift in the relative power of different groups within the communities. In both of these cases, forests were located in what the community had defined as parceled areas (rather than common property) within the *ejido* boundaries. The outcome of this division was that it gave the owners of these forests the ability to make a credible threat to cut them down if their demands were not met. In the first case, the forest holders were receiving payments but requested they be adjusted to reflect the proportion of forest located in their parcels. This proposal was voted upon and accepted by the assembly and will be put into effect in the next round of payments. In the second, the members with forested parcels were not receiving payments and threatened to cut down their forest if they did not receive some proportional compensation in the next round. An additional result of the program in this community was that participation in conservation activities was reduced. This phenomenon was a direct consequence of the way in which the payments were divided up – only those with rights to the commons received them. It is somewhat unusual that only a small part of the membership of an *ejido* would have rights to the commons. In this case, the decision had been made to give commons rights to those with very small private parcels. Prior to the program, all *ejido* members had participated in forest surveillance and maintenance. Once the payments were received, non-recipients withdrew from these activities.

In two happy cases, we observed that the allocation of the PES funds resulted in an increased environmental awareness and participation of a greater number of community members in conservation activities. Although the authors of the cases noted that one of these communities clearly had higher levels of social capital than some of the other participants, it also received a much larger payment, both in total and on a *per capita* basis. In addition, the payments were not divided up equally among members, but rather were distributed according to the level of participation in the activities they deemed necessary to fulfill the program requirements.

8.4.2.2 Other Case Study Findings

Misunderstanding of the Program

One of the most discouraging findings was that in none of the communities visited were the objectives and rules of the program clear to the members. This was not surprising given the time restrictions on program promotion. Interestingly, the majority of *ejidos* were able to identify the cities that benefited from the hydrological services provided by conservation of their forests, but none of them realized that the payments they were receiving were meant to be in compensation for these services. In several cases, interviewees stated that they thought the payments were a poverty-alleviation mechanism somehow linked with forests.

Corruption

Another unsettling finding was that, in at least one case, the intermediary responsible for helping the communities fill out the paperwork for the program covered a "fee" equivalent to some percentage of the final payment.

Slippage

The term slippage, coined by Wu (2000), refers to the bringing into production of other land as a result of removing land from production and putting it into a conservation program. Although in most of our cases this was not a risk, given that the forest integrated into the program was not slated for any use by the *ejidos* in the first place, we did observe slippage in two cases in the *ejidos* in Northern Mexico. In one case, cattle were removed from the forest to be entered into the program and placed in another area not previously used for grazing, although it is not clear if this area was forested or not and whether it was located in the commons or in a private parcel. In the second instance, the community put in the program forested land that they had programmed for harvest, and instead harvested another area of the forest.

Another phenomenon which is related to slippage is the use of the program as a way of receiving payments for land which the *ejido* intends to use productively in the future. We saw this in the forestry *ejidos* that decided to enroll into the program hectares that are part of their 10-year harvesting rotation. With permission of the forestry authorities, these *ejidos* then modified their forestry plan to put different hectares into production.

8.5 Learning from the Mexican Experience

8.5.1 Lessons in Political Economy

There are multiple lessons to be derived from the administrative and political processes to which the Mexican PES program was subjected. Many of the forces that modified the program's objectives were not foreseeable and could not have been circumvented. In this section we focus on aspects of policy design and implementation, which could provide useful guidance for the continuation of Mexico's program and for programs in other countries.

8.5.1.1 Program Design and Promotion

The first important lesson for policy designers is the need to establish clear objectives and criteria for the program before promotional activities take place. The nationwide tour caused confusion regarding the purpose, rules, and financial mechanisms of the program. This resulted in bad blood at the local government level in places and an overabundance of unqualified applications, which merely exacerbated CONAFOR's staffing and time constraints. Clearly defined criteria and objectives could also have helped minimize the problems that occurred during the implementation phase and increase transparency of the program. A well-defined program may even have facilitated the early stage negotiations with Hacienda and improved the quality of the "rules of operation." Finally, clear objectives would have aided in the promotion of the program and the understanding of it among the participants.

Participants in the early phases of the Mexican program emphasize the importance of forming an advisory group of both national and international experts to aid in the policy design process. The combination of expertise from outside the country and experts aware of the realities of implementing programs in Mexico expedited the design of the program and allowed recommendations to be made quickly and effectively. Whether or not these recommendations get implemented depends very much on the relationship between the policymakers and the intermediate provider or implementing agencies, which leads us to the next point.

8.5.1.2 Choice of Implementing Agency

In the Mexican case, the choice of CONAFOR as an intermediate service provider had very important impacts on both payment structure and targeting. Many of these changes resulted from the fact that CONAFOR's traditional program objectives and constituency – owners of commercial forests – differed from the program's objectives and target population. It was easier for CONAFOR to communicate and negotiate with this group since they had already established relationships through other programs. As we saw in the summary statistics section, the result of this relationship was that 63% of participating *ejidos* extract wood for sale; 79% of the PES hectares in 2003 and 85% in 2004 were in watersheds that were categorized as "not overexploited."

An additional objective that influenced CONAFOR's implementation of the program in later rounds was the desire to support another of its programs – the Priority Mountains Program. This program's budget shared the PES objective of preserving water production through forest conservation. These two features made it logical to funnel the PES funds towards these mountains, with the useful result of reducing administrative costs by concentrating the two programs in the same geographical areas. Forty-six percent of the 527, 515 hectares enrolled in the PES program in 2004 were located in areas within the Priority Mountains Program.

It is extremely important to reiterate that CONAFOR brought to the program two essential benefits: the desire to implement the program and the political clout necessary to obtain a budget for it from Congress. Without CONAFOR, it is very likely that Hacienda would have blocked the allocation of money to a PES program indefinitely. The trade-off here is an important one: One chooses an intermediate service provider whose incentives are partially misaligned with the objectives of the policymakers in exchange for obtaining a budget for the program.

8.5.1.3 PES Contracts

The contracts between the intermediate provider, CONAFOR, and the final service providers, the *ejidos*, must give the forest communities sufficient incentives to cease their extractive activities in favor of conservation. This requires that payments be high enough to compensate for the loss of forest extraction, agriculture, or cattle grazing and that there be sufficient monitoring and enforcement of program rules. The case studies and the statistics regarding deforestation risk show us that much of the forest currently under contract is likely to have a very low opportunity cost – that is to say, one would not have to pay very much in order to compensate for the loss of income from activities currently taking place in these forests. In this sense, the magnitude of the payments appears to be high enough given the forests that are enrolled in the program – whether or not these payments would be high enough to preserve all of the water services at risk of being lost is another issue, and one we will return to in the discussion on targeting.

One way to eliminate the guesswork in the magnitude of payment design is to use an auction process to induce potential participants to reveal the minimum payment, which they would accept in exchange for conservation of their forest. Although we have yet to hear of such an approach being applied in an ES scheme, the Conservation Reserve Program in the United States did take a step in this direction by allowing potential participants to place a bid that can affect the probability that they will be included in the program. Another option would be to conduct rigorous contingent valuation studies in areas targeted by the program.

One feature of the contracts, which is important and easily replicable in other situations, is that contracts should be made over the entire forested area. In order to avoid the movement of productive activities from PES hectares to other previously unused forests within the *ejidos*, it is very important that contracts for payments specified that there should be no change in the entire forested area. This does not imply that payments should be given for all of the hectares of forest within the *ejido*, but rather that the contracts should eliminate the possibility that deforestation be reallocated from one spot in the community to another. Agreements can allow for some pre-specified amount of forest conversion. Were the program not to have followed this strategy, an *ejido* receiving payments for 10 out of 100 forested hectares within its boundaries might then deforest with impunity the remaining 90 hectares not included in the program. Obviously, the choice of which hectares to pay for should not be arbitrary. The logical option is to pay for those hectares of land, which are at risk of being deforested. Such an approach is described in more detail in Sect. 8.5.3.

The timing of the payments of the PSA-H is sensible and easy to replicate. Payments are given at the end of each period, after verification of the conserved forest cover. In effect, the payments, since they are made on a yearly basis, are a rental contract for the ES provided by the forest over the year. This arrangement is logical since it is much easier to withhold a payment than to request its return, and there is a clear moral hazard problem with paying before the receipt of a service. As a result of the need to spend the initial budget quickly, the first year's payments were given for forest conserved in the previous year. This is not a method we would recommend for other programs, although it was politically expedient.

An important part of being able to give or withhold payments relies on having an objective measure of change in forest cover. Here we find another positive lesson from the Mexican strategy. The monitoring scheme consists in choosing communities at random and assessing the quality of their forest cover using satellite images, which are both transparent and difficult to manipulate.

8.5.1.4 Within Community Contracts

Mexico is unique in having most of its forest held under common property. However, many other countries also have substantial tracts of forest under similar institutional arrangements, and it is important to mention lessons learned within this context. The case studies indicate that members of most communities did not know why they were receiving the payments. This could be quite detrimental to achieving forest conservation. This is because the payments must provide an incentive for individuals within a community to cease their deforestation activities or, in cases where deforestation pressures come from outside the communities, to increase conservation activities like forest monitoring for encroachment. These changes must either come through an income effect that is large enough to remove the need to extract goods from the forest, or through a price effect in the form of a transfer – be it in cash or kind – conditional on ceasing extraction or participating in conservation. The PES program belongs in the category of a conditional cash transfer (CCT) that creates a price effect on forest conservation. Per unit of payment received, a CCT should have a larger incentive effect on forest conservation than a non-conditional payment would have.[4]

8.5.1.5 Give Voice to Water Service Providers

One final political lesson is to allow water service providers to participate in the design and management of the program. Although bureaucratic limitations did not allow CONAFOR to funnel the payments through the municipalities, the early participation of the water service providers could have worked in several ways to bolster the success of the program. First, because they know their localities, these service providers could have helped to target properties, which were particularly important for the provision of water. There is a general sense among water providers in Mexico that problems with water supply are simply normal seasonal or yearly fluctuations, which are not directly linked with overall management of the watershed. Where water providers are concerned about this link, as is the case in southern Veracruz and some states in Northern Mexico, this awareness was triggered by severe drought years and crippling water shortages. This knowledge is absolutely essential for the development of markets for hydrological services, which is the final goal of the program. Water service providers are an important link in the accountability circle. They can generate awareness among consumers of water, who can then pressure policymakers either directly or indirectly. The providers themselves are directly linked with the government because they are either municipalities or private providers operating under the supervision of municipal authorities who are in communication with the government. They are, therefore, in a position to demand results from the program – increasing the accountability of the ES providers, and helping provide another source of pressure on policymakers to continue allocating a budget for the program.

8.5.2 Financing Lessons

8.5.2.1 Sustainability of the Funding Source

As was described above, the current financing from the program, consisting in an annual budget of $20 million, has been approved by Congress. This does probably

[4] See by analogy the discussion on incentive effects from cash transfers versus. CCT for education under Progresa (de Janvry & Sadoulet, 2006).

not qualify as a sustainable financial arrangement since, though it has been written into law, it is decoupled from the intentions of the program and subject to the political process. The Mexican program is seeking sustainability through the development of local markets for ES, a criterion that led to the selection of *ejidos* with downstream populations of over 5,000. As we saw in the section describing the current participants in the program, the distribution of the enrolled hectares is widely dispersed – in 2003, the program enrolled 271 properties in 15 states. The small number of participants per large population area may make it difficult to establish markets for two reasons. First, there may not be a sufficient number of hectares enrolled to actually make a substantial impact on the downstream water quality and quantity. Second, dispersion of the participants may make it costly to organize such markets. Although the development of markets would be sustainable as long as demand for ES is strong, it is currently unclear how the transition from subsidy to market will occur.

In some Mexican cities, with Coatepec in Veracruz serving as an outstanding example, markets for ES have developed in the absence of the payment program (though the program has been used to support Coatepec in the past 2 years). It is important to note, however, that initiation of the program in Coatepec followed a water crisis in the city, which raised the local demand for water services from forests in the mountains above the city.

8.5.2.2 Mechanisms to Guarantee Long-Term Contracts

Despite the potential tenuousness of the program's budget, we do extract one very positive lesson from the financing of Mexico's program: the usefulness of creating a trust fund which guarantees the ability to provide payments to recipient communities over an extended period. For ES programs to be taken seriously, funding must be guaranteed over a substantial period of time. The FFM is a clever mechanism that circumvents the political budgeting process by allowing money allocated in 1 year to be used in subsequent years. This security, however, comes at the cost of a substantial part of the budgeting money sitting idle each year. In the case of the FFM, as each year's new budget comes in, four fifths of it is put into the fund, where it is paid in equal installments over the next 4 years. If one could rely on the yearly financing of the scheme, considerably more hectares could be enrolled in the first 5 years of the process than are actually in it. However, in the face of insecure political outcomes, the trust fund mechanism plays an important role in enabling long-term contracts with service providers in spite of reliance on politically uncertain annual budget appropriations.

8.5.2.3 Use Bankable Certificates

An alternative to the FFM approach is to use the strategy undertaken by PROCAMPO, an agricultural subsidy program introduced in 1994 to compensate farmers for the negative price effects of NAFTA. Under PROCAMPO, farmers are given payment certificates against which they can borrow money from the bank. Using such a strategy would allow for all of the money granted by Congress each year to be used to pay communities. Had such an approach been taken, CONAFOR would have been able to contract 5 times as many hectares in 2003. The key to making such a system work is a guarantee of funding from the federal budget for the length of the contracts, in this case, 5 years.

8.5.3 Targeting Lessons

8.5.3.1 Target Public Goods Important Within National Boundaries

Mexico was wise in its choice of hydrological services provided by forests as the focus of the program. Because the water quality and quantity associated with forests is a good that is solely consumed within watersheds, almost all of which are entirely within national boundaries, it was much easier to seek financing from Congress and to look towards the development of local markets. Despite the fact that the targeting of the payments was skewed by the choice of the implementing agency (see Sect. 8.5.1.2), the current targeting scheme is moving towards achieving the goals of the program, with big improvements in the 2004 implementation, where communities located in priority watersheds are given preference. It is also a scheme that would be simple to apply in various contexts, although it does require having sufficient information to prioritize the watersheds that are key to preserve the environmental service of concern.

8.5.3.2 Take into Account Risk of Service Loss

The most efficient way of allocating payments in environmental programs is to pay the lowest cost possible for those hectares of land containing benefits that are at risk of being lost. If the risk of service loss (in this case, deforestation) is not taken into account, then large amounts of money will be spent paying for ES that were never at risk of being lost in the first place. Our summary statistics on program participants in Mexico (see Sect. 8.4.1) showed that a large number of hectares enrolled (72% in 2003 and 52% in 2004) had either low or very low risk of deforestation. This implies that the current targeting strategy is inefficient. The efficiency of the current scheme could be enormously improved by taking into account both the risk of losing these benefits and the cost of conserving them. In a 2004 paper simulating the effects of different payment targeting schemes to *ejidos* in Mexico, Alix-Garcia et al. (2008) showed that for the same budget, payments allocated to maximize expected benefits per dollar led to a fourfold increase in efficiency over a scheme analogous to the current one which offers a flat payment per hectare with a cap on the number of allowable hectares.

Operationalizing this scheme means developing some measure of environmental benefits, weighting these benefits by the deforestation risk, and creating a ratio of expected benefits to opportunity cost that allows the ordering of properties. One then begins to pay those with the highest expected benefit/cost ratio and proceeds down the line until the budget is exhausted. Clearly, the three elements necessary to implement this scheme are: (1) some measure of the environmental benefits offered by land in each *ejido*, (2) a measure of the opportunity cost per hectare, and (3) an estimate of the risk of forest loss. It would be possible to implement this approach by using a bidding process (as suggested above) to reveal the opportunity cost of hectares in forest, and then use the ratio of the expected environmental benefits to the bid made as the targeting criterion.

Payments in the second round of the program begin to fill one of the criteria of a targeting strategy that maximizes environmental benefits per dollar, that is to say, payments are broadly allocated to regions where water resources are over-utilized. They are, therefore, directed towards communities where the environmental benefits are relatively higher. Another related lesson is that, in general, where forestry projects are very profitable, forests will be conserved because it is in the interest of the owners of this forest that it keeps supplying lumber over the long term. Payments should therefore be directed away from these communities towards those with unprofitable forestry projects or to non-commercial forests.

8.6 Concluding Remarks

We began this chapter with an overview of the state of the Mexican forest, whose considerable riches are at risk of being lost due to a combination of perverse incentives, one of which is the lack of a market for the ES that it provides. This forest is a prime example of a natural resource which supplies services in addition to its extractive value; it sequesters carbon dioxide, shelters biodiversity, prevents erosion, provides a destination for local and international tourists, and plays an important role in regulating a complex hydrological system. Mexico chose to pay for hydrological services through a PES program. The possibilities and pitfalls of this experience have taught us lessons that will both help improve the Mexican program and assist in designing PES programs for other services and in other countries.

Our analysis showed that payments as they were distributed in 2003 and 2004 did not necessarily achieve the goals of the program – they were largely allocated to hectares of land that were not within critical watersheds. They are also so fragmented in their distribution that they are unlikely to be providing measurable services to downstream water providers. In addition, they were not targeted at forests that were at risk of being lost. Our case studies showed that there was little pressure to deforest in the communities chosen to receive payments and that, as a result, there were very few behavioral changes induced by the program payments. In some cases, however, the payments did serve to increase participation in conservation activities. One serendipitous effect of the targeting was that the majority of payments went to poor and very poor forest-holders.

The sources of this bias in the program are various, many of them stemming from factors beyond the program designer's control. One of the most important was the choice of service provider, CONAFOR, whose objectives were not directly aligned with those of the proposed policy. The trade-off in this choice, however, was a large one. CONAFOR was very successful in lobbying for a budget for the program and in ensuring that its first phase was implemented within a very restrictive political timetable. Other important factors that affected and will continue to affect the program's success are related to accountability. Without awareness on the part of water providers and consumers, accountability of forest-holders to provide ES will be very limited, and it is unlikely that local markets for hydrological services will emerge. In addition, without pressure from these groups, it will become increasingly difficult for the program to continue to receive budgetary support from Congress. There are also problems of accountability within communities – if the program continues to be misunderstood by recipients, the contracts are likely to be broken and market formation hindered.

The program in Mexico is still quite young, and will surely have many future lessons to teach us. A thorough evaluation of the actual impact of the program on forest cover will eventually need to be done, and this will require considerable GIS work as well as further surveying of the participant (and some non-participating) communities. In addition, the question of the optimal design of payment contracts in the context of forest common properties has yet to be answered.

Furthermore, we do not know whether the payment level that is currently being used is appropriately set. It is clear that the payment level was high enough to attract a substantial number of participants, but it would seem that often those who chose to participate had no intention of cutting down the forest in the first place. As we saw in the case studies, some communities showed absolutely no change in behavior upon receiving program payments. Part of this may have been because the overall payment amounts were not very high; they could probably have been lower with the same result. Calibration of the payments must take into account the fact that forests at higher risk of deforestation, i.e., with a greater opportunity cost, will require larger payments. The logical conclusion is that payments must be differentiated according to the level of risk associated with a given forest. The design of such a differentiated scheme, however, requires considerably more research.

Finally, recalling that the goal of the program is to develop markets for ES within Mexico, an essential part of future research should include a rigorous assessment of where these markets can truly be developed. This requires knowing which forests are essential to which watersheds, if they are at risk or not, and the quality and quantity of the demand for services by downstream users. Integral to such an analysis is identifying forests that could be preserved through means other than PES – i.e., through changes in the incentive structure created by forest policy. The portrait of forests at risk could then be completed by those that cannot be saved by markets or through changes in forest policy. It is these that will require either mandated protection or continuous payments from federal or international funding sources.

Acknowledgments This document is based upon a report prepared for Randy Stringer formerly at the Comparative Studies Service Agricultural and Development Economics Division, United Nations Food and Agriculture Organization (FAO). The work was coordinated by Alain de Janvry and Elisabeth Sadoulet, professors at the Department of Agricultural and Resource Economics, University of California at Berkeley. The Political Economy section is largely based on the contribution by Josefina Braña Varela and Maria Zorilla Ramos. Fieldwork for the study cases was coordinated by Jaime Sainz, Adán Martínez, and Josefina Braña.

References

Alix-Garcia J, de Janvry A, Sadoulet E (2008) The role of deforestation risk and calibrated compensation in designing payments for environmental services. Environ Dev Econ 13 (3):375–394.

Central Intelligence Agency (CIA) (2005) World Fact Book 2004. Available online at http://www.cia.gov/ cia/publications/factbook/fields/2004.html.

Comisión Nacional Forestal (CONAFOR) (2001). Semarnat Programa Nacional Forestal 2001–2006.

Comisión Nacional Forestal (CONAFOR) (2004). www.conafor.gob.mx.

Colegio de Posgrados (COLPOS) (2004) Valuación de programa de pago por servicios ambientales hidrológicos. Working paper.

National Population Council (CONAPO) (2000), Indices de Marginación, 2000. Mexico City, Mexico.

de Janvry A, Sadoulet E (2006) Making conditional cash transfer programs more efficient: Designing for maximum effect of the conditionality. World Bank Econ Rev 20:1–29.

García Coll I (2002) Potencial de recarga de acuíferos y estabilización de ciclos hídricos de areas forestadas. Reporte de Investigación. México: INE-DGIPEA (processed).

INE, Instituto Nacional de Ecología (2004) Encuesta de los ejidos participantes en el programa de PSAH. Survey of ejido participants in the PSA-H.

Mayrand K, Paquin M (2004) Payments for environmental services: A survey and assessment of current schemes. Unisfera International Center (for the Commission for Environmental Cooperation of North America). Available at http://www.cec.org/files/PDF/ECONOMY/PES-Unisfera_en.pdf.

Muñoz C, Guevara A, Bulás JM, Torres JM, Braña J (2005) Pagar por los servicios hidrológicos del bosque en México. Draft book chapter.

National Institute of Geography, Statistics and Information Systems (INEGI) (2005), Mexico City, Mexico.

Secretariat of Environmental and Natural Resources (SEMARNAT) (2000) National forest resources, Mexico City, Mexico.

SEMARNAT (2004) La experiencia de México en el pago por servicios ambientales hidrológicos y el fondo forestal Mexicano. Powerpoint presentation.

The Nature Conservancy (2003) Home page. Available at http://www.nature.org.

Velázquez A, Mas J, Palacio J (2003) Análisis del cambio de uso del suelo Convenio INE-IGg (UNAM). (Oficio de autorización de inversión 312.A.-00215), Instituto de Geografía, UNAM.

World Bank (2005) Current World Bank work on payments for ecological services. Available at http://lnweb18.worldbank.org/ESSD/envext.nsf/44ByDocName/PaymentsforEcologicalServicesCurrentProjects

Wu J (2000) Slippage effects of the conservation reserve program. Am J Agric Econ 82 (4):979–992.

Chapter 9
Agricultural Landscape Externalities, Agro-Tourism, and Rural Poverty Reduction in Morocco

Khalil Allali

Abstract This chapter presents an empirical analysis of agricultural landscape externalities and an economic valuation of their potential impacts to reduce rural poverty in Morocco's Western High Atlas Mountains. The externality concept is applied at the farm level to inventory landscape externalities, to analyze their internalization forms, and to evaluate their economic benefits on rural households. Of 134 farms studied, the attributes of agricultural landscape externalities are identified and characterized. Then, internalization forms of landscape externalities and their economic contributions to the economy of the rural households are examined and evaluated. The development of agro-tourism requires adaptation of local agricultural practices to internalize landscape externalities. The substantial benefit provided by agro-tourism farms, as compared with returns of the traditional farms, reveals an important approach to reduce rural poverty. Various forms of agricultural policy interventions are introduced and interpreted as market solutions to compensate farmers for provision of environmental services.

9.1 Introduction

This chapter attempts to provide new insights into the environmental role of agriculture, particularly those pertaining to landscape externalities with the potential to promote sustainable development. A better understanding of the beneficial linkages between marketed agricultural production and its positive landscape externalities should help create a more favorable environment for a regional dynamic, leading to greater economic diversification and potential poverty reduction in rural areas.

This study considers scenic agricultural landscape with multiple environmental amenities an input to rural tourism in Morocco, an essential component of the tourism sector. The economic benefits are evaluated in relation to income, employment, investment, and welfare of local population and their contribution to rural poverty reduction.

L. Lipper et al. (eds.), *Payment for Environmental Services in Agricultural Landscapes*, Natural Resource Management and Policy.
doi: 10.1007/978-0-387-72971-8_9;

The specific objectives of the study are to: (1) identify the main landscape externalities generated by the farming practices specific to the agro-ecosystem of Morocco's Western High Atlas region; (2) assess the main agricultural landscape externalities generated by those farming practices, identifying the markets concerned and the various categories of beneficiaries; and (3) evaluate the economic benefits and relationship to rural poverty reduction.

9.2 The Methodological Approach

9.2.1 Concepts and Theoretical Framework

In Morocco, agricultural landscapes are considered a non-market output and consequently not recorded by the country's statistical department (Allali, 2003a, 2003b). Landscape amenities are viewed as a positive unintentional environmental service, generated jointly by farming practices without specific supplementary costs. Likewise, farmers who provide these environmental externalities are not directly remunerated by beneficiaries, who maybe either other rural residents or tourists. The analytical model adopted in this study is based on the theory of externalities, particularly production externalities (Hill, 1999; Delaunay & Gadrey, 1987; Gadrey, 1996, 2000; Aznar, 2004).

Landscape externalities are generated by the production space (natural patrimony, farmland, and livestock buildings); the consumption space (farm residential area); and the related infrastructure (paths, dirt roads, boundaries, fenced-off areas). Types of landscape externalities considered are all spatially localized and correspond to the attributes of the rural countryside that make it visually pleasing. This study considers only agricultural landscape externalities that have a visual and aesthetic dimension and a positive impact on rural landscape quality. The following three externality categories are assessed to evaluate their economic contribution to rural poverty reduction (see Fig. 9.1):

- *Landscape,* referring to the agricultural contribution to landscape upkeep, landscape structuring, and rural sightseeing (recreation, sports, and relaxation);
- *Natural patrimony,* referring solely to the positive effects of farming practices on the basic natural resources (soil conservation and water management) in a context of ecological fragility;
- *Biodiversity,* referring to the contribution of farming practices to the conservation in situ of local genetic diversity, particularly animal resources.

The study was carried out in Morocco's Western High Atlas region, focusing on three major agro-ecological systems: the high mountains, mid-mountain areas, and the foothills (see map, Appendix 9.1). These systems were chosen to represent the diversity both of the natural environment and agricultural practices. Within each of these three agro-ecological systems, representative zones were selected according

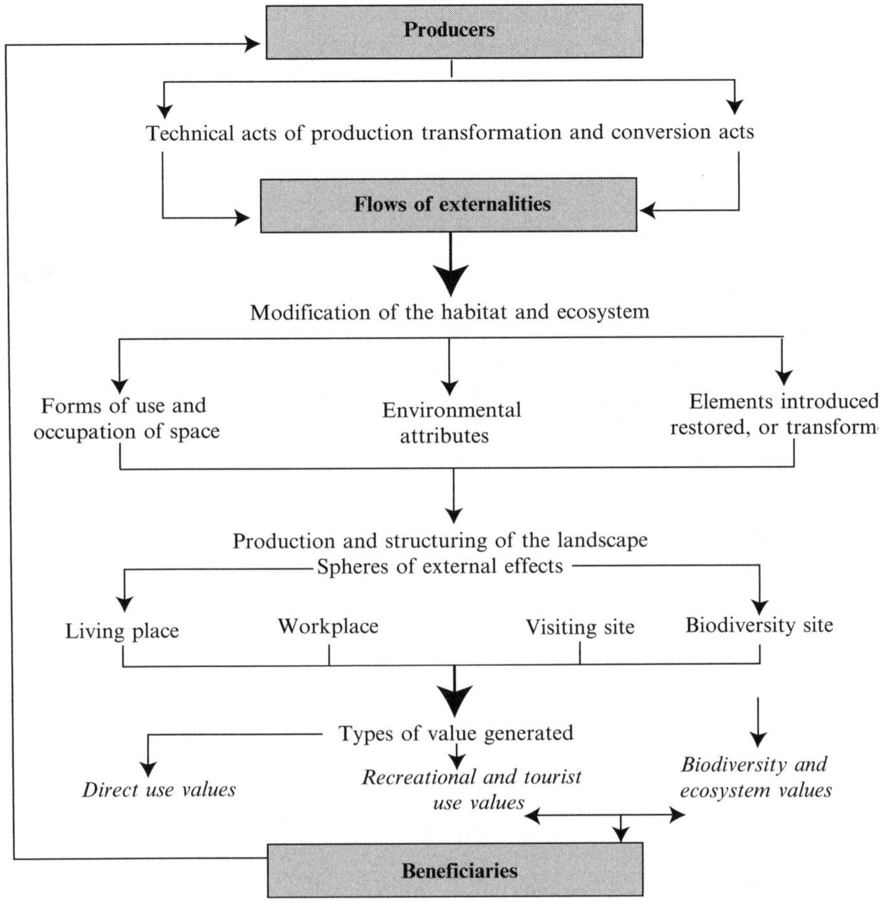

Fig. 9.1 Overview of the analytical model used for agricultural landscape externalities. *Source*: Author's survey

to the following criteria: (1) the importance of agriculture and livestock activities, as potential sources of landscape externalities, and (2) the importance of rural tourism, as an indicator of the potential and real opportunities of internalization of the externalities.

This typology covers all typical cases as well as the local diversity found in mountain zones. Four representative zones of the variation in the Western High Atlas Mountains were chosen (see Appendix 9.2). Within each of the zones considered, a simple random sampling approach was used to select farms. The sample selected consists of 134 farms, spatially distributed according to the weight of the three agro-ecological zones, and the four representative areas studied (see Appendix 9.3).

Field surveys were carried out in two stages. The first survey focused on an inventory and analysis of landscape externalities in relation to farming practices and interventions. The second survey focused on existing forms of capturing or internalizing externalities for food security and poverty reduction.

The farm-level surveys produced a description of externalities by documenting: (1) the relationship between cause (farming practices and processing activities) and positive external effect, (2) the nature of the farming practices and processing activities in generating the external positive effect, (3) the existence of indicators to support and measure the relationship, and (4) the spatial dimension of the external positive effect generated. The inventory and survey exercise coincided with the principal growing season, which facilitated the assessment of visible elements of the agricultural landscapes.

Following the collection of the farm level inventory data, a list of priority landscape externalities was established. The three main criteria used to draw up this list included: (1) the consistency of the collected data with the definition of the three characteristics of the externality (Fig. 9.2), (2) the relative size of the external effects generated within the area (physical presence and spatial and temporal dimensions of the effect), and (3) the possibility of evaluating their impact on household economies in monetary terms.

9.2.2 Methods Used to Evaluate Impact on Household Economies

Evaluation of the contribution of agricultural landscape externalities to rural poverty reduction took place in two complementary stages (see Fig. 9.3). The first consisted of the establishment of linkages between each of the three categories

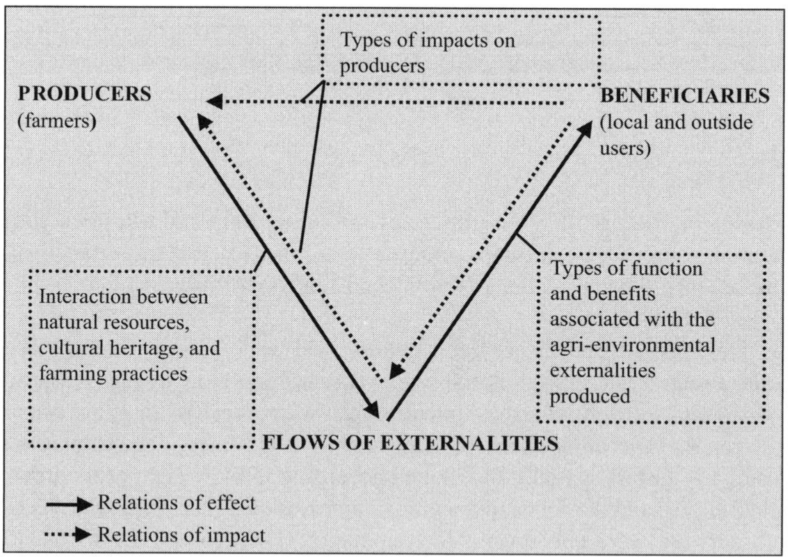

Fig. 9.2 Main poles of the triangle of agricultural landscape externalities. *Source*: Author's survey

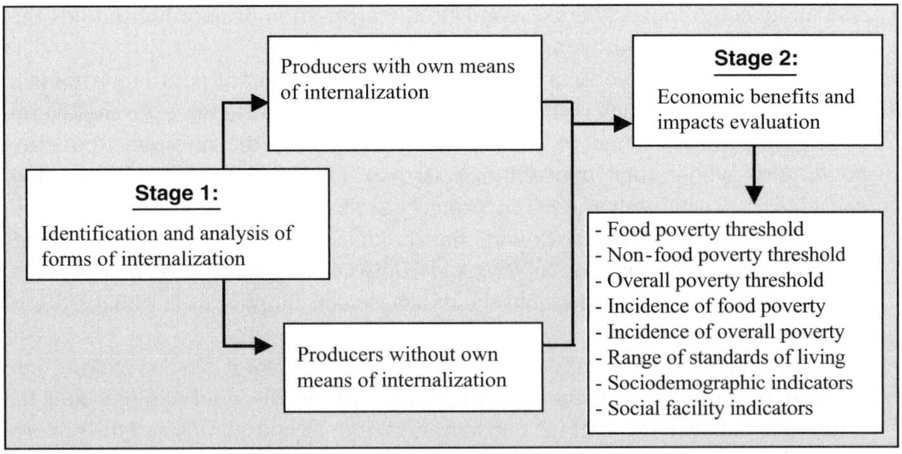

Fig. 9.3 Evaluation of landscape externalities and impact indicators used. *Source*: Author's survey

of externality considered and the positive effects identified. Each form identified was then analyzed in terms of its direct effects on job creation, income generation, investment stimulation, and introduction of farms into the market. Two distinct types of farmers-producers were distinguished, based on the output of this first stage: (1) producers of agricultural landscape externalities without their own means of internalization and (2) producers of agricultural landscape externalities with their own means of internalization. The latter can in turn be divided into three groups according to the volume and nature of their supply of tourist services.

The second stage of evaluation concerned the relationship between supplying various agricultural landscape externalities and rural poverty reduction. This evaluation focused exclusively on the producers and was concerned with identifying the level of poverty for each of the representative areas studied and each of the identified producers (with or without internalization).

Monetary poverty was used as the criterion to establish the poverty level of the households, determined by the standard of living, defined in terms of the effective consumption of the household per adult-equivalent. Three poverty thresholds were used: (1) the food poverty threshold, corresponding to the minimum expenditure needed for an individual or household to obtain a basket of food items that meets both nutritional norms for a balanced diet and the consumption habits of the rural population in question; (2) the non-food poverty threshold, corresponding to the minimum expenditure needed to obtain the non-food goods and public services essential to a household; and (3) the overall poverty threshold, corresponding to the sum of the food and non-food poverty thresholds.

Using average poverty thresholds established for rural areas in Morocco for 1999 (Directorate of Statistics, High Commissariat for Planning, 1999) and updated for 2005, the incidence, distribution, and depth of poverty were determined for each of the groups of externality producers per zone defined as follows:

- The incidence of poverty corresponds to the proportion of poor households that fall below the food and overall poverty thresholds.
- The distribution of poverty corresponds to the proportion of poor households in the area falling into each of four situations defined by distance from the overall poverty threshold (range of standards of living): (1) the non-poor, covering households whose total consumption is over 150% of the poverty threshold considered; (2) the vulnerable, covering households whose consumption is between 100% and 150% of the poverty threshold; (3) the poor, covering households whose consumption is between 75% and 100% of the poverty threshold; and (4) the extremely poor, covering households whose consumption is less than 75% of the poverty threshold.
- The consumption deficit ratio is used to define the depth of poverty, taking into account both the proportion of poor households in the total sample, and the difference between the average consumption of poor households and the poverty threshold.

9.3 Study Results

The results of the study are presented in five sections. First, the conditions for generating agricultural landscape externalities are considered, presenting the results of the inventory of sources of production, differences in these sources depending on area, and the factors influencing them. The constituent elements of the various agricultural landscape externalities identified are then considered, with their major landscape attributes and their functions, values, and spheres of trade-offs. This leads to an examination of the various forms of internalization recorded on the farms studied and an evaluation of their various economic contributions to the economy of the households considered. Last comes a review of the results of the evaluation of the potential impact of agricultural landscape externalities on rural poverty reduction, once the value of the externalities have been internalized through rural tourism revenues.

9.3.1 Principal Sources of Agricultural Landscape Externalities

Of 134 farms studied, 13 main sources of landscape externalities were identified on the basis of an assessment of the physical, visual, and location-specific elements of the landscape. Only sources potentially capable of generating landscape externalities that meet criteria for the three ingredients for positive landscape externalities were taken into account. These sources are linked mainly to technical acts of production and operations to modify the farm practices to conserve local natural resources (see Table 9.1).

Technical production activities were found to be the most frequent means of generating landscape externalities, particularly stone clearing and putting uncultivated land into cultivation (10%), irrigation of farmland (8%), terracing (11%), fruit

Table 9.1 Main sources of agricultural landscape externalities

Sources of agricultural landscape externality	Cases recorded	Percentage of total	Percentage of sample
Stone clearing and land reclamation	75	10	56
Water rehabilitation of dryland	62	8	46
Terracing and terraced plots	84	11	63
Fruit tree plantations on slopes	98	12	73
Construction of stone edges around plots	72	9	54
Planting of ornamental trees around houses	29	4	22
Growing of flowers and gardening around houses	49	6	37
Creation and upkeep of earth irrigation channels	73	9	55
Creation and upkeep of dirt road around the farm	65	8	48
Rehabilitation of dirt road outside the farm	19	2	14
Conservation of the typical architecture of houses	85	11	63
Renovation and landscape integration of farm buildings	15	2	11
Conservation of local mule breeds	63	8	47
Total	789	100	–

Source: Author's survey

tree planting on slopes (12%), construction of borders around plots (9%), and the creation and upkeep of earth irrigation channels (9%).

The most common activities to modify the farm's constructed elements were: growing flowers (6%) and planting ornamental trees (4%) around the home, conservation of locally characteristic home architecture (11%), extension and integration of livestock buildings into the landscape (2%), creation and upkeep of trails around the farm (8%), and rehabilitation of trails outside the farm (2%). The only activity observed in the category of the conservation of local resources was the rearing of local mule breeds (8%).

Examination of the nature of the activities and interventions that generate landscape externalities indicates that sources were much more likely to be found in the *production space* (77%) as compared with *circulation and consumption spaces* (10% and 13%, respectively) (see Table 9.2). These modifications involved natural resources, particularly soil and water (69%), natural patrimony, particularly local animal genetic resources (8%), farm buildings (13%), and circulation spaces (10%). This result confirms the importance of agricultural space as a source of landscape externalities and the considerable contribution of diversified rural areas highlighting the position of the farm as the place of agricultural landscape externality production.

The most important farm activities generating landscape externalities jointly with agricultural production are land improvement and irrigation operations, the

Table 9.2 Classifications of main sources of agricultural landscape externalities (as a percentage)

Types of space concerned	Sources of landscape externalities	Modified habitat and ecosystem	Acts of intervention	Percentage of total
Production	Stone clearing and land reclamation	Soil	Rehabilitation	
	Water rehabilitation of dryland	Soil and water	Irrigation	
	Terracing and terraced plots	Soil and water	Rehabilitation	
	Fruit tree plantations on slopes	Soil	Planting	
	Construction of stone edges around plots	Soil	Upkeep	69
	Planting of ornamental trees around houses	Soil	Planting	
	Growing of flowers and gardening around houses	Soil	Planting	
	Creation and upkeep of earth irrigation channels	Soil and water	Rehabilitation Upkeep	
	In situ conservation of local mule breeds	Local genetic resources	Animal husbandry	8
Circulation	Creation and upkeep of dirt road around farms	Soil	Rehabilitation	10
	Rehabilitation of dirt road outside farms		Upkeep	
Consumption	Conservation of the typical architecture of houses	Residential context	Renovation	13
	Landscape integration of farm buildings		Extension	

Source: Author's survey

gardening and planting of fruit or ornamental trees, animal husbandry, and upkeep and rehabilitation work, as well as renovation and extension operations. The breadth of each of these sources of landscape externality depends to a large extent on the nature of the modified habitat and ecosystem and the how these habitats are modified.

In view of the mountainous nature and rough terrain of the four areas considered and the relative scarcity of arable land, the total externality supply potential from all sources is small (see Table 9.3). This is also the case for externalities generated from other resources and interventions. One of the main causes of the lack of supply potential lies in the small size of farms in Morocco's Western High Atlas region.

The distribution of externality sources across the four study zones provides a clearer idea of the differences in landscape externalities according to area. Using the total number of cases recorded per zone as an indicator, Zone 3 producers show much stronger involvement compared with farmers in Zones 2 and 4, and even more as compared with those in Zone 1 (see Table 9.4). Zone 3 shows a much higher share of

Table 9.3 Indicators of the magnitude of supply potential from various sources of agricultural landscape externalities

Sources of agricultural landscape externalities	Kinds of resource and habitat concerned	Unit[a]	Average size (survey sample)
Stone clearing and land reclamation	Soil	ha	17.5
Water rehabilitation of dryland	Soil and water	ha	12
Terracing and terraced plots	Soil and water	ha	10.5
Fruit tree plantations on slopes	Soil	tree	4,620
Construction of stone edges around plots	Soil	km	1.7
Planting of ornamental trees around houses	Soil	tree	160
Growing of flowers and gardening around houses	Soil	m^2	1,660
Creation and upkeep of earth irrigation channels	Soil and water	m	4,470
Creation and upkeep of dirt road around the farm	Soil	m	5,480
Rehabilitation of dirt road outside the farm	Soil	m	3,600
Conservation of the typical architecture of houses	Residential context	m^2	14,400
Renovation and landscape integration of farm buildings	Residential context	m^2	1,760
Conservation of local mule breeds	Animals	Animal	73

Source: Author's survey
[a]Units: hectares (ha), kilometers (km), meters (m), and number of trees and animals

the total number of landscape externality-generating practices at 40%, compared with 23% and 22%, respectively, in Zones 2 and 4, and 15% in Zone 1.

If we consider the average number of cases recorded per farm, the ranking changes somewhat. Zone 4 moves into last place with an average of four sources per farm, while Zones 2 and 3 show the highest numbers with an average of seven cases in each.

Overall, it appears that farms in medium and high mountain zones generate more landscape externalities and produce higher average levels per farm than those in the foothills. The higher altitude farms require producers to adapt their farming practices and natural resource management methods to the difficult conditions resulting in more externalities being generated.

Apart from their numbers, the area managed also provides information on location-related conditions for landscape externality generation. Here again, the data show that the levels in Zone 3 are higher than those in the other zones. Zones 2 and 4 have similar levels, which are higher than those in Zone 1 (see Table 9.5).

The results concerning the inventory, magnitude, and distribution of sources of agricultural landscape externalities show considerable variations among the four

Table 9.4 Distributions of sources of externality in terms of representative area studied

Sources of agricultural landscape externalities	Number of cases recorded per representative area				
	Zone 1	Zone 2	Zone 3	Zone 4	Total
Stone clearing and land reclamation	7	18	28	22	75
Water rehabilitation of dryland	9	15	23	15	62
Terracing and terraced plots	15	20	43	6	84
Fruit tree plantations on slopes	18	21	37	22	98
Construction of stone edges around plots	11	15	27	19	72
Planting of ornamental trees around houses	3	6	13	7	29
Growing of flowers and gardening around houses	3	13	22	11	49
Creation and upkeep of earth irrigation channels	12	13	33	15	73
Creation and upkeep of dirt road around the farm	10	16	23	16	65
Rehabilitation of dirt road outside the farm	0	7	12	0	19
Conservation of the typical architecture of houses	17	20	28	20	85
Renovation and landscape integration of farm buildings	0	8	7	0	15
Conservation of local mule breeds	11	10	22	20	63
Total cases recorded	116	182	318	173	789
Percentage of total cases recorded	15	23	40	22	100
Average number per farm	6	7	7	4	6

Source: Author's survey

study zones. The main factors explaining these differences, especially the ranking with regard to agricultural landscape externalities are considered next.

9.3.2 *Factors Affecting the Sources of Landscape Externalities*

The factors examined here are linked to the combined features of location, structure, and/or functioning of farms. Four types of factors were identified as sources of variability among the four study zones: geographical position, the environment of the farm, productive natural resources, and the levels of diversification and intensification of agricultural production.

9.3.2.1 Effects of the Farm's Geographical Position

The main finding is that the further the farm is from the foothills region (Zone 4) toward the mid- and high-mountain regions, the more favorable the conditions become for generating landscape externalities (see Table 9.6). An altitude of between 1,000 m and 1,200 m appears more favorable for externalities than altitudes of more than 1,200 m, a fact linked mainly to the use of farmland, bearing in mind the agro-ecological demands of cultivated plant species and the status of the available natural resources.

Table 9.5 Magnitude of supply sources identified by zone

Sources of agricultural landscape externalities	Importance by zone					
	Unit[a]	Zone 1	Zone 2	Zone 3	Zone 4	Total
Stone's clearing and land reclamation	ha	2	4	9.5	2	17.5
Water rehabilitation of dryland	ha	2	3	6	1	12
Terracing and terraced plots	ha	1	6	3	0.5	10.5
Fruit tree plantations on slopes	tree	250	590	1,970	1,810	4,620
Construction of stone edges around plots	km	0.25	0.25	0.7	0.5	1.7
Planting of ornamental trees around houses	tree	15	28	77	40	160
Growing of flowers and gardening around houses	m^2	50	750	575	285	1,660
Creation and upkeep of earth irrigation channels	m	630	1,050	1,740	1,050	4,470
Creation and upkeep of dirt road around the farm	m	600	1,350	2,100	1,430	5,480
Rehabilitation of dirt road outside the farm	m	0	800	2,800	0	3,600
Conservation of the typical architecture of houses	m^2	2,480	3,950	4,270	3,700	14,400
Renovation and landscape integration of farm buildings	m^2	0	530	1,230	0	1,760
Conservation of local mule breeds	animal	11	11	29	22	73

Source: Author's survey
[a]Units: hectares (ha), kilometers (km), meters (m), and number of trees and animals

Farms distant from the *douar* (village), lying in a valley bottom or on valley slopes and located close to a forest, provide more landscape externalities than those in other geographical positions. This finding cannot be generalized to all regions of the country, but it does suggest that in high-mountain zones, the location of farms is one of the major factors affecting agricultural landscape externality generation.

9.3.2.2 Effects of the Features of Productive Natural Resources

Marked differences amongst the four study zones were found for three of the indicators selected to measure the natural resources used in agricultural production. These are the total arable land per farm, the irrigated arable land per farm, and the numbers of sheep and goats (see Table 9.7).

The data suggest that landscape externalities are lower in areas with higher average amounts of cultivated and irrigated lands. The further the farm is from the foothills (Zone 4) toward the mid (Zones 2 and 3) and high mountains (Zone 1), soil resources become very limited, forcing farmers to adopt integrated farming and conservation management practices, and the potential to produce externalities increases. The findings indicate that in high-mountain zones, the relationship between farm size and landscape externality is not linear. The same applies to the numbers of sheep and goats, which would appear to have no positive correlation with the farm's landscape externalities.

Table 9.6 Geographic positions of farms and average number of externality sources

Feature	Zone 1	Zone 2	Zone 3	Zone 4
Mean number of sources per farm	6	7	7	4
Agro-ecological zone	High mountain	Mid mountain	Mid mountain	Foothills
Average altitude (in m)	>1,200	1,000–1,200	1,000–1,200	<1,000
Location of farm in relation to the *douar* (%):				
In the *douar*	90	44	39	39
Near the *douar*	5	26	33	46
Isolated from the *douar*	5	30	28	15
Location of farm in relation to the valley (%):				
In the valley bottom	0	11	48	10
On a slope	85	82	35	85
Near a water course	15	7	17	5
Location of farm in relation to the forest (%):				
Near the forest (less than 1 km)	70	52	59	37
Far from the forest (more than 1 km)	30	48	41	63

Source: Author's survey

Table 9.7 Importance of natural resources used and levels of externalities

Feature	Zone 1	Zone 2	Zone 3	Zone 4
Mean values	6	7	7	4
Number of farmers interviewed	20	27	46	41
Total arable land (ha)	9.5	28	16	161
Average total arable land per farm (ha)	0.4	1	0.4	4
Average irrigated arable land per farm (ha)	0.4	0.7	0.4	1.6
Number of sheep per farm	23	12	15	4
Number of goats per farm	10	5	6	3
Number of cattle per farm	1	2	2	1
Number of mules per zone	11	10	22	20
Farms of less than 1 ha per zone (%)	90	48	85	17

Source: Author's survey

9.3.2.3 Effects of Agricultural Diversification and Intensification

The effect of intensification on landscape externality production is not always linear but depends on various other factors, particularly the nature of the technologies adopted, equipment and tools used, and the state of natural resources. The results indicate that the rate of crop intensification does not have a clear negative correlation with the level of landscape externality produced. The average levels of

Table 9.8 Crop intensification rate and levels eternality source

Feature considered	Zone 1	Zone 2	Zone 3	Zone 4
Mean of sources per farm	6	7	7	4
Crop intensification rate (%)	197	105	102	115
Share of cereals in cropping pattern (%)	118	69	73	84
Share of fruit tree plantations in cropping pattern (%)	75	26	22	21
Share of pulses, horticultural crops, and fodder crops (%)	4	10	7	2
Share of fallow (%)	0	0	0	8

Source: Author's survey

externality produced is higher in Zones 2 and 3, where crop intensification rates are lower, but similar levels are also found in Zone 1, where the crop intensification rate is the highest (see Table 9.8).

The diversification of agricultural production, measured by the relative shares of different crops in the cropping pattern, appears to be positively correlated with the levels of landscape externality produced (Zones 2 and 3). Crop diversification in mountain zones, particularly terracing and multi-cropping on one plot, are attributes and amenities typical of agricultural landscapes.

9.3.3 *Types of Agricultural Landscape Externalities*

The study identified the relationship between each practice and the landscape externality produced (see Table 9.9). For example, cultivated fields and fruit or ornamental tree plantations (green, shade, layering of vegetation) account for 41% of the total number of externality sources accounted for in this study. The other relatively important share comes from landscaping around the home (23%), improving the physical beauty of the house, and farm buildings. Of lesser importance are passages and open-air recreational spaces (19%) (tracks, tourist trails), the upkeep and shaping of plots (9%) (borders and rows), and aspects of animal biodiversity (8%).

9.3.4 *Agricultural Landscape Externalities and Their Positive Effects*

Identification of positive effects requires first a definition of the main functions associated with the various externalities identified and the values associated with them in relation to the types of beneficiaries (see Table 9.10).

The functions associated with agricultural landscape externalities depend to a large degree on the nature of the landscape and on the users' perception of its quality. From an aesthetic, recreational, and cultural point of view, three major functions can be linked to the landscape externalities identified: (1) as a residence and workplace for local inhabitants; (2) as a place for holidays, relaxation, and open-air recreational and sporting activities for visitors, both national and foreign; and (3) as a place of biodiversity and ecosystem for informed visitors

Table 9.9 Relationship between sources and type of externality

Source	Landscape element(s) concerned	Details of externalities	Relative share (%)
Stone clearing and land reclamation	Cultivated fields	Green	41
Water rehabilitation of dryland	Plantations	Shade	
Terracing and terraced plots		Layers of vegetation	
Fruit tree plantations on slopes			
Construction of stone edges around plots	Stone walls	Borders, rows	9
Planting of ornamental trees around houses	Constructed sector	Environment of the residential context and farm buildings	23
Growing of flowers and gardening around houses	Local heritage		
Conservation of the typical architecture of houses			
Renovation and landscape integration of farm buildings			
Creation and upkeep of earth irrigation channels	Tracks	Passages	19
Creation and upkeep of dirt road around the farm	Irrigation channels	Spaces for open-air recreational and sporting activities	
Rehabilitation of dirt road outside the farm			
Conservation of local mule breeds	Animals	Local breeds and biodiversity	8

Source: Author's survey

who are sensitive to environmental issues (ecotourists, ecologists, and others). The values associated with the externalities identified and the spheres of positive trade-offs can be defined in terms of these three functions. Since much of these values are not captured in markets and are external to the farm, the challenge is to identify how farmers, the main producers of landscape externalities, can obtain remuneration for the environmental services they supply jointly with agricultural production.

9.3.5 Ways of Internalizing Agricultural Landscape Externalities

In the Moroccan context, the integration of agriculture and rural tourism seems to offer interesting possibilities for internalizing agricultural landscape externalities. The various functions associated with agricultural landscape externalities can thus be capitalized on through the reception and lodging of tourists, the support of

Table 9.10 Functions, values, and beneficiaries of the identified agricultural landscape externalities

Element of the agricultural landscape	Details of the landscape externality	Functions associated with the externality	Values associated with the externality	Potential beneficiaries
Cultivated fields	Green	Places for work and visit	Direct use values	Producers
Plantations	Shade			
Stone walls	Layering of vegetation			
	Borders and rows			
Buildings	Environment of the residential context and farm buildings	Places for work and visit	Values of recreational and tourist use	Residents and tourists
Tracks irrigation channels	Passages and environment for open-air recreational and sporting activities	Places for work and visit	Values of recreational and tourist use	Ecotourists
Natural patrimony (animals)	Local breeds and biodiversity	Place of biodiversity	Biodiversity and ecosystem values	Biodiversity users

Source: Author's survey

tourists during excursions, and the practice of open-air recreational and sporting activities and services as guides and interpreters.

An inventory of all the tourist services offered by farms in the four representative areas was made to identify forms of internalization. This inventory revealed a considerable proportion of producers lacking the means to internalize the landscape externalities they produce (see Table 9.11). Over half the farms in Zone 1 (55%) and over one third in Zones 2 and 4 (33% and 39%, respectively) fall into this category, while in Zone 3 only a minority of producers (11%) is not involved in tourist activities. Examination of the nature of the tourist services supplied shows an unequal distribution across the four zones. Board and lodging services are more frequently supplied by farms in Zone 2 (33%), while support services (mule drivers and cooks) are more common among farms in Zones 1 and 4 (35% and 37%, respectively), and a combination of tourist services is found among farms in Zone 3, with 44% of them combining board and lodging as well as support and guide services.

These findings reveal a certain specialization in the tourist services supplied by farms as a function of their geographical location with respect to the main tourist circuits. In the high mountain and foothill areas (Zones 1 and 4), board and lodging structures are less widespread than in the mid-mountain areas (Zones 2 and 3). These differences are a result both of the strong tourist attraction of the mid-mountain areas because of the wealth and diversity of its agricultural and mountain landscapes, and also its proximity to the summit of Toubkal and trekking and

Table 9.11 Distribution of internalization types recorded per representative area (as a percentage)

Nature of tourist services supplied	Zone 1	Zone 2	Zone 3	Zone 4
Supply of tourist services (%):	45	67	89	61
Board and lodging	10	33	17	19
Support (mule drivers and cooks)	35	19	28	37
Board, lodging, support, and guiding	0	15	44	5
Without tourist services (%)	55	33	11	39
Total	100	100	100	100

Source: Author's survey

Table 9.12 Types of producers according to the nature of their supply of tourist services

Producer group	Features of the group	Number	Percentage of total
Group 1	Producers supplying board and lodging services	25	19
Group 2	Producers supplying only support services	40	30
Group 3	Producers combining both board and lodging services and support and guide services	28	21
Group 4	Producers supplying no tourist services	41	30
Total		134	100

Source: Author's survey

horseback riding trails. The advantages and amenities of the agricultural landscapes supplied by foothill and high mountain farms are thus outside tourist circuits and cannot be directly optimized by their producers.

A typology of four producer groups was established on the basis of the nature of tourist services supplied (see Table 9.12). Almost one third (30%) of farms (Group 4) have no possibility of directly internalizing the landscape externalities they produce. This considerable proportion of farms is not able to obtain remuneration for the landscape services they supply jointly with their agricultural production. Amongst farms with means of internalization, board and lodging services are offered by 19%, solely support services (mule drivers and cooks) by 30%, or both by 21%.

At this stage in the analysis, one of the major questions is that of whether there is a correlation between the level of externalities produced and the means of internalization. The distribution of the various producer groups according to type and source of the externality generated helps provide an answer to this question. Groups 2 and 3 come first in terms of the number of sources of externality recorded, each with 28% of the total cases recorded (see Table 9.13). Group 1 comes next, with 24%, while Group 4 has 20%. This trend is slightly modified when the number of cases recorded per farm is considered. Group 3 again comes in first place, with an average of eight externality sources per farm, then Group 1, with an average of seven, and Group 2, with an average of six, while Group 4 remains in last place, with an average of only four. These findings suggest that farms involved in supplying tourist services tend to produce more landscape externalities than those confined solely to agricultural production.

Table 9.13 Distribution of sources of externality according to producer group

Source of agricultural landscape externality	Number of cases recorded per representative area				
	Group 1	Group 2	Group 3	Group 4	Total
Stone clearing and land reclamation	22	17	24	12	75
Water rehabilitation of dryland	14	19	18	11	62
Terracing and terraced plots	22	20	28	14	84
Fruit tree plantations on slopes	20	29	19	30	98
Construction of stone edges around plots	19	19	17	17	72
Planting of ornamental trees around houses	12	2	14	1	29
Growing of flowers and gardening around houses	19	4	22	4	49
Creation and upkeep of earth irrigation channels	20	20	22	11	73
Creation and upkeep of dirt road around the farm	12	20	13	20	65
Rehabilitation of dirt road outside the farm	2	10	6	1	19
Conservation of the typical architecture of houses	18	24	21	22	85
Renovation and landscape integration of farm buildings	4	3	6	2	15
Conservation of local mule breeds	3	39	11	11	64
All cases recorded	187	225	221	156	789
Percentage of all cases recorded	24	28	28	20	100
Average number per farm	7	6	8	4	6

Source: Author's survey

The results also show that farms in Groups 2 and 3 are more involved in generating landscape elements associated with the constructed sphere, with 35% and 30%, respectively, of the cases recorded (see Table 9.14). This result should be seen in relation to the amount of the board and lodging services supplied by these two groups of farms. On the other hand, most (62%) of the landscape externalities that involve the rearing of mules are generated by farms in Group 2, which are the main suppliers of support services (mule drivers and cooks).

9.3.6 *Economic Benefits Associated with Landscape Externalities*

The typology of producers allows for an evaluation of the contribution of each form of internalization to employment, income, investment, and food security, and a comparative analysis of all the situations existing on the ground.

9.3.6.1 Employment and Income Effects

The tourist services supplied by farms create employment opportunities both for family members and for other local inhabitants. The survey data show that 181 of

Table 9.14 Amount of the externalities identified according to group (as a percentage)

Landscape element	Details of the externality	Group 1	Group 2	Group 3	Group 4
Cultivated fields	Green	24	27	28	21
Plantations	Shade				
	Layers of vegetation				
Stone walls	Delimitation and edging	26	26	24	24
Constructed areas	Environment of the residential context and farm buildings	30	19	35	16
Local heritage					
Tracks	Passages and spaces for open-air recreation and sporting activities	22	32	26	20
Irrigation channels					
Living natural heritage	Local breeds and biodiversity	4	62	17	17
Animals					

Source: Author's survey

the 815 people in the sample are involved in tourist services of some kind. This gives a rate of involvement of all the population surveyed of 22% (see Table 9.15). The number of jobs created increases with the number of tourist services supplied by farms. On average, two people per household are employed full-time in board and lodging services during the tourist season. Support services (mule drivers and cooks) for excursionists employ on average one person per household, although this average masks situations in which there are more than three people per household involved in support services. When a farm provides a combination of tourist services, the employment it can create is on average three people per household. In most cases (85%), the employments created by tourist services benefit family members or neighbors.

Supplying board, lodging, food supplies, and support to tourists provides farmers with the opportunity to receive remuneration, even if only partial, for the various efforts they make in terms of the upkeep and shaping of agricultural landscapes. Apart from direct job creation, the supply of tourist services from farms can also generate substantial, diversified income. The survey data allowed an assessment of the proceeds from the various tourist services supplied by farms. These are on average DH (Dirham) 14,000 per year for lodging, 9,000 per year for board, DH 5,000 per year for support services, and DH 12,000 for guide services (DH 10 $\approx$ US \$1.00). The impact of each of these forms of tourist income on the economy of the farm can be assessed by its correlation with income diversification and household income. The more the farm diversifies its supply of tourist services, the greater the contribution of tourist income to overall income (see Table 9.16).

In Group 3, those supplying several tourist services, the share of tourist income represents an average of over two-thirds (73%) of the annual overall income (DH 52,000 per household per year). This level of contribution becomes close to half (49%) in Group 1 and close to one third (29%) in Group 2. One of the effects of this

Table 9.15 Employment potential of various forms of internalization

Producer group	Inhabitants surveyed	Average size of household	Number involved in tourism	Level of involvement in tourist activities (%)	Average number of household members involved in tourist activities
Group 1	149	5.96	50	34	2
Group 2	253	6.32	42	17	1
Group 3	171	6.11	89	52	3
Group 4	241	5.88	0	0	0
Total	815	6.1	181	22	1

Source: Author's survey

Table 9.16 Income diversification potential of various forms of internalization

Producer group	Average overall income (DH/year)	Share of agricultural income (%)	Share of tourist income (%)	Share of supplementary income (%)
Group 1	30,450	34	49	17
Group 2	18,870	62	29	9
Group 3	51,880	21	73	6
Group 4	16,250	68	0	32
Overall	26,830	40	46	14

Source: Author's survey

diversification of income sources appears to be a lower dependence of farms on supplementary income. In Group 4, producers without any tourist activity, the share of supplementary income is very large – over one third (34%) of the household's overall annual income. Examination of the income categories of the four producer groups indicates that Group 3 has the highest share of households earning more than DH 30,000 per year (see Table 9.17). This result indicates the considerable potential of tourist-based forms of internalization for improving the income of farms.

9.3.6.2 Effects on Investment

Adapting farms for tourism generally requires modification of buildings (extension, consolidation, renovation, etc.) and material resources to ensure tourists' comfort (sanitary installations, furniture, cooking and eating utensils, etc.). Such operations require major investments, especially for farms hoping to supply full, high-quality services.

Survey data concerning the size of investments made during the five previous years show that one third (33%) of farms had invested more than DH 20,000, with an average of about DH 45,000 for the period over the whole sample (see Table 9.18). However, the distribution of sums invested varied by producer group, showing large differences depending on the level of integration of tourist services.

Table 9.17 Distribution of producer groups according to income category (as a percentage)

Producer group	Average overall income (DH/year)	Income category (DH/Household/Year)			
		<1,000	10,000–20,000	20,000–30,000	>30,000
Group 1	30,450	0	32	36	32
Group 2	18,870	3	72	20	5
Group 3	51,880	0	5	25	70
Group 4	16,250	24	56	15	5
Overall	26,830	8	45	23	24

Source: Author's survey

Table 9.18 Sums invested in tourist activities according to producer group (as a percentage)

Producer group	Average sum invested (in DH)	Categories of sums invested over the past 5 years			
		Less than 5,000	5,000–10,000	10,000–20,000	More than 20,000
Group 1	49,400	4	16	8	72
Group 2	9,500	47	35	10	7
Group 3	150,000	7	7	14	72
Group 4	6,400	41	34	19	5
Overall	44,500	29	25	13	33

Source: Author's survey

In Group 3, the sums invested were much higher than those in Groups 1 and 2, exceeding DH 20,000 in most cases (72%).

Investment associated with the supply of tourist services is also associated with improved residential living conditions on farms (see Table 9.19). All the indicators of residential living conditions, especially drinking water supplies, electricity connections, and the presence of showers and toilets, are much more favorable in Groups 1 and 3 than Groups 2 and 4. In other words, the greater the investments linked to board and lodging services, the greater the improvement in living conditions on farms.

9.3.6.3 Others Effects on the Household Economy

Apart from the economic effects discussed above, the supply of tourist services is also an activity that generates contacts and links with the outside world. Board, lodging, and support services in their various forms also affect the behavior and strategies of the suppliers. One of the most important effects is linked to the supply of the goods needed to house and feed tourists, particularly the purchase of goods from nearby towns. This type of market integration leads to modifications in dietary habits and hence to changes in the structure of food expenditure.

Table 9.19 Indicators of the residential living conditions by producer group (as a percentage)

Producer group	Access to drinking water	Electricity connection	Presence of showers	Presence of toilets
Group 1	80	85	75	95
Group 2	58	75	10	70
Group 3	96	96	88	100
Group 4	41	63	5	63
Overall	63	77	34	78

Source: Author's survey

Table 9.20 Share of domestic production in the food expenditure of producer groups

Producer group	Average share (%)	Category of share in domestic production				
		Less than 20%	20–40%	40–60%	60–80%	More than 80%
Group 1	24	56	24	16	0	4
Group 2	34	20	42	25	10	3
Group 3	15	68	32	0	0	0
Group 4	33	20	61	15	2	2
Overall	28	37	42	15	4	2

Source: Author's survey

Examination of the part played by domestic production in the food expenditure of the various producer groups threw light on the issue (see Table 9.20). The results indicate that the more involved a farm is in supplying tourist services, the more sharply the share of domestic production in food expenditure falls. While this share is over one third of food expenditure for Groups 2 and 4 (33% and 34%, respectively), it is only 15% and 24%, respectively, for Groups 3 and 1, and in more than half of these latter groups (68% and 56%, respectively) the share of domestic production in food expenditure is 20% or less. These figures indicate a link between the supply of integrated tourist services with market integration of the farm.

The results examined so far suggest two important facets of agricultural multifunctionality in the context of agricultural development. On the one hand, the integration of tourist activities into agricultural production provides considerable support to the revitalization of farms and the strengthening of the capacity of diversifying livelihoods in rural areas. On the other hand, the contact that households establish with tourists encourages farms to open up to the market environment and hence to adopt new types of behavior and approaches helpful to their modernization.

9.3.7 Impacts on the Rural Poverty Reduction

The approach adopted to measure poverty focuses mainly on comparisons of "economic well-being" through the concept of monetary poverty (Ravallion, 1994). Consumption, rather than income, was used to measure poverty, bearing in mind its direct links with an individual's or household's "well-being" (Coudouel et al., 2002). Food and overall household expenditures were estimated and expressed in per adult equivalents, in order to measure monetary poverty through indicators concerning the incidence of poverty or poverty in numbers of inhabitants, the range of standards of living, and the consumption deficit ratio (expenditure deficit ratio).

9.3.7.1 Level and Structure of Food and Overall Expenditure

Survey data concerning the food and non-food expenditures of households indicates that average annual overall expenditure is about DH 6,000 per adult equivalent (see Table 9.21). The share of food expenditure is almost half (47%) of overall expenditure, or DH 2,800 per adult equivalent per year, while non-food expenditure is estimated at about DH 3,200 per adult equivalent per year, or 53% of total expenditure.

Examination of the expenditure levels of the various groups shows that, overall, households in Groups 1 and 3 spend more than those in Groups 2 and 4. In terms of non-food expenditure, the levels of Groups 1 and 3 are on average more than double those of Groups 2 and 4, but the differences are less in the case of food expenditure, where the multiplier does not exceed 1.6 points between Groups 1 and 4.

Almost half (47%) the households in Group 3 spend more than DH 7,000 per adult equivalent per year (see Table 9.22). This category is less than one third (28%) for households in Group 1 and almost non-existent (2%) for those in Groups 2 and

Table 9.21 Levels of food and overall expenditure according to producer group, in DH

Producer group	Food expenditure	Non-food expenditure	Overall expenditure
Group 1	3,500	4,640	8,200
Group 2	2,420	1,960	4,380
Group 3	3,390	5,910	9,300
Group 4	2,210	1,590	3,800
Overall	2,770	3,180	5,950

Source: Author's survey

Table 9.22 Structure of annual overall expenditure according to producer group (as a percentage)

Producer group	Average overall expenditure	Category of expenditure in DH/adult equivalent/year			
		Less than 3,000	3,000–5,000	5,000–7,000	More than 7,000
Group 1	8,200	4	28	40	28
Group 2	4,380	15	50	32	2
Group 3	9,300	0	18	36	46
Group 4	3,800	24	59	15	2
Overall	5,950	13	42	29	16

Source: Author's survey

4. In these two latter groups, the proportion of households spending less than DH 3,000 per adult equivalent is relatively high, or 15% and 24%, respectively.

Sharp differences in overall expenditures between Groups 1 and 3 on the one hand and Groups 2 and 4 on the other are apparent. This difference not only reflects a major income effect as the source of variations in expenditure, but can also indicate the existence of major differences in the living conditions of the various groups.

9.3.7.2 Incidence of Food Poverty and Overall Poverty

The poverty thresholds used to measure the incidence of food and overall poverty in the households considered are DH 2,377 and 3,842 per adult equivalent per year. The results show that more than one third (37%) of the 134 households studied are unable to obtain the necessary food basket, that is, the equivalent of 2,400 Kcal per day per adult equivalent (see Table 9.23).

About four households out of ten are food poor, although there are considerable differences in the incidence of food poverty amongst the producer groups. It is particularly high in Group 4, producers without means of internalization, with six households out of ten in a situation of food poverty. The proportion falls as farms diversify their activities by supplying tourist services. The lowest incidence of food poverty is found in Group 3, where only one household in ten lives below the threshold, while Groups 1 and 2 fall midway with, respectively, two and four households out of ten in a situation of food poverty. So far, as the overall poverty threshold is concerned, households in Groups 2 and 4 are the poorest in comparison with the average recorded for the whole sample. In Group 4, the incidence of overall poverty is very high, with 63% of households living below the overall poverty threshold. In Group 2, the incidence is close to the average for the sample, with 47% of households living below the threshold. On the other hand, the incidence in Groups 1 and 3 is well below the average for the sample with, respectively, 28% and 10% of households living below the overall poverty threshold.

These results indicate a strong correlation between the level of involvement of farms in tourist activities and the incidence of food and overall poverty. The more involved a farm is in supplying tourist services, and hence the more diversified its sources of income, the less affected it is by food and overall poverty. Comparing farms without any form of internalization to others with board and lodging structures,

Table 9.23 Incidence of food poverty and overall poverty per group (as a percentage)

Producer group	Incidence of food poverty	Incidence of overall poverty
Group 1	24	28
Group 2	45	47
Group 3	8	10
Group 4	61	63
Overall	37	40

Source: Author's survey

the incidence of food and overall poverty falls by an average of 16%, and this difference is even greater (more than 50%) in the case of farms that combine a number of tourist services (Group 3). However, in the case of Group 2, the supply of support services to tourists does not seem to improve the households' wealth status, inasmuch as the incidence of food and overall poverty remains above the average for the sample.

9.3.7.3 Situation of Households in Relation to Food and Overall Poverty Thresholds

In order to assess relative poverty amongst groups, a classification of households was established, distinguishing four levels of "living standards":

- The "non-poor," encompassing households whose total consumption is over 150% of the poverty threshold considered.
- The "vulnerable," encompassing households whose total consumption is between 100% and 150% of the poverty threshold considered.
- The "poor," encompassing households whose total consumption is between 75% and 100% of the poverty threshold considered.
- The "extremely poor," encompassing households whose total consumption is below 75% of the poverty threshold considered.

The results of this ranking showed that about 20% of the households considered fall into the extremely poor category in terms of the food poverty threshold; in other words, they are unable to cover 75% of the expenditure required to obtain the required minimum daily calorie intake (see Table 9.24). The distribution of households in terms of their food poverty status shows major differences among the four groups studied. The proportion of extremely poor households is very high in Groups 2 and 4, with 25% and 39%, respectively, classified as extremely poor in dietary terms. On the other hand, in Groups 1 and 3, extremely poor households are in a minority or non-existent (4% and 0%, respectively). At the other extreme, Groups 1 and 3 have the highest proportions of non-poor households (28% and 25%, respectively). Vulnerable households are the largest category, with almost half (47%) of all households falling into it, a finding that is also valid for the individual groups, except Group 4 where the proportion of vulnerable households is smaller (29%).

Table 9.24 Poverty status of the various producer groups in terms of the food poverty threshold (as a percentage)

Producer group	Non-poor	Vulnerable	Poor	Extremely poor
Group 1	28	48	20	4
Group 2	5	50	20	25
Group 3	25	67	8	0
Group 4	10	29	22	39
Overall	15	47	18	20

Source: Author's survey

Table 9.25 Poverty status of the various producer groups in terms of the overall poverty threshold (as a percentage)

Producer group	Non-poor	Vulnerable	Poor	Extremely poor
Group 1	24	48	20	8
Group 2	13	40	32	15
Group 3	63	29	8	0
Group 4	13	24	46	17
Overall	30	31	28	11

Source: Author's survey

There are similar proportions of households classified as poor in all the groups considered (about 20%) except Group 3, which has only 8%.

Two essential elements stand out in the profiling of households according to their poverty status measured against the food poverty threshold: (1) the considerable proportion of vulnerable households in all groups, which would indicate that a number of households are potential candidates for joining the ranks of the poor; and (2) the extreme poverty that particularly affects households dedicated to agriculture without any supplementary activities.

The distribution of households according to their poverty status measured against the overall poverty threshold shows that while the proportion of non-poor households measured against the overall poverty threshold is doubled, that of extremely poor households has been halved (see Table 9.25). Under these criteria, the poor now represent almost one third (28%) of the whole sample.

Vulnerable and poor households measured against the overall poverty threshold make up 90% of Group 4, 72% of Group 2, and 68% of Group 1. In Group 3 the proportion of vulnerable or poor households is only 37%, with the non-poor accounting for 63%.

9.3.7.4 Depth of Poverty Among Households

The last analytical component of the poverty profile of the four groups of households concerns the depth of poverty, determined through the consumption deficit ratio. This ratio allows the proportion of poor households in the whole sample and also the difference between the average consumption of the poor and the overall poverty threshold to be taken into account.

The results of the survey reveal an expenditure deficit ratio with regard to the overall poverty threshold of 16% for all household groups (see Table 9.26). In other words, taken together, the poor households of the four groups have an average consumption per adult equivalent of 84% of the overall poverty threshold (DH 3,842 per adult equivalent). To move out of poverty, average consumption should increase by at least 19% to make up the deficit. However, this situation does not affect all the groups of households in the same way.

Table 9.26 Consumption deficit ratio for the various producer groups

Producer group	Average consumption of the poor (in DH)	Consumption deficit ratio (%)	Rate of growth in consumption to make up the deficit (%)
Group 1	3,487	9	10
Group 2	3,237	16	19
Group 3	3,642	5	5.5
Group 4	3,129	19	23.5
Overall	3,230	16	19

Source: Author's survey

The greatest depths of poverty are found in Groups 2 and 4, which are also the two groups most affected by both food and overall poverty. Given their consumption deficit ratios of 16% and 19%, respectively, these two groups need to increase their average consumption by 19% and 23.54%, respectively, in order to overcome their poverty. On the other hand, average consumption levels are relatively high in Groups 1 and 3, rising above the average for the sample, as is reflected in generally small consumption deficit ratios.

The foregoing analysis shows considerable differences between Groups 1 and 3 on the one hand and Groups 2 and 4 on the other in terms of the incidence and depth of food poverty and overall poverty. Comparing farms without any form of internalization to others offering tourist services, all the standard-of-living indicators improve markedly. These results suggest that tourist activities integrated into agricultural production could have significant potential for reducing poverty. However, it should not be forgotten that the possibilities of establishing tourist activities are also limited by a whole range of factors, such as location, the environmental features of the area, and the nature of the structural elements of agricultural and natural landscapes, as well as tourists' present and future preferences.

9.4 Conclusions and Policy Implications

This chapter presents an empirical analysis of landscape externalities of agricultural origin and an assessment of their relationship to food security and poverty reduction in Morocco's Western High Atlas region. In attempting to establish linkages among the environmental "services" of agriculture and rural poverty reduction, a number of methodological as well as empirical results were produced.

Methodologically, the use of the concept links between the three main ingredients involved in the generation of agricultural landscape externalities – producers, externality flows, and beneficiaries – made it possible to distinguish the differences and particular features of agricultural landscape externalities. Second, the inventory of externality production sources, applied on the farm level, led to the formulation of a procedure to identify, classify, and analyze the factors that play a part in generating agricultural landscape externalities. The nomenclature and typology adopted in recording sources of production can then be used in empirical studies of agricultural landscape externalities in other contexts and regions. Third, the linkage between production sources and the constituent and structural elements of agricultural landscapes made it easier to determine the nature of farms' involvement in the generation of landscape externalities. Differentiation between agricultural landscape externalities according to area, definition of the functions and values associated with them, and identification of beneficiaries can all help in defining a procedure for empirical analysis of agriculture's contribution to the production of agricultural and rural landscapes. Fourth, the choice of rural tourism as the main source of value for landscape externalities sheds light on the various means of internalization at the farm level. It also provides insights into the spheres of economic activity relevant for evaluation.

Although this research was a case study specific to Morocco's high-mountain zones, it generates some data and indicators useful not only in gaining an accurate picture of the environmental role of agriculture in this region, but also in recognizing its present and potential contributions to rural poverty reduction.

The first result highlighted the multiplicity of agricultural landscape externalities and the variability in their breadth. In all the zones considered and regardless of predominant land use, a minimal generation of landscape externalities was found. This is mainly a result of the interaction between activities, natural resources, and the environment and varies by the nature and degree of the modification of agricultural-use support goods. The main factors showing a positive correlation with the level of landscape externality produced were: location-linked features of farms, the ways in which natural resources, particularly soil and water, are used, especially when combined with conservation management techniques and crop diversification in relation to land-use patterns. There are few landscape externalities generated outside farms. However, on-farm benefits are realized from several forms of externality production: the rehabilitation of valley bottoms and lower slopes by shaping soil into terraces, the planting of fruit trees along water courses, the construction of irrigation channels and the development of an irrigation system all not only play environmental and landscape roles but also enhance local assets. These results also provide insights into the farming practices that boost agriculture's environmental performance in mountain zones.

The second major result comes from the systematic inventory of agricultural landscape externalities in the four representative areas studied. Assessment of landscape attributes and their distribution according to their origin revealed major trends in the means of generating agricultural landscape externalities. From an aesthetic, recreational, and cultural point of view, three main functions are associated with the landscape externalities identified: as residences and workplaces for the producers (farmers) and local inhabitants, as places for holidays and open-air recreational and sporting activities for visitors, whether national or foreign, and as places of biodiversity and ecosystem for those interested in discovering nature.

The third result concerned the proportion of farms that internalized landscape externalities jointly with agricultural production through tourist activities. Results presented here suggest that the combined supply of several tourist services offers the greatest remuneration potential for farmers. The provision of hospitality, and board and lodging is a means of capitalizing on agricultural landscapes as is the provision services such as mule drivers, interpreters, and guides.

The economic contributions of various forms of internalization were categorized in terms of job creation, income generation, investment stimulation, market introduction, and improved living conditions. Evaluation of the impact of the various internalization forms showed major differences in the incidence of food and overall poverty and in the depth of poverty. All these standard-of-living indicators improve considerably when the farm combines the supply of tourist services with its other functions, a fact that highlights the very great potential of tourist activities, once they have been integrated into the farm as means of capitalizing on agricultural landscapes, to reduce poverty in its various forms.

Nevertheless, this trend should not be interpreted as a recommendation for the generalized extension of tourist services within farms, but as a possible approach to be promoted where local conditions permit. It should not be forgotten that the possibilities of establishing tourist activities are also limited by a whole range of factors, connected for example with location, environmental features of the area, and the nature of the structural elements of agricultural and natural landscapes, as well as tourists' present and future preferences.

In the context of developing countries where financial resources are limited, the establishment of measures to integrate agriculture and rural tourism under a program to promote a multi-activities approach in rural areas would surely be more effective and practical than a transfer of cash as a payment for environmental services. This also means that debate on the issue of environmental externalities must broaden its focus from agriculture and the environment to encompass rural tourism as well. In other words, any system to compensate farmers for the environmental services they supply, whatever its nature (agri-environmental measures, payment for environmental services, etc.), cannot be really effective or sustainable without an integrated territorial approach, allowing organizational and operational foundations to be laid in order to facilitate the emergence of multifunctional farms. The development of agro-tourism farms would then be a model to be promoted as a way of capitalizing on both agricultural resources and landscape and tourist potential.

Appendix

Appendix 9.1 Geographic localization and administrative limits of study zones

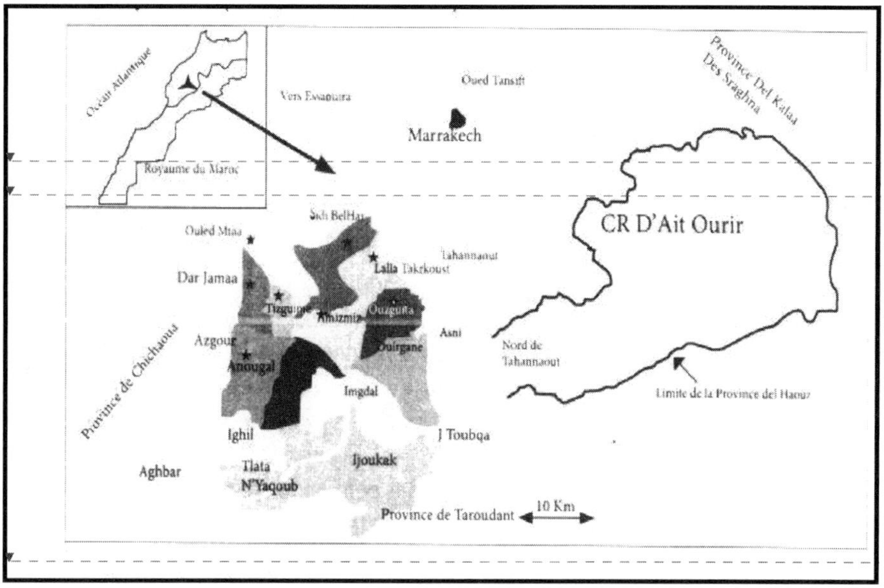

Appendix 9.2 Main features of the four representative areas studied

Representative area	Agro-ecological zone	Dominant use	Amount of natural resources	Production system(s) practiced	Development state of rural tourism	Rural communes selected
Zone 1	High mountain	Agrosylvo-pastoral	Abundant water resources	Very little crop diversification	Very little or no tourist activity	Imgdal
			Limited soil resources	Tree growing predominant, goat rearing preponderant		
Zone 2	Medium mountain	Agro-pastoral	Limited water resources	Diversified crops	Expanding tourist activity	Anougal
			Relatively large soil resources	Intensive tree growing		
				Considerable dairy farming		
Zone 3	Medium mountain	Agro-pastoral	Limited water resources	Cereal crops predominant	Major and well-established tourist activity	Asni
			Limited soil resources	Intensive tree growing		
				Extensive sheep rearing		
Zone 4	Foothills	Agro-pastoral	Limited water resources	Less diversified crops	Marginal and localized tourist activity	Amzmiz
			Greater soil resources	Major cereal-, olive- and almond-growing		Amghras
				Extensive animal husbandry		Lala Takarkoust

Source: Author's survey

Appendix 9.3 Composition of the sample of farms studied

Representative area	Agro-ecological zone	Rural commune(s) concerned	Total number of farmers	Farmers surveyed	Percentage of all farmers	Percentage of the sample
Zone 1	High mountain	Imgdal	500	20	4	15
Zone 2	Medium mountain	Anougal	579	27	5	20
Zone 3	Medium mountain	Asni	1,600	46	3	34
Zone 4	Foothills	Amzmiz	421	41	3	31
		Amghras	595			
		Lala Takarkoust	379			
Total			4,074	134	3.3	100

Source: Author's survey

Appendix 9.4 Spheres of trade-offs in terms of categories of externality studied

Category of externality	Spheres of positive trade-offs
Landscape	Rural tourism – jobs – poverty
	Rural tourism – income – poverty
	Rural tourism – investment – poverty – standard of living
	Rural tourism – sociocultural heritage – income–poverty
Plant and animal biodiversity	Domestic production potential – food security
	Rural tourism – income – poverty
Natural resources: soil and water	Domestic production potential – food security

Source: Author's survey

References

Allali K (2003a) Valeur sociale de la conservation du paysage agricole au Maroc: Application de la méthode d'évaluation contingente au paysage agricole de la Chaouia. Projet Rôles Socio-économiques de l'agriculture et Implications politiques, ROA-FAO/INRA-Maroc.

Allali K (2003b) Externalités pécuniaires positives de l'agriculture sur le tourisme de montagne au Maroc. Projet Rôles Socio-économiques de l'agriculture et Implications politiques, ROA-FAO/INRA-Maroc.

Aznar O (2004) Services environnementaux et espaces ruraux. Une approche par l'économie des services. Thèse de Doctorat d'Economie, Université de Bourgogne- Faculté des Sciences Economiques et de Gestion, Dijon.

Coudouel A, Hentschel JS, Wodon QT (2002) Poverty measurement and analysis, in the PRSP Sourcebook. World Bank, Washington, DC.

Delaunay J-C, Gadrey J (1987) Les enjeux de la société de service. Presses de la Fondation nationale des Sciences Politiques, Paris.

Direction de la Statistique, Haut Commissariat au Plan, Ministère de la prévision économique et du plan (1999) Enquête nationale sur les niveaux de vie des ménages 1998-1999. Royaume du Maroc.

Gadrey J (1996) L'économie des services, première édition 1992. la Découverte, Paris.
Gadrey J (2000) The characterization of goods and services: An alternative approach. Rev Income Wealth 46(3):369–387.
Hill P (1999) Tangibles, intangibles and services: A new taxonomy for the classification of output. Can J Econ 32(2):426–446.
Ravallion M (1994) Fundamentals of pure and applied economics (Vol. 56). Poverty comparisons. Harwood Academic Publishers, Chur, Switzerland.

Chapter 10
Exploring Environmental Services Incentive Policies for the Philippine Rice Sector: The Case of Intra-Species Agrobiodiversity Conservation

Nobuhiko Fuwa and Asa Jose U. Sajise

Abstract This chapter considers a hypothetical scheme of green payments to induce intra-specific agrobiodiversity in the context of Philippine rice farming. We empirically estimate a model of farmer behavior and then simulate the consequences of alternative (hypothetical) PES schemes under a fixed budget constraint. We find that, under this particular application, there is a clear trade-off between the two policy goals of enhancing agrobiodiversity and poverty reduction. Even the totally untargeted lump-sum subsidy would have a larger poverty reduction impact than would the first-best conservation subsidy payment scheme. Therefore, policymakers would be required to strike a delicate balance between the two competing policy objectives. In addition, there is also a clear trade-off between the efficiency of targeted conservation payment and the information requirement for implementing subsidy schemes.

10.1 Introduction

There has been an increasing recognition that agriculture or agricultural activity produces not only food and fibers but it also produces as joint products environmental services that are not traded in markets. These environmental services include climate regulation, carbon sequestration, waste absorption and breakdown, biodiversity and wildlife conservation, soil and water conservation, and a host of others. The recognition of such positive externalities has led to the attempts to correct the underprovision of these services through payments for environmental services (PES) or "green" subsidies. This market-based instrument has been used extensively in developed countries. For instance, the United States has a land retirement program under the Conservation Reserve Program (CRP) and

L. Lipper et al. (eds.), *Payment for Environmental Services in Agricultural Landscapes*. Natural Resource Management and Policy.
doi: 10.1007/978-0-387-72971-8_10;

Environmental Quality Incentives scheme aimed at providing incentives for sustainable agricultural practices, while countries like Canada (National Farm Stewardship Program) and the United Kingdom (Country Stewardship and Organic Farming Scheme) have similar incentive systems. It is ironic, however, that in agriculture- dependent developing countries like the Philippines these policy instruments have not yet been explored.

This chapter considers a hypothetical scheme of green payments to induce intraspecific agrobiodiversity in the context of Philippine rice farming. While most of the existing studies focus on efficiency aspects of agricultural environmental services payments (see Kurkalova et al., 2003; Feng et al., 2004, 2005; Lankoski & Ollikainen, 2003), this study explores potential trade-offs between biodiversity conservation and poverty reduction goals. We attempt to quantify the magnitude of such trade-offs by empirically estimating a model of farmer behavior and then simulating the consequences of alternative (hypothetical) PES schemes under a fixed budget constraint. PES schemes have been primarily designed with efficiency objectives in mind. However, a review by Pagiola et al. (2005) points to the possibility of synergies between poverty reduction and efficiency goals. They conclude that poverty impacts of these schemes depend on a number of technical and economic factors notably the population composition of target areas, targeting schemes, tenure security, and the size of the payments itself.

Casual reference to the poverty impacts of PES schemes abound in the literature,[1] but there have been relatively few empirical studies that examine PES for agriculture and its poverty alleviation implications. The intent of this study is similar to Alix-Garcia et al. (2004) who empirically addressed the conservation-poverty link in a different context, i.e., that of PES for watershed management. Antle and Stoorvogel (2006), on the other hand, looked at agricultural "green subsidies" and poverty, but the focus is on the carbon sequestration function of agriculture. They used a simulation model to explore the potential impacts of payments for agricultural soil carbon sequestration on poverty and farm households and the sustainability of agricultural systems. Their results support the claim that carbon payment contracts provide sufficient incentives for farmers to shift to sustainable systems while reducing poverty.

Using a nationwide data set from the Philippines, we focus on the farmer behavior of planting traditional rice varieties alongside modern rice varieties, and examine policy instruments that could potentially induce farmers to adopt this "environmentally friendly technology." This chapter addresses three issues: (1) How much would it cost to induce rice farmers to plant traditional varieties, i.e. implementation cost of an intra-species conservation payments scheme? (2) What would be the most effective form of payment scheme as an environmental policy instrument? (3) What are the poverty implications of these payment

[1] See for instance the literature in PES for watershed management and biodiversity conservation. The article by Wu et al. (2001), on the other hand, is a good theoretical paper on the distributional consequences of different conservation-targeting strategies.

schemes? In addressing these issues, we pay particular attention to the potential trade-offs involved between the higher farm profit from *not* planting traditional rice varieties (since modern rice varieties tend to allow farmers to obtain higher profit through their higher yields) and the potential benefits of maintaining biodiversity in rice farming that may not be captured (entirely) by individual farmers. Such trade-offs could be particularly acute for relatively poorer farmers. From policymakers' point of view, the potentially efficient (optimal) policies for the goal of environmental preservation may not be fully consistent with poverty reduction goals. Such potential trade-offs from a policymaking point of view is our major focus in the following analysis.

The rest of this chapter is organized as follows. The next section briefly introduces the issue of biodiversity conservation in the context of rice farming, in general, and the issue of traditional rice variety, in particular. Section 10.3 presents the empirical model to be used for the analysis. Section 10.4 is a short description of the dataset used. The next three sections present our results; Sect. 10.5 presents our results on the determinants of the adoption of traditional variety cultivation, while Sect. 10.6 discusses our results on the determinants of farm profit and the effects of traditional variety cultivation on farm profit. Section 10.7 presents the results of our policy simulations, with a focus on the impact of environmental service payment schemes on poverty outcomes. The final section concludes.

10.2 Biodiversity Benefits of In Situ Conservation of Traditional Rice Varieties

Any loss of biodiversity is irreversible, and such losses have been increasingly recognized as a major policy issue in developing countries. Genetic diversity is an important component for the continuous improvements of rice crops, as cultivars need to be invigorated every 5–15 years to better protect them against diseases and pests (International Rice Research Institute, IRRI, 2004). Furthermore, the recent advances in biotechnology have led to a renewed recognition of the importance of maintaining biodiversity as the basis for technological breakthroughs. Commercial rice production also relies heavily on the genetic diversity of rice as a source of material for plant breeding and improvement (IRRI, 2004, p. 25). In addition to the potential roles of traditional rice varieties as raw materials for genetic improvements, the use of traditional varieties has been found to be potentially effective in controlling certain types of pests. For example, recent experiments conducted in the southwestern province of Yunnan, China, have found that intercropping rows of different rice varieties can control the rice blast disease "that costs the rice industry millions of dollars annually." The cropping practice allows blast-susceptible traditional varieties to be conserved *in situ* and also reduces the cost of pesticides (IRRI, 2004, p. 27).

While there exist some estimated 140,000 rice varieties, it is widely recognized that the number of rice varieties has declined dramatically, especially since the introduction of the high-yielding rice varieties (HYVs) in the 1960s. In the Philippines alone, there were "more than a few thousand" rice varieties grown in the 1950s. Today, only two varieties cover 98% of the land planted with rice (IRRI, 2004, pp. 24–25).

In the following analysis, we focus on the practice of growing "traditional" rice varieties, i.e., in situ on-farm conservation of traditional varieties, as an environmentally friendly agricultural technology that the government might consider encouraging farmers to "adopt." The potential advantages of on-farm (in situ) conservation of biodiversity, in contrast with ex situ conservation – such as a gene bank – can be summarized as follows (Tuan et al., 2003):

- On-farm conservation conserves the evolutionary processes of local adaptation of crops to their environments.
- It conserves diversity at all levels – the ecosystem, the species, and the genetic diversity within species.
- It conserves ecosystem services critical to the functioning of the Earth's life-support system, thus improving the livelihoods for resource-poor farmers through economic and social development.
- It maintains or increases farmers' control over and access to crop genetic resources.
- It ensures farmers' efforts are an integral part of national Plant Genetic Resources (PGR) systems and involves farmers directly in developing options for adding benefits of local crop diversity.
- It links the farming community to gene banks for conservation and utilization.

However, due to the absence of sufficient information that would allow us to estimate potential values of biodiversity conservation from paddy rice cultivation in the Philippine context, our focus here is exclusively on the cost side (i.e., the amount it would cost to induce farmers to adopt farming practices that provide certain environmental services as externalities) and not on the benefit side (e.g., valuation of environmental services). Needless to say, policy decisions would need to be based on both the cost (as pursued here) and the benefit (not pursued here) sides of alternative policy instruments.

10.3 The Empirical Model: Treatment Effects and the Choice of Cultivating Traditional Rice Variety

In light of the potential trade-offs between farm profit and conservation, we first estimate the likely losses in farm profits due to the adoption of traditional rice variety cultivation, and then discuss the potential amount of subsidies that need to be provided to the farmers as an environmental service payment under alternative policy scenarios. The general model that we use in this case study is the following endogenous switching model:

$$\pi_i^a = X_i\beta^a + u_i^a \tag{10.1a}$$

if TV cultivation adopted,

$$\pi_i^{na} = X_i\beta^{na} + u_i^{na} \tag{10.1b}$$

if TV cultivation not adopted,

$$I_i^* = Z_i\gamma + \varepsilon_i \tag{10.1c}$$

$$I_i = 1(\text{TV cultivation adopted})\text{ if } I_i^* > 0$$

$$= 0(\text{TV cultivation not adopted})\text{ if } I_i^* < 0$$

where ${\pi^a}_i$ is the profit of parcel i adopting traditional varieties, while ${\pi^{na}}_i$ is the profit of parcel i not adopting traditional varieties. X_i is the respective matrices of independent variables. I_i is the indicator variable representing the adoption decision of the farm household on parcel i. Households adopt traditional varieties ($I = 1$) if and only if $I^* > 0$, otherwise the farmers plant modern varieties only ($I = 0$). The endogenous switching regression model is appropriate if the participation or adoption decision is an endogenous choice. Simple OLS estimation is likely to yield inconsistent estimates.

The approach used in this chapter draws from the literature on microeconometric evaluation of programs and policies (see the work of Heckman, 1974, 1976; Heckman & Robb, 1985). Existing studies have used alternative methods to estimate the value of green subsidies. For example, Kurkalova et al. (2003) estimated the incentive payments in the form of an irreversibility and risk premium needed to induce the adoption of conservation tillage. They estimate this premium as one that is over and above the compensation for expected profit losses. Other studies have resorted to direct questioning or CVM type of techniques to estimate adoption subsidies (see Lohr & Park, 1995). Unlike the Antle and Stoorvogel (2006) study that used a simulation model to study carbon soil sequestration contracts, we use a revealed preference approach in the estimation of green subsidies for rice intra-specific agrobiodiversity. We employ similar concepts as with Kurkalova et al. (2003), but limited only to compensation for expected profit loss.

The first step in calculating incentive payments for technology adoption is to identify factors that affect the level of rice farming profits, i.e., estimation of Eqs. (10.1a)–(10.1c) through a two-stage estimation. We initially estimate Eq. (10.1c) using the probit maximum likelihood method. We then use the estimated coefficient vector γ to calculate the inverse Mills ratios:

$$E(u_i^a|\varepsilon_i \le Z_i\gamma) = -\sigma_u^a \frac{\phi(Z_i\gamma)}{\Phi(Z_i\gamma)}$$

and

$$E(u_i^{na}|\varepsilon_i Z_i\gamma) = \sigma_u^{na}\frac{\phi(Z_i\gamma)}{1-\Phi(Z_i\gamma)},$$

which are added to estimate Eqs. (10.1a) and (10.1b), respectively, to estimate β^a and β^{na} by ordinary least squares:

$$\pi_i^a = X_i\beta^a - \sigma_u^a\frac{\phi(Z_i\gamma)}{\Phi(Z_i\gamma)} + u_i^a \quad \text{for } I_i = 1 \qquad (10.1a')$$

$$\pi_i^{na} = X_i\beta^{na} + \sigma_u^{na}\frac{\phi(Z_i\gamma)}{1-\Phi(Z_i\gamma)} + u_i^{na} \quad \text{for } I_i = 0 \qquad (10.1b')$$

The vector of the determinants of profit X_i includes: the age of the household head, its square, years of schooling of the head, household size, demographic composition of the household members, the distance from the nearest market, and the size of landholding and regional dummy variables. In addition to the variables included in the vector X_i, the determinants of technology adoption (Z_i) include, as identifying instruments, dummy variables for access to drying facilities, access to storage facilities, and access to extension services. The underlying assumption is that access to those postharvest facilities and access to extension services affect the decision to plant traditional varieties but do not directly affect farm profit.

The net benefits from planting traditional varieties then are obtained by calculating the counterfactual profit. The counterfactual profit is the expected income if, for instance, a non-adopting or pure modern variety farmer is forced to plant traditional varieties on their farm. In equation form the subsidy or the net benefit required to compensate a farmer for technology shifts can be obtained by:

$$\Delta = E\left[\pi_{na}|I_i^* < 0\right]\pi_{na} - E\left[\pi_a|I_i^* < 0\right] \qquad (10.2)$$

Since there is the possibility of having negative profits, i.e., the actual profit being less than the counterfactual profit, then the required subsidy or conservation payments to promote agrobiodiversity in the farm is simply:

$$\text{subsidy} = \min(0, \Delta).$$

The next step in our analysis is to assess the likely impact of conservation payments on the levels of poverty. The headcount poverty ratio is used to assess the changes in the poverty levels with and without the conservation payment scheme. The official provincial poverty lines constructed by the National Statistical Coordination Board are used as the basis for computing the headcount poverty ratio.

10.4 The Data Set

The data set for our analysis is taken from the DAR-UPLB[2] Comprehensive Agrarian Reform Program Impact Assessment Project. This data set came from a nationwide survey of 1,855 household beneficiaries of the Comprehensive Agrarian Reform Program. It contains detailed demographic, socioeconomic, and farm production data. A total subsample of 1,041 *rice*-farming households was used.

Tables 10.1 and 10.2 are cross-tabulations that describe the data set in terms of the number of households and parcels under traditional and modern variety cultivation. Around 42% of all households planted only modern varieties while 25% were pure traditional variety cultivators. The same percentages are observed for the parcels. This means that modern varieties are more widely cultivated by households and that more plots are planted solely for modern varieties. On the other hand, households who cultivate both traditional and modern varieties account for only 23% of the sample. In terms of parcels, only 20% of all parcels are planted with both modern and traditional varieties. This means that there is a relatively lower level of agrobiodiversity within parcels and geographically.

Table 10.3 summarizes household characteristics by household types of rice variety adoption. We can see that pure traditional variety cultivators tend to have lower incomes, lower level of education, fewer productive assets, less access to

Table 10.1 Number of households, by type of rice variety cultivation

	No. of households not planting modern varieties	No. of households planting modern varieties	Total
No. of households not planting traditional varieties	108	436	544
No. of households planting traditional varieties	262	235	497
Total	370	671	1,041

Source: Calculation by Nobuhiko Fuwa and Asa J. Sajise using DAR-UPLB Comprehensive Agrarian Reform Program Impact Assessment data

Table 10.2 Number of parcels, by type of rice variety cultivation

	No. of parcels not planted with modern varieties	No. of parcels planted with modern varieties	Total
No. of parcels not planted with traditional varieties	258	1,075	1,333
No. of parcels planted with traditional varieties	569	485	1,054
Total	827	1,560	2,387

Source: Calculation by Nobuhiko Fuwa and Asa J. Sajise using DAR-UPLB Comprehensive Agrarian Reform Program Impact Assessment data

[2] Department of Agrarian Reform -University of the Philippines at Los Baños.

Table 10.3 Mean values of relevant household characteristics, by type of rice variety cultivation

Variable	Pure traditional rice farming household (*N* = 262)	Pure modern rice farming household (*N* = 436)	Both modern and traditional varieties farming household (*N* = 235)
Total income (pesos)	77,182	131,632	101,970
Age of household head (years)	55.6	55.9	56.7
Education level of household head (years)	2.1	2.7	2.4
Household size	5.3	5.4	5.2
Productive assets (pesos)	15,245	23,047	26,640
Distance to market (km)	0.44	0.34	0.42
Access to drying facilities (dummy)	0.21	0.69	0.72
Access to storage facilities (dummy)	0.05	0.14	0.08
Extension services (dummy)	0.67	0.82	0.75
Male household members (0–15 years old)	0.85	0.84	0.81
Female household members (0–15 years old)	0.79	0.70	0.66
Male household members (15–60 years old)	1.46	1.57	1.50
Female household members (15–60 years old)	1.39	1.53	1.43
Male household members above 60 years old	0.35	0.32	0.34
Total farm area (hectares)	6.33	3.38	2.12

Source: Calculation by Nobuhiko Fuwa and Asa J. Sajise using DAR-UPLB Comprehensive Agrarian Reform Program Impact Assessment data

postharvest facilities, and are farther away from markets but have larger farms compared to both pure modern variety cultivator. In terms of these same characteristics, agrodiverse rice farming households fall in between pure modern variety and traditional cultivators. The overall trend is that for most of the mentioned variables, agrodiverse farming households are better than pure traditional cultivators but are relatively worse off compared to pure modern variety cultivators. These observations suggest that there would be potential opportunity costs in any scheme that attempts to induce pure modern variety users to adopt traditional varieties in their farms.

10.5 Factors Affecting Rice Variety Choice Among Farmers

The results of the probit estimation of adopting traditional rice variety cultivation are shown in Table 10.4. Households with better-educated household heads tend to have lower probability of adopting traditional rice varieties in their parcels. House-

Table 10.4 Probit estimation of the choice of planting traditional rice varieties

Variable	Coefficient	*P*-value	Marginal effects
Age of household head (year)	−0.006	0.745	0.003
Age of household head squared (year)	0.000	0.573	−0.000
Education of household head (year)	−0.048*	0.004	−0.019
Household size	0.028	0.463	0.011
Assets (pesos)	1.21e − 06*	0.048	0.0483 (per 100,000)
Male household members (0–15 years old)	−0.033	0.473	−0.013
Female household members (0–15 years old)	0.030	0.524	0.012
Male household members (15–60 years old)	−0.043	0.318	−0.017
Female household members (15–60 years old)	−0.111*	0.010	−0.044
Male household members above 60 years old	0.068	0.419	0.027
Distance to market (km)	0.017**	0.317	0.007
Access to drying facilities (dummy)	−0.165*	0.008	−0.066
Access to storage facilities (dummy)	−0.284*	0.004	−0.113
Extension services (dummy)	−0.221*	0.001	−0.088
Land allocation (hectare)	−0.022	0.369	0.009
Region 0 dummy[a]	−1.446*	0.005	−0.575
Region 1 dummy	0.010	0.936	0.004
Region 2 dummy	−0.171*	0.045	−0.068
Region 4 dummy	0.054	0.628	0.022
Region 5 dummy	−0.174	0.132	−0.069
Region 6 dummy	−0.457*	0.000	−0.182
Region 7 dummy	0.456*	0.013	0.181
Region 8 dummy	0.330*	0.023	0.132
Region 9 dummy	−0.618*	0.001	−0.246
Region 10 dummy	−0.158	0.693	−0.063
Region 11 dummy	0.431*	0.006	0.172
Region 12 dummy	−0.021	0.932	−0.008
Region 13 dummy	−0.627*	0.002	−0.249
Constant	0.421	0.455	–
Log likelihood	−1487.44		

Source: Calculation by Nobuhiko Fuwa and Asa J. Sajise using DAR-UPLB Comprehensive Agrarian Reform Program Impact Assessment data

*Significant at 5% level; **significant at 10% level

[a]The Central Luzon (Region 3), which is often called the "Rice Bowl of the Philippines," is set as the reference region

holds with larger amounts of productive assets are also more likely to adopt traditional rice variety, which is rather surprising. Demographic composition of the household also has some effects on the decision to adopt traditional rice varieties. In particular, households with more female members between the working ages of 15–60 are less likely to adopt traditional variety. Exposure to extension services also reduces the probability of traditional variety adoption. This is not surprising since most extension agents have encouraged adoption of modern rice varieties. Furthermore, private seed suppliers and input dealers often provide extension services that also promote modern varieties through various contractual arrangements. Access to storage facilities also reduces the probability of adoption of traditional varieties. This probably just captures the fact that postharvest facilities in the Philippines are not very efficient. Lastly, regional locations also have significant effects. Households in Regions 7, 8, and 11 are more predisposed to planting traditional varieties, compared to those in the Central Luzon region (Region 3), which has traditionally been regarded as "the rice bowl of the Philippines." Households in Regions 0, 2, 6, 9, and 13, on the other hand, have lower adoption compared to Central Luzon (Region 3).

Also shown in Table 10.4 are the computed marginal effects of each of the variables. The dummy variables for the regions seem to have relatively large effects on the probability of adoption. These effects reflect the combined effects of geographical/location specific variations in natural environment (climate, topographic, soil, etc.) and in socioeconomic conditions (e.g., infrastructure access, opportunities in non-agricultural economic activities, distance from large cities, etc.). The largest marginal effect among regional dummy variables is found for Region 0. This implies that the farmers living in Region 0 (Cordillera region) have the probability of adoption of 58% points higher, on average, than that of the farmers living in Central Luzon, after controlling for the household-level characteristics. Similarly, the farmers living in Region 7 have the adoption probability of 18% points higher than do the farmers in Central Luzon. Among the household characteristics, having an additional 100,000 peso worth of productive assets is associated with 5% point increase in the probability of adopting traditional varieties, while additional years of schooling lower the adoption probability by 2% points. Exposure to extension services is associated with a 9%-point increase in the probability of adoption.

10.6 Rice Farming Profits and Traditional Varieties

Tables 10.5 and 10.6 show the estimation results of the determinants of farm profit using endogenous switching regression model, i.e., Eqs. (10.1a') and (10.1b'), respectively. The signs of the coefficients are mostly the same between the two "regimes." One contrasting point estimate is the education of household head, where the estimated coefficient is negative for TV parcels while it is positive for non-TV parcels although neither is statistically significant. Also the negative coefficient on the size of land, under both "regimes," suggests diminishing returns to scale, in line with the often-found empirical regularity in developing agriculture of the "inverse relationship between land size and productivity." The point estimate of the magnitude of the inverse relations, however, is about twice as large on TV parcels as it is on non-TV parcels. Since the Central Luzon region, the base region, is among the wealthiest regions in the country, with favorable agricultural conditions, most of the regional dummies are negative and significant. As we can also

Table 10.5 Determinants of rice farm profit: Traditional variety adopters

Variable	Coefficient	*P*-value
Age of household head (year)	8.013	0.98
Age of household head (squared, year)	0.677	0.78
Education of household head (year)	−195.678	0.50
Household size	−316.8929	0.53
Assets (pesos)	0.0157*	0.045
Male household members (0–15 years old)	418.285	0.50
Female household members (0–15 years old)	1522.289*	0.02
Male household members (15–60 years old)	296.150	0.64
Female household members (15–60 years old)	516.191	0.43
Male household members above 60 years old	507.846	0.67
Distance to market (km)	−360.684**	0.08
Land allocation (hectare)	−3602.293*	0.00
Region 0 dummy[a]	−7466.403	0.16
Region 1 dummy	855.239	0.58
Region 2 dummy	−5793.535*	0.00
Region 4 dummy	−4471.86*	0.00
Region 5 dummy	−4169.953*	0.01
Region 6 dummy	−8389.748*	0.00
Region 7 dummy	−5316.260*	0.02
Region 8 dummy	−5364.069*	0.00
Region 9 dummy	−12867.700*	0.00
Region 10 dummy	−12892.040*	0.04
Region 11 dummy	−503.421	0.81
Region 12 dummy	−430.866	0.90
Region 13 dummy	−10419.610*	0.01
Mills ratio	9109.964*	0.02
Constant	8159.567	0.33

Source: Calculation by Nobuhiko Fuwa and Asa J. Sajise using DAR-UPLB Comprehensive Agrarian Reform Program Impact Assessment data

*Significant at 5% level; **significant at 10% level

[a]The Central Luzon (Region 3), which is often called the "Rice Bowl of the Philippines," is set as the reference region

see, coefficients on the Mill's ratio are statistically significant in both "regimes," implying that the error terms of the profit determination functions, i.e., Eqs. (10.1a') and (10.1b'), are both correlated with the error term of the determinants of the traditional variety adoption, i.e., Eq. (10.1c).

10.7 Conservation Payments and Their Impacts on Poverty Levels

The counterfactual rice profit based on Eq. (10.1a') above can provide the necessary conservation payment that would compensate households for shifting to more agrodiverse rice farms. Under the hypothetical (first-best) subsidy for the traditional variety introduction scheme, each household currently *not* planting traditional

Table 10.6 Determinants of rice farm profit: Traditional variety non-adopters

Variable	Coefficient	*P*-value
Age of household head (year)	518.489*	0.05
Age of household head (squared, year)	−4.513**	0.05
Education of household head (year)	256.921	0.31
Household size	−9.592	0.99
Assets (pesos)	0.051*	0.00
Male household members (0–15 years old)	94.119	0.88
Female household members (0–15 years old)	1,198.578**	0.07
Male household members (15–60 years old)	190.131	0.75
Female household members (15–60 years old)	−891.254	0.15
Male household members above 60 years old	1420.093	0.21
Distance to market (km)	−434.288**	0.09
Land allocation (hectares)	−1,613.869*	0.00
Region 0 dummy[a]	−4,382.6	0.32
Region 1 dummy	−3,093.919**	0.07
Region 2 dummy	−3,348.368*	0.01
Region 4 dummy	−3,047.287*	0.05
Region 5 dummy	−5,013.542*	0.00
Region 6 dummy	−8,047.464*	0.00
Region 7 dummy	−4,668.445	0.16
Region 8 dummy	−5,171.136*	0.04
Region 9 dummy	−4,768.957*	0.05
Region 10 dummy	−1,898.011	0.35
Region 11 dummy	−4,677.608	0.48
Region 12 dummy	−6,924.852	0.19
Region 13 dummy	−6,924.852*	0.01
Mills ratio	6,856.129**	0.08
Constant	6,098.135	0.47

Source: Calculation by Nobuhiko Fuwa and Asa J. Sajise using DAR-UPLB Comprehensive Agrarian Reform Program Impact Assessment data

*Significant at 5% level; **significant at 10% level

[a]The Central Luzon (Region 3), which is often called the "Rice Bowl of the Philippines," is set as the reference region

varieties is assumed to be paid a subsidy to compensate for the losses due to the adoption of traditional varieties. The estimated subsidy needed for each household is calculated based on the counterfactual profit obtained as the fitted value using the regression equation in Table 10.5 applied to the plots currently not planted with traditional varieties, i.e., those observations with $I = 0$, which are the observations used to estimate Eq. (10.1b') as reported in Table 10.6. The mean subsidy payment based on the scheme is estimated to be 15,601 pesos per parcel. This direct payment scheme would cost the total of around 18,767,923 pesos to implement in total.

Under the hypothetical policy scheme of providing subsidies to convert farms exclusively planted with modern rice varieties to plant (at least partially) traditional varieties, a total of 544 or 52% of the sample (of 1,041) households in our data set would be eligible to receive such subsidies. Most of these households, on average,

Table 10.7 Mean values of characteristics of eligible and non-eligible farmers

Variable	Eligible household (*N* = 544)	Non-eligible household (*N* = 497)
Total income (pesos)	123,231	88,902
Total rice profit (pesos)	29,918	22,682
Age of household head (years)	56.2	56.1
Education level of household head (years)	2.5	2.3
Household size	5.3	5.3
Productive assets (pesos)	21,395	20,633
Distance to market (km)	0.48	0.43
Access to drying facilities (dummy)	0.40	0.30
Access to storage facilities (dummy)	0.12	0.06
Extension services (dummy)	0.80	0.71
Total area (hectares)	2.97	2.65

Source: Calculation by Nobuhiko Fuwa and Asa J. Sajise using DAR-UPLB Comprehensive Agrarian Reform Program Impact Assessment data

have significantly higher pre-subsidy incomes, and slightly larger farms than their non-eligible counterparts as shown in Table 10.7. Other household characteristics, such as schooling, age, the value of productive assets, and household size, are roughly the same between the two groups. These comparisons again emanates from the fact that most of pure modern variety cultivators are found in the low lands. Here government support for agriculture tends to be more intense than that in less favorable upper lands, with its emphasis on modern agriculture. In addition, lowland rice farmers likewise have more access to extension agents and thus are more knowledgeable in productivity-increasing technologies.

Under this subsidy scheme, the total of 18,767,923 pesos is distributed among 544 eligible households (first column in Table 10.8). Since some of the beneficiary households live below the poverty line, this hypothetical subsidy scheme contributes to a modest decline in the headcount poverty ratio from 39.0% to 32.2%, a 17% decline in the headcount poverty ratio (the first row of the second and third columns in Table 10.9). As we have seen, however, those households that are not currently planting traditional varieties tend to be slightly better educated and to have higher profit and income, thus those households who are likely to be the subsidy recipients tend to be relatively better-off households. This suggests a likely trade-off between the policy goals of pursuing biodiversity and that of poverty reduction, in this particular context. As a benchmark to see such a trade-off, we could consider an alternative hypothetical subsidy scheme where the same total amount of 18,767,923 pesos would be distributed equally among all households (18,029 pesos each), a totally untargeted lump-sum subsidy scheme (second column in Table 10.8).

Such a subsidy scheme would reduce the headcount poverty ratio to 24.0%, leading to a roughly 40% decline, compared to the 17% decline under the conservation subsidy scheme, in the headcount ratio (the first row of the fourth and fifth columns in Table 10.9). Under this scheme, however, traditional varieties would be

Table 10.8 Alternative policy scenarios for conservation/poverty subsidy payment

	(1) Household-specific payment	(2) Untargeted lump-sum subsidy	(3) Uniform poverty subsidy	(4) Uniform conservation payment
Total subsidy cost (pesos)	18,767,923			
Eligibility criterion	Non-TV cultivators expected to incur losses from TV adoption	None	Below poverty line	Currently not planting traditional varieties
Number of beneficiaries	544	1,041	406	544
Subsidy amount	Parcel specific	Uniform among households	Uniform among poor households	Uniform among MV households
Amount per beneficiary	34,499 (average)	18,029	46,226	34,499
Leakage (land areas not planted TV) (hectares)	0	82.2	567.3	126.8
(Percentage of eligible land)	0	(23.6)	(35.1)	(7.8)

Source: Calculation by Nobuhiko Fuwa and Asa J. Sajise using DAR-UPLB Comprehensive Agrarian Reform Program Impact Assessment data

introduced only a fraction of the lands; there would be an estimated "leakage" of 382 hectares or 24% of the land that would not be converted (at least partially) to traditional rice varieties, while 100% of the eligible parcels, by design, would be planted (at least partially) with traditional varieties under the first-best subsidy scheme. Thus, even the totally untargeted subsidy payment is much more "pro-poor" than the hypothetical conservation payment scheme considered here.

In order to assess the potential opportunity costs of the conservation payment scheme in terms of poverty reduction, we can alternatively consider a poverty-focused uniform payment scheme, holding the total subsidy budget constant at 18,767,923 pesos, where all the households living below the poverty line would receive a uniform amount of 46,226 pesos. This would obviously be much more preferred from poverty reduction standpoint compared to the totally untargeted subsidy. Under this payment scheme, the headcount poverty ratio would decline to 9.7%, a 75% decline compared to the pre-subsidy poverty incidence (the first row of the sixth and seventh columns in Table 10.9). Comparing the headcount poverty ratio under the first-best subsidy scheme, 32.2% (found in the second column of the first row of Table 10.9), and the poverty ratio under the "uniform poverty subsidy," 9.7% (found in the sixth column of the first row of Table 10.9), the difference between the two poverty ratios (i.e., 22.5% points) can roughly be seen as the

Table 10.9 Headcount poverty ratio under alternative policy scenarios

Region	Status quo	House hold-specific payment	Change (%)	Untargeted lump-sum subsidy	Change (%)	Uniform poverty subsidy	Change (%)	Uniform conservation payment	Change (%)
All regions	39.0	32.2	−17.4	24.0	−38.5	9.7	−75.1	28.1	−28.0
Region 0	50.0	37.5	−25.0	29.2	−41.6	12.5	−75.0	37.5	−25.00
Region 1	50.6	42.3	−16.4	32.9	−35.0	16.5	−67.39	38.8	−23.32
Region 2	30.7	26.4	−14.0	18.9	−38.4	7.1	−76.9	25.0	−18.6
Region 3	37.1	34.0	−8.4	27.8	−25.1	16.5	−55.5	32.0	−13.9
Region 4	39.8	34.4	−13.6	28.0	−29.7	16.1	−59.6	31.2	−21.6
Region 5	47.6	44.0	−7.6	28.6	−39.9	13.3	−72.1	35.2	−26.1
Region 6	37.1	29.7	−20.0	20.6	−44.6	5.7	−84.6	22.3	−39.9
Region 7	46.9	34.7	−26.0	26.5	−43.4	4.1	−91.3	28.6	−39.1
Region 8	48.3	40.5	−16.3	29.2	−39.5	7.9	−83.7	37.1	−23.3
Region 9	40.5	32.6	−19.5	18.9	−53.3	8.1	−80.0	16.2	−60.0
Region 10	42.9	42.9	−0.1	28.6	42.9	0	−100.0	28.6	−33.4
Region 11	17.6	8.8	−49.9	8.8	−49.9	2.9	−83.3	8.8	−49.9
Region 12	25.0	25.0	0.00	16.7	−33.3	8.3	−66.7	16.7	−33.3
Region 13	22.7	22.7	0.00	9.1	−60.0	0	−100.0	9.1	−60.0

Source: Calculation by Nobuhiko Fuwa and Asa J. Sajise using DAR-UPLB Comprehensive Agrarian Reform Program Impact Assessment data

opportunity costs *in terms of poverty reduction (forgone)* for policymakers associated with the conservation subsidy payment (a PES) scheme under consideration.

At the same time, however, the likely "leakage" in land conversion to traditional rice varieties would increase to 35% of the eligible parcels from 24% under the totally untargeted subsidy scheme. Our example thus illustrates a case of direct trade-offs between the policy goals of biodiversity conservation and poverty reduction. This is essentially because (1) the kind of biodiversity we are considering here involves the adoption of a technology that would typically lead to loss in farm profit, (2) those households who are already practicing this ("environmentally friendly") technology tend to be less wealthy farmers while better-off farmers tend not be using the technology, and, therefore, (3) the environmental service payment would need to be targeted to those non-adopter farmers, who happen to be better-off farmers. As a result, given the same amount of budget, a subsidy scheme that is more efficient in inducing the adoption of traditional rice varieties is less pro-poor, while more pro-poor subsidy schemes tend to be less efficient as conservation payment schemes. In this particular application, therefore, policymakers would need to strike a balance between the two competing policy objectives.

Apart from the possible trade-offs between the environment and poverty reduction goals, another potential trade-off that policymakers are likely to face is the possible trade-off between the efficiency of payment scheme and the increase in the cost of information required for implementing subsidy schemes. The first-best subsidy scheme we considered above (i.e., first column of Table 10.8) assumes that the government is able to elicit the information on both the current and the counterfactual profit (where currently non-adopters of traditional varieties adopt such a technology) from each household. Since this is rather unrealistic, we could consider some other subsidy schemes that are less stringent in information requirement. One alternative is to distribute a uniform amount among all the farmers who are currently not adopting traditional varieties. Such a subsidy, holding the total subsidy amount constant at 18,767,923 pesos, would amount to distributing a subsidy of 34,499 pesos (in lieu of parcel-specific subsidy corresponding prospective profit loss) to each eligible household (where the farmers are not currently planting traditional varieties). This subsidy scheme, not surprisingly, is less efficient than the first-best conservation subsidy scheme (where the expected leakage is zero by design), leading to a leakage in land conversion of 8% (fourth column of Table 10.8). The poverty reduction impact under this scheme, however, is larger than that of the first-best conservation scheme considered above; this scheme would lead to a 28% reduction in poverty incidence, compared to the 17% reduction under the first-best scenario (the first row of the eighth and ninth columns in Table 10.9).

The leakage share of land conversion under this subsidy scheme (i.e., 8%), however, is still much lower compared to the 24% and 35% under the untargeted lump-sum subsidy and the poverty-targeted subsidy, respectively. At the same time, however, the poverty reduction impact under this subsidy scheme is smaller; the headcount poverty ratio after this subsidy scheme is implemented would be 28% compared to 8% under the poverty-focused subsidy scheme. This last scheme, therefore, might be seen as a middle-ground option among the alternative payment schemes we have considered here, with a moderate leakage in terms of biodiversity conservation, a relatively modest information requirement, and a better poverty reduction performance (compared to the first-best conservation payment scheme).[3]

10.8 Concluding Remarks

This case study has shown the poverty implications and the cost of promoting agrobiodiversity in rice farming. Poverty effects of a direct conservation scheme appear to be quite sensitive to how the specific subsidy scheme is designed. Under this particular application of preserving traditional rice varieties in the Philippines, there is a clear trade-off between the two policy goals of enhancing biodiversity and poverty reduction. Even the totally untargeted lump-sum subsidy would have a larger poverty reduction impact than would the first-best conservation subsidy payment scheme. There is also a clear trade-off between the efficiency of targeted conservation payment and the information requirement for implementing subsidy schemes. While compensating the exact amount of profit losses due to technology adoption is obviously more efficient in terms of eliminating possible "leakages," the information requirement for such scheme is perhaps unrealistically high. One interesting result of our analysis is that a less informationally stringent, thus less efficient from a conservation point of view, subsidy scheme is more pro-poor than the efficient subsidy scheme. Under this particular policy example, therefore, policymakers are likely to be required to strike a delicate balance between the two competing policy objectives.

Acknowledgments The authors would thank Ms. Aileen Lapitan for her insights on the data set; Ms. Marjorie Ann L. Dator, Ms. Dieldre S. Harder, and Ms. Jocelyn T. Tabali for their research assistance; and the DAR-UPLB Comprehensive Agrarian Reform Program Impact Assessment Project for the data set used in this study.

References

Alix-Garcia J, de Janvry A, Sadoulet E (2004) Payments for environmental services: To whom, for what, and how much? Department of Agricultural and Resource Economics Working Paper, University of California, Berkeley.

Antle JM, Stoorvogel JJ (2006) Agricultural carbon sequestration, poverty and sustainability. Available at http://www.tradeoffs.montana.edu.

Feng H, Kling CL, Gassman PW (2004) Carbon sequestration, co-benefits, and conservation. CARD Working Paper No. 04-WP379.

[3] In fact, there would be another issue of potentially perverse incentive effects; the farmers currently planting traditional varieties may shift to modern varieties in order to (appear to) be "eligible" for the subsidy scheme, which would lead to even larger leakages. While this is a real possibility, this issue is not pursued further here.

Feng H, Kurkalova LA, Kling CL, Gassman PW (2005) Economic and environmental co-benefits of carbon sequestration in agricultural soils: Retiring agricultural land in the upper Mississippi River Basin. CARD Working Paper No. 05-WP384.

Heckman J (1974) Shadow prices, market wages, and labor supply. Econometrica 42:679–694.

Heckman J (1976) Varieties of selection bias. Am Econ Rev 80:313–318.

Heckman J, Robb R (1985) Alternative methods for estimating the impact of interventions. J Econom 30:239–267.

International Rice Research Institute (IRRI) (2004) IRRI's environmental agenda: An approach toward sustainable development. International Rice Research Institute, Los Baños, Philippines.

Kurkalova L, Kling C, Zhao J (2003) Green subsidies in agriculture: Estimating the adoption costs of conservation tillage from observed behavior. CARD Working Paper No. 01-WP 286.

Lankoski J, Ollikainen M (2003) Agri-environmental externalities: A framework for designing targeted policies. Eur Rev Agric Econ 30:51–75.

Lohr L, Park TA (1995) Utility-consistent discrete-continuous choices in soil conservation. Land Econ 71:474–490.

Pagiola S, Arcenas A, Platais G (2005) Can payments for environmental services help reduce poverty? An exploration of the issues and evidence to date from Latin America. World Dev 33:237–253.

Tuan HD, Hue NN, Sthapit BR, Jarvis DI (eds.) (2003) On-farm management of agricultural biodiversity in Vietnam. Proceedings of a Symposium December 6-12, 2001, Hanoi, Vietnam. International Plant Genetic Resources Institute, Rome, Italy.

Wu J, Zilberman D, Babcock BA (2001) Environmental and distributional impacts of conservation targeting strategies. J Environ Econ Manag 41:333–338.

Chapter 11
Assessing the Feasibility of Wetlands Conservation: Using Payments for Ecosystem Services in Pallisa, Uganda

Imelda Nalukenge, John M. Antle, and Jetse Stoorvogel

Abstract This chapter reports on a study of the potential for payments for ecosystem services to encourage the adoption of more sustainable agricultural practices in the Pallisa district in southeastern Uganda. Due to low productivity and population pressure, the subsistence agriculture that dominates the upland areas is increasingly encroaching on wetland areas critical to a many ecosystem services. While encroachment is illegal, enforcement has not been effective, raising the possibility that a positive incentive mechanism might be a more effective approach to wetlands protection. This study began with a workshop designed to learn about the potential importance of wetlands and their services from local and national stakeholders, and to assess the legal and institutional setting in which environmental policy is being implemented. The next step was to implement a quantitative analysis of ecosystem service supply, to estimate the possible rates of participation by farmers in contracts for wetlands conservation and the impact on farmers' incomes. The analysis suggests that payments for ecosystem services could be a viable alternative to conventional environmental regulation if local institutions can manage contracts with farmers at a reasonable cost, and if national and international beneficiaries are willing to pay for wetlands protection.

11.1 Introduction

This chapter reports on a study of the potential for payments for ecosystem services (PES) to encourage the adoption of more sustainable agricultural land use and management practices in the Pallisa district in southeastern Uganda, one of the poorest parts of Uganda. Due to low productivity and population pressure, the subsistence agriculture that dominates the upland areas is increasingly encroaching on wetland areas that occupy about one third of the district. These wetlands are

L. Lipper et al. (eds.), *Payment for Environmental Services in Agricultural Landscapes*, Natural Resource Management and Policy.
doi: 10.1007/978-0-387-72971-8_11,

considered to be critical to a number of ecosystem services on which life depends in this region. While encroachment is illegal according to Uganda's national environmental laws, enforcement has not been effective. Our goal in this study is to assess the potential for an alternative approach to environmental protection based on creating positive incentives for farmers in the form of PES.

This study began with a workshop designed to learn about the potential importance of wetlands and their services from local and national stakeholders, as well as the legal and institutional setting in which environmental policy is being implemented. The next step was to conduct a quantitative analysis of the economic feasibility of PES. This economic analysis was used to estimate the possible rates of participation by farmers in contracts for wetlands conservation and the impact on farmers' incomes.

In the next section we provide background about Pallisa's agriculture, its impacts on ecosystem services, and Uganda's environmental policies. We then discuss PES as an alternative approach to agricultural and environmental policy. Next we describe the results of the stakeholder meeting and the quantitative analysis of the economic feasibility of PES. Finally, we discuss how these results could be utilized to support implementation of a PES policy for wetlands protection in Uganda, either through a government program or through a market for ecosystem services.

11.2 Agriculture, Ecosystem Services, and Policy

Like elsewhere in the country, Pallisa district's major economic activity is farming where small-scale, semi-subsistence agriculture dominates. Farming in the district is characterized by growing a mixture of crops and rearing livestock on small landholdings averaging 1.8 acres per household. The landscape is characterized by rain-fed upland farming interspersed with about one third of wetland areas (Fig. 11.1). The typical upland small-scale mixed farming systems comprise cotton, cassava, and finger millet. Other crops include sweet potatoes, bananas, sorghum, and legumes like groundnuts, cowpeas, and green gram. These crops are grown mainly as intercrops, with cassava the most widely grown crop for food security reasons. Rice is grown in the wetlands, sometimes intercropped with maize.

A challenge facing farmers in Pallisa district is the general decline of crop yields resulting from declining soil fertility levels. Due to low external input use, farmers depend heavily on land and water resources for their survival. The declining soil fertility together with drought has led to increasing encroachment of wetlands by the farming and pastoral communities. The increasing use of natural resources to support agricultural activities makes this part of the country environmentally vulnerable. According to the participants in the project workshop (discussed below), local people are concerned that if wetland encroachment progresses, farmers will use increasing amounts of nutrients and pesticides that will have impacts on water quality and wildlife as well as on human health. According to the National

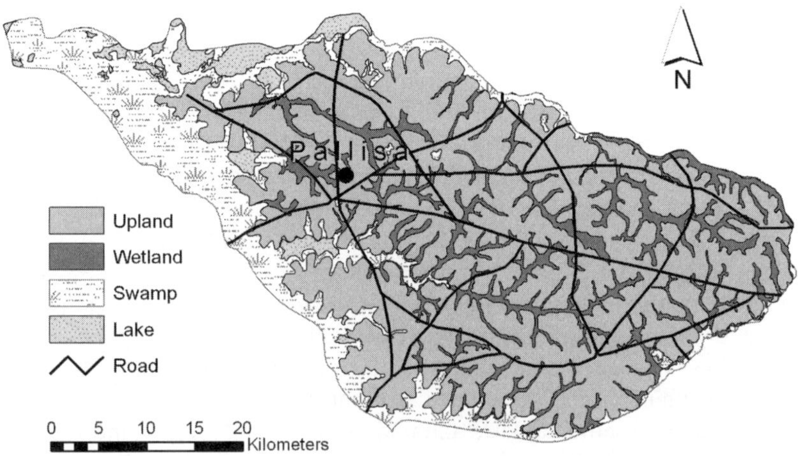

Fig. 11.1 Pallisa district map showing upland and wetland areas. Courtesy of Imelda Nalukenge, John Antle, and Jetse Stoorvogel

Environment Management Authority (National Environmental Management Authority, NEMA, 2003), the use of agro-chemicals in the district is increasing in response to declining levels of yields in cotton growing areas as well as introduction of horticultural farming.

The legal and institutional framework that defines and protects wetlands in Uganda appears to have important limitations (NEMA, 2001). Until recently, government policy encouraged the drainage of swamps for agricultural and other economic activities. Currently, an estimated 2,376.4 km^2 of wetland area have been converted representing a 99.3% increase in area since the 1960s. The rate of conversion is estimated at about 2.9% per annum in Uganda. According to NEMA (2003), the situation of wetland degradation has reached critical levels in the Eastern part of the country inclusive of the Pallisa district. In this region, 20% of the virgin wetlands have been converted to agricultural and other uses compared to 2.8%, 2.4%, and 3.6% in the Central, Northern, and Western regions, respectively. Pallisa district is reported as one of the most affected sites in terms of wetland degradation. Pallisa district occupies an area of 1,992 km^2 out of which 711 km^2 is wetland area that amounts to 35.7% of district area (NEMA, 2001). In 1992, 553 km^2 of the total cropland area was in wetlands, and this comprises about 44% of the total cropland area. The converted area is expected to be increasing due to agriculture as well as activities that are closely linked with developments in urban centers where there is an increasingly high demand for bricks, sand, gravel, and other building raw materials as a result of the current construction boom.

There are two major environmental concerns in Pallisa district – wetland conversion to open up land for rice growing and other economic activities, and deforestation caused by clearing of swamp forests for domestic fuelwood supplies and commercial timber provision. The wetland resources in Pallisa have associated primary roles and functions that include sediment, nutrient and toxin retention, stabilization of the hydrological cycle and micro-climate, biological diversity and species richness, and biomass production (e.g., papyrus and reeds).

Wetlands in Pallisa are used for hunting and fishing, shifting cultivation including rice growing, cattle grazing, brick making, and harvesting raw material (sand, gravel, clay, papyrus, poles) for building houses. The shallower water part of the wetlands has been put under intensive cultivation mainly for rice growing. Rice growing currently occupies 68% of the reclaimed wetland area in Pallisa district. According to NEMA (2001), the impact of such massive wetland utilization has been a reduction in the number of permanent streams, disappearance of permanent springs, and reduced groundwater yield in wells. A further impact of wetland conversion has been associated with the shift from perennial crops such as bananas and coffee to annual crops such as rice and maize.

The current environmental laws and policies in Uganda follow the 1995 constitution and are part of the National Environment Management Policy (NEMP). The NEMP sets out the overall policy goals, objectives, and principles for environmental management. Under the NEMP, the overall policy goal is to achieve sustainable, social, and economic development.

Enforcement issues associated with the management of ecological resources are guided by the 1995 constitution, which is the supreme law that provides regulations for environmental protection and conservation. The constitution provides for sustainable use of natural resources of Uganda and the protection of the right to environment. According to the NEMP, the specified aim of preserving natural resources such as wetlands is to: enhance the quality of life; ensure integration of environmental issues in developmental objectives; conserve, preserve, and restore agro-ecosystems; and to optimize resource use and sustainable consumption of resources. The current law provides for sustainable use of wetlands through a stipulation of the types of activities to be regulated. The laws allow communities to undertake activities such as brick making, fish farming, recreation, drainage, and cultivation of crops in the wetlands under regulation. Further, traditional activities such as medicinal and papyrus harvesting, hunting, and collection of water are allowed, as well as cultivation of up to 25% of the wetland area. In addition, all activities that were carried out before 1995 when the law was enacted are allowed. In undertaking the various activities in the wetlands, it is important to note that preservation activities are highly influenced by the property rights that communities have over the wetlands. Communities' beliefs are that wetlands belong to them as well as the government, and so both parties must be held accountable. The law, however, provides that wetland areas are held in trust by government or local governments for the common good of all the people of Uganda and should not be alienated through individual ownership. This conflict of landownership implies that although the national policy has established wetlands as public lands to be managed in the public interest, the local communities have and continue to use wetlands to meet their needs for subsistence production of livestock and crops, particularly rice. While the rural communities can meet their immediate needs through the use of wetlands, there is a general observation and consensus that the use of wetlands for agriculture has the potential to significantly degrade water quality and quantity and to cause other adverse environmental effects.

The conclusion is that the present system of environmental laws is not being successful in achieving protection of the wetlands. A key reason for this failure is that the current approach is based on regulations that tend to lead to conflicts between local people and the regulatory agency, NEMA, and its local representatives. These regulations are difficult to enforce, and the resulting open-access utilization of the wetlands is leading to a classic "tragedy of the commons" outcome. Currently, the Pallisa district government spends ~50 million Ush per month to undertake monitoring activities in the wetlands. This cost is high because farms are small and distributed throughout the district. One possible advantage of resource management based on PES is that it replaces an adversarial regulatory approach with positive incentives for farmers to preserve environmental resources. While a PES system would also require monitoring and verification of compliance with contracts to provide ecosystem services, the PES approach could provide a more positive relationship between farmers and the government.

There are three key challenges to implementing a PES system. First, the wetland services must be identified and quantified, and the economic feasibility of implementing PES must be assessed at the farm level. Second, institutional arrangements for the implementation of a PES approach need to be identified and assessed. This includes both supply-side and demand-side components. On the supply side, the costs of negotiating contracts with farmers and the costs of monitoring and verifying compliance with contracts for provision of ecosystem services. On the demand side, the beneficiaries of ecosystem services must be identified, and mechanisms for assessing their willingness to pay for the services and mechanisms to link demanders to suppliers must be identified and implemented.

11.3 PES as an Alternative Approach to Environmental Policy

Experience with environmental regulation has shown that command-and-control regulations often do not produce desired environmental outcomes. Most of the environmental programs that have been used in industrialized countries have been based on paying farmers to adopt certain "best management practices" deemed to reduce soil erosion, improve water quality, or have other environmental benefits. However, until very recently, these programs have not been designed to pay farmers in direct relation to the environmental benefits they produce. This is analogous to the experience with environmental regulation of industrial firms, wherein the regulatory agencies typically used "command-and-control" regulation to achieve improvements in air and water quality. Experience has shown that a system based on performance standards and incentives is a far more efficient way to achieve environmental quality improvements. Similarly, the typical approach to agricultural conservation policies based on adoption of government-prescribed "best management practices" is an inefficient way to achieve environmental objectives.

Fundamentally, "one-size-fits-all" solutions are not efficient when conditions vary greatly across the landscape as they do in agriculture. Moreover, prescriptive policies are costly because they fail to create incentives for participation by those farmers who can provide the ecosystem service at the lowest cost per unit of service. In developing countries, experience with regulatory interventions has been even less successful because of weak regulatory and enforcement institutions (Blackman & Harrington, 2000).

There is a large and growing body of science underpinning the concept of ecosystem services and demonstrating that agricultural activities have impacts on ecosystem function and the provision of those services. A recent study by Zhang et al. (2007) provides an overview of these impacts, which include soil nutrient cycling, water provision and climate regulation, and disservices such as pest damage and competition for water and other resources. The scientific literature establishes that farmers' land-use and management decisions may affect biological and physical systems through a number of mechanisms. These effects may be limited to the land owned by the farmer, such as a change in soil productivity, or may have off-site effects such as chemical runoff into surface waters. Without policies that affect farmers' incentives, there is ample evidence showing that most farmers make land-use and management decisions to maximize their perceived economic well-being. These decisions result in a supply of ecosystem services that is determined by farmers' economic incentives to supply market goods (crops and livestock) but does not take into account society's valuation of the ecosystem services. To increase the supply of ecosystem services beyond this private equilibrium, demanders must provide farmers with incentives to change their management decisions in ways that increase those services. In most cases, ecosystem services are public goods, so some form of government intervention or assignment of property rights is needed to create these incentives.

11.3.1 Economic Analysis of Participation in Ecosystem Service Contracts

In this section we use a simple static model to analyze farmers' participation in ecosystem service contracts. This model demonstrates that participation generally depends on several key factors: the spatial distribution of returns for competing land use and management activities; the spatial distribution of associated with the competing practices; the design of the incentive mechanism embodied in the contract; and other factors influencing decisions such as risk and transaction costs (Antle & Stoorvogel, 2006).

We consider a farmer's choice between two competing land uses or management practices, *a* and *b*, in a geographic region. The different land uses are expected to yield different combinations of marketable product and ecosystem services. The land-use decision in each time period is based on the farmer's goal to maximize expected economic returns to the land. We initially assume there are no costs to switching from practice a to practice b, and we discuss relaxing that assumption later. Under these assumptions, activity a is chosen if it yields higher expected returns than activity b, otherwise b is chosen. Let the difference in returns per hectare between the two practices (returns to a minus returns to b) be denoted as $\omega(p, s)$, where p represents output and input prices and s denotes the site. Thus, the farmer adopts practice a if $\omega(p, s)$ is positive, and adopts b otherwise. We can interpret $\omega(p, s)$ as the opportunity cost per hectare, in terms of forgone returns, for adopting practice b.

We assume that an ecosystem service of $e(s)$ units per hectare per time period is produced at each site s when practice b is in use, and that zero services are produced if practice a is in use at the site. Here we treat e as a scalar for convenience, but in the case of wetlands protection, multiple ecosystem services are usually produced at each site. In principle, each of the services can be priced and quantified, although in practice it is often difficult to do so. For implementation of an ecosystem service payment system, an approach that has been used successfully in a number of countries is to approximate ecosystem services with an index based on site characteristics and land management practices (e.g., the Environmental Benefits Index used in the United States, see Cattaneo et al., 2005). It should be noted that some processes determining $e(s)$, such as soil carbon sequestration, are spatially independent, whereas in other cases such as habitat preservation or water quality protection, there may be spatial dependencies. These spatial dependencies will need to be taken into account in designing an efficient mechanism for provision of the service. Similarly, there may be spatial dependencies in opportunity costs if, for example, there are positive learning externalities associated with the adoption of alternative management practices.

In the private equilibrium that occurs without incentives to supply e, we assume that farmers allocate land to the use with the highest returns. To increase the supply of ecosystem services above the quantity provided in this private equilibrium, a payment is offered to the land managers for increasing the quantity of the ecosystem service.[1] The farmer can choose practice a and receive the expected returns to that activity, or can choose practice b and receive the expected returns to b plus the ecosystem service payment $g(s)$. When payments are per unit of ecosystem service, $g(s) = p_e e(s)$, where e is interpreted as the expected amount of services produced with practice b, whereas if payments are made for adoption of practices, $g(s) = g_b$. The farmer will choose activity b if the net benefit n of changing practices is positive,

$$n = g(s) - \omega(p, s) > 0 \tag{11.1}$$

and will choose practice a otherwise.

Farmers' site-specific, land-use decisions generate a regional supply of ES that is determined by the joint spatial distribution of ES and opportunity cost. Define the spatial distribution of net benefit as $\phi(n)$. For $g = 0$, the area under the positive tail of $\phi(n)$ represents those land units where farmers use practice b without environmental service payments. Define the quantity of ecosystem service supplied in this initial equilibrium as $S(p)$. As p_e increases, n becomes positive for additional land units, farmers adopt practice b on those land units, and the quantity of ES supplied becomes greater than $S(p)$. The total amount of ecosystem service supplied, $S(p, p_e)$, is then calculated by summing all the quantities of services produced on the additional land units where farmers are willing to enter into contracts. The properties of the supply curve $S(p, p_e)$ are also determined by the form of $\phi(n)$, which is derived from the distributions of ω and e. When $g > 0$, $\phi(n)$ is a convolution of the distributions of ω and e. The particular form of $\phi(n)$ will depend on the processes generating the site-specific quantities of e and ω. For example, if there are no spatial dependencies between these processes, the distribution of e can be defined independently of the incentives provided to farmers. If payments are made per unit of ES, the mean and higher moments of n can be derived from the moments of e and ω using standard formulas for linear combinations of random variables. When payments are based on practices, g is constant, $E(n) = g_b - E(\omega)$, and the higher moments of n and ω are equal.

In some cases, there may be costs of adjustment associated with changing practices. These costs of adjustment may involve capital investments or learning about alternative management practices. However, in the case of payments to protect wetlands by stopping to use them for agricultural production, there are no significant costs of adjustment. In addition to adjustment costs, there may be behavioral and institutional factors that influence farmers' willingness to change land use and management practices. There is a sizeable literature on the adoption of conservation practices in agriculture that is closely related to the problem of ecosystem service supply. For example, studies in the United States show that characteristics of farm decision makers affect their willingness to adopt conservation tillage, although how they impact decisions appears to depend on their geographic location and other factors (e.g., Fuglie & Kascak, 2001). In addition, the literature on technology adoption shows that risk and uncertainty effectively raise the perceived costs of changing practices (Sunding & Zilberman, 2001). Risk is most likely to impact decision making when there is a substantial difference in risk associated with the land use options, e.g., as would be the case when farmers are choosing between risky crop production and a riskless government payment for idling land. In the case of switching from wetlands rice production to upland crops, farmers may perceive uplands crops that depend on rainfall to be riskier than rice production, so they may require an additional financial incentive to compensate for this perceived risk. However, the PES may be perceived as essentially riskless, so it is unclear which option is less risky.

Another factor that is likely to affect farmers' willingness to participate in ecosystem service contracts is transaction costs. These costs include the time and other resources farmers spend learning about the ecosystem service contract, as well as costs of verifying compliance with the contract. In addition, when the processes governing the provision of ecosystem services are spatially dependent, as is likely to be the case for ecosystem services from wetlands, efficient provision may require cooperation among groups of farmers within an agro-ecological zone.

These coordination costs are likely to depend on factors such as the number of farms participating, the number of hectares under contract, and the number and frequency of verification measurements required for contracts. If these costs are allocated to participants according to the number of hectares under contract, then the net benefit of contract participation, Eq. (11.1), is modified by subtracting transaction costs. If these transaction costs do not vary spatially, they simply shift the mean of the spatial distribution of net benefits in the negative direction.

11.4 Pallisa Stakeholder Workshop

The first part of the Pallisa study involved a participatory workshop held in the district with local and national stakeholders (Antle et al., 2006). The objective of this workshop was to consult the various stakeholders involved in the management of agro-ecosystems about the feasibility of adoption of the PES as an alternative to conventional agricultural and environmental policy tools. The workshop utilized the Tradeoff Analysis methodology in which stakeholders are asked to identify key sustainability indicators of concern, as well as policy and technology scenarios that might be used to improve the sustainability of the production systems in the region (Stoorvogel et al., 2004). The sustainability indicators identified are those that influenced the desired and undesired changes in the social, economic, environmental, and health conditions of farming communities around the wetlands and other private and public stakeholders. In the meetings, the stakeholders identified the following variables as issues related to economic activities and wetland conservation in Pallisa district:

(a) Economic and social indicators:

- Agricultural productivity
- Agricultural production
- Income
- Food security
- Poverty
- Conflict over use
- Soil productivity and degradation

The stakeholders described negative changes in production, productivity, incomes, food security, and soil productivity currently observed due to wetland degradation. The current situation has also led to rising conflicts and poverty levels in the district.

(b) Environmental indicators:

- Water quality (eutrophication, contamination, sedimentation)
- Water quantity (less water in the wells and boreholes)
- Biodiversity (fish, birds, plants)
- Microclimate

(c) Health indicators

- Malaria
- Malnutrition
- Bilharzia (due to more snails)

Indicators in categories b and c above were described as worsening.

Workshop participants identified the following options to deal with perceived threats to ecosystem services in the district:

- Create awareness and provide training on improved soil and water conservation (for both uplands and lowlands)
- Restore wetlands and water catchments (e.g., agro-forestry)
- Improve access and affordability of fertilizers
- Establish cover crops and green manure
- Train and involve communities in the development of environmental action plans
- Enact environmental byelaws (and enforcement) by local governments
- Encourage fish farming
- Promote Integrated Pest Management (IPM)
- Dispose of waste properly
- Improve rice varieties for uplands
- Improve markets for produce

All participants agreed that the organizational structure for implementing wetlands protection should be a combination of community-based and non-governmental organizations and local governments. One group suggested including special interest groups, specifically focused on wetlands management. A general remark was that the implementation should be consistent with environmental laws and policies at the national level. There was also concern that political interference could hamper attempts to reduce encroachment in the wetlands.

Presentations and discussions by participants at the workshop made clear that while national policy has established wetland areas as public lands to be managed in the public interest, local people have long used the wetlands and are continuing to use them to meet their needs for subsistence and for production of livestock and crops, particularly rice. While this use of the wetlands is helping rural households to meet their immediate needs, there was a consensus among workshop participants that agricultural use of the wetlands was leading to potentially significant degradation of water quantity and quality, which in turn has other adverse environmental effects.

A major conclusion that the research team took from thc workshop is that the present system of national environmental laws is not being successful in achieving protection of the wetlands. A key reason for this failure is that the current approach is based on regulations that tend to lead to conflicts between local people and the regulatory agency, NEMA, and its local representatives. These regulations are difficult to enforce, and the resulting open-access utilization of the wetlands is leading to a classic "tragedy of the commons" outcome.

Based on the workshop, there appear to be two key challenges to implementing a PES system. First, the wetland services must be identified and quantified, and the economic feasibility and institutional feasibility of implementing PES must be assessed. The workshop provided a list of services that local stakeholders value, and this list needs to be supplemented with other scientific information, such as the extent of wetlands utilization for agriculture and other activities. Second, the question of who would pay for the ecosystem services must be addressed, and mechanisms for linking demander to suppliers must be identified. The beneficiaries would range from local people and communities, to national policy organizations (e.g., NEMA) that represent the national interest, to people downstream in other countries in the Nile watershed, and to global organizations and individuals interested in environmental conservation.

11.5 Economic Feasibility of PES for Wetlands Protection in Pallisa

Using the analytical framework discussed above, an empirical analysis was carried out utilizing the minimum-data (MD) methods described by Antle and Valdivia (2006). The MD method exploits the structure of the PES supply problem by recognizing that the analysis hinges on the spatial distribution of opportunity cost for farmers to change practices, Eq. (11.1). The MD method is based on deriving the mean and variance of opportunity cost from means, variances, and covariances of the net returns of the base and alternative practices.

For this preliminary study of Pallisa, data from an existing detailed farm survey were utilized in combination with some supplemental secondary data. The primary data source was a survey carried out by Wageningen University in collaboration with Makerere University, using the Nutrient Monitoring methodology (see www.nutmon.org). These highly detailed, field- and farm-level data were collected from 40 farms in the late 1990s. The data contained 472 useable observations for 36 farms, providing estimates of costs and returns for the upland production system, defined as subsistence crops plus maize. However, at the time these data were collected, relatively little rice was being grown. Therefore, secondary data on rice yields and cost of production were obtained from the Africa Rice Center and used to estimate costs and returns for rice in lowland and upland areas. The data used in the MD analysis are summarized in Table 11.1, stratified into a small farm size group (averaging 3.2 acres) and a large farm size group (averaging 8.4 acres). In the case of rice, measures of yield variability were not available, so the variability of upland rice was assumed to be similar to other upland crops, and yield variability of lowland rice was assumed to be 50% of upland rice.

Table 11.1 Summary statistics for minimum data analysis of PES in Pallisa[a]

	Mean revenue (Ush/acre)	Revenue coefficient of variation	Mean production cost (Ush/acre)
Subsistence crops	110,931/104,317	108/126	20,515/13,414
Maize	97,232/59,910	103/136	23,846/12,690
Lowland rice	640,000/640,000	50/50	384,000/384,000
Upland rice	320,000/320,000	100/100	167,000/167,000

Source: Nutrient monitoring data for subsistence crops and maize, Pallisa district data for rice yields and cost of production.
[a]First number is small farms, second number is large farms. Ush/acre = Uganda shillings per acre

Figure 11.1 presents results for participation in a contract that would pay farmers to stop using the wetlands areas for crop production for five scenarios:

- Base: Rice grown in lowlands plus subsistence crops and maize in the uplands.
- Base + Upland: Rice grown in lowlands and uplands, plus subsistence crops and maize in the uplands.
- Base + Improved Upland: Same as Base + Upland but upland rice yields increased 50%.
- Improved Lowland: Same as Base but lowland rice yields increased 50%.
- Base + TC: Same as Base with 18,000 Ush per acre per year transaction cost.

In the base case with zero PES, the model indicates that about 17% of farms would utilize a cropping system that did not include wetlands, implying about 83% would utilize wetlands on their farms. We do not have data on the average share of cropland in wetlands, so we have assumed 30% based on anecdotal information (the complete model is available from the authors). This assumption plus the 83% simulated rate of utilization of the base system implies that about 25% of the cropland area would be wetlands rice. We do not have data to indicate the current proportion of farms that do utilize the wetlands for crops, but the model baseline is consistent with the data showing a relatively large and increasing encroachment in the wetlands for crop production. With better data, the model could be re-configured to fit the actual situation as well as to simulate a scenario of increasing wetlands utilization.

Examination of the base scenario shows that farmers' participation is quite elastic at low payment levels. To interpret these results, note from Table 11.1 that mean returns to subsistence crops and maize are in the range of 50,000–80,000 Ush per acre, whereas mean returns to lowland rice are about 260,000 Ush, thus to induce a high rate of contract participation, a payment in excess of 100,000 Ush is required (note, the exchange rate is about 1800 Ush per US dollar). For example, to reduce wetlands use by 50%, a payment of about 125,000 Ush per acre would be required. Assuming that there are about 1,000 km^2 of wetlands and about 20% converted to rice production, this implies about 50,000 acres of wetlands in rice production. The cost of returning 50% of lowland rice to wetlands would be about 3.125 billion Ush, or about US $1.7 million per season (there are two growing seasons per year). Alternatively, we can ask how much wetlands protection could be purchased with the existing regulatory budget.

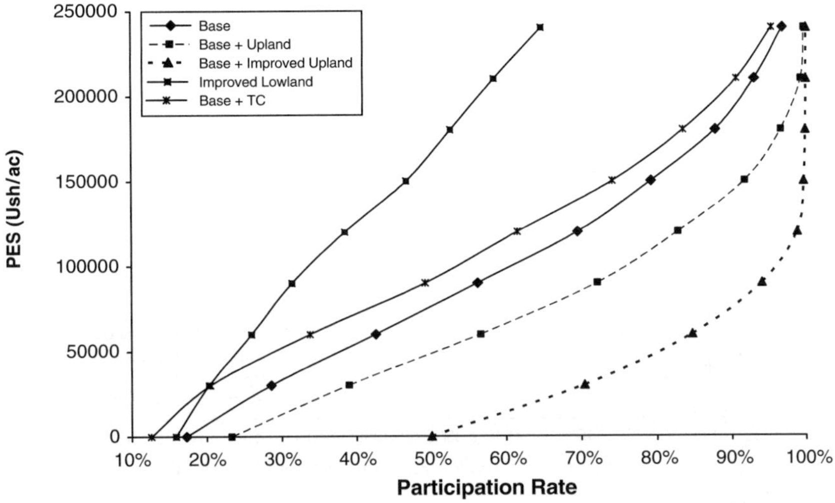

Fig. 11.2 Participation rates in contracts for wetlands protection in Pallisa district, Uganda. Courtesy of Imelda Nalukenge, John Antle, and Jetse Stoorvogel

The other scenario results show the importance of the returns to the alternative crops to the willingness of farmers to participate in wetlands conservation contracts at any given payment level. The addition of upland rice to the crop mix improves the returns to the uplands system and thus shifts the participation curve to the right, but if substantial improvements could be made in uplands rice productivity as part of the contract participation, say, through the availability of improved seed and fertilizer, this causes the participation curve to shift dramatically (note that the baseline point where the payment is zero increases from 17% to 50%). Conversely, an increase in the productivity of the lowlands rice would have the opposite effect, increasing the opportunity cost of abandoning the wetlands. Figure 11.2 also shows that transaction costs could also have an effect on participation. When transaction costs become significant, the participation rate is reduced. Note, however, that the magnitude of this effect declines as the payment level increases.

The potential impact of PES on improvement of incomes and reduced poverty levels is portrayed in Fig. 11.3. Under all the study scenarios, the results show that PES has great potential of increasing the farm incomes of the poor rural agricultural communities of the district.

11.6 Conclusions

The stakeholders' workshop provided an informative overview of the issues related to wetlands management in the Pallisa district, from the perspective of the local stakeholders, and in terms of national policy. Discussions and presentations at the

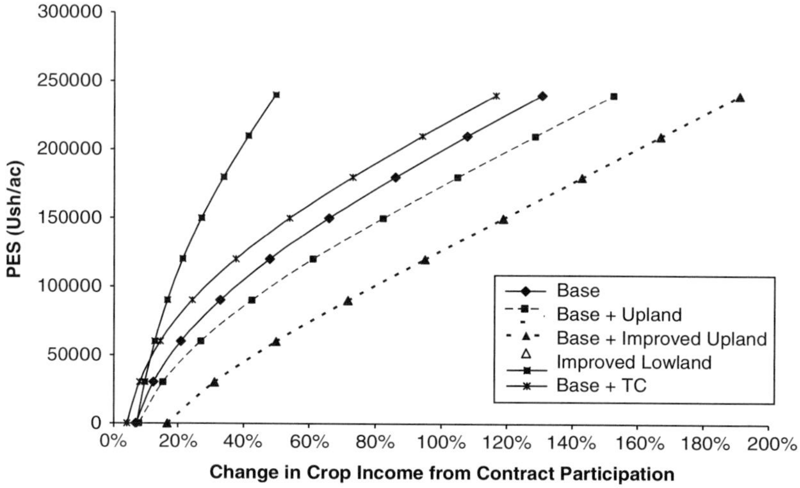

Fig. 11.3 Impact of contract participation on crop income in Pallisa district, Uganda. Courtesy of Imelda Nalukenge, John Antle, and Jetse Stoorvogel

workshop made clear that while national policy has established wetland areas as public lands to be managed in the public interest, local people have long used the wetlands and are continuing to use them to meet their needs for subsistence and for production of livestock and crops, particularly rice. While this use of the wetlands is helping rural households to meet their immediate needs, the consensus is that agricultural use of the wetlands was leading to potentially significant degradation of water quantity and quality, which in turn has other adverse environmental effects.

To undertake the assessment of the economic feasibility of PES, this study utilized several methodologies and, using data from the Pallisa district, MD simulation methods were employed to model the supply of ecosystem services that could be induced by providing farmers payments for reducing utilization of wetlands. The analysis confirms that farmers would be willing to participate in PES contracts to protect wetlands, but the cost could be substantial and would depend critically on the production alternatives available to farmers. Under the baseline technology, a US $50 per hectare payment would be required to induce 60% of farms not to utilize wetlands. However, if viable upland rice varieties were available to farmers, a US $50 per hectare payment could induce from 80% to 100% of farms to stop using wetlands for rice cultivation. The key challenges to implementing a PES system in Pallisa would be to establish institutions to manage contracts with farmers, and to determine how much beneficiaries would be willing to pay for wetlands protection. The beneficiaries would range from local people and communities, to national policy organizations (e.g., NEMA) that represent the national interest, to people downstream in other countries in the Nile watershed, and to global organizations and individuals interested in environmental conservation.

References

Antle JM, Stoorvogel JJ (2006) Predicting the supply of ecosystem services from agriculture. Am J Agric Econ 88(5):1174–1180.

Antle JM, Valdivia RO (2006) Modeling the supply of ecosystem services from agriculture: A minimum data approach. Aust J Agric Resour Econ 50:1–15.

Antle JM, Nalukenge I, Stoorvogel JJ, Claessens L (2006) Payments for ecosystem services as a new approach to tackle land degradation in Pallisa district: Workshop report. Available at http://www.tradeoffs.montana.edu.

Blackman A, Harrington W (2000) The use of economic incentives in developing countries: International experience with industrial air pollution. J Environ Dev 9:5–44.

Cattaneo A, Claassen R, Johansson R, Weinberg M (2005) Flexible conservation measures on working land: What challenges lie ahead? U.S. Department of Agriculture. Economic Research Service, Economic Research Report No. 5. Available at http://www.ers.usda.gov/publications/ERR5.

Fuglie KO, Kascak CA (2001) Adoption and diffusion of natural-resource-conserving agricultural technology. Rev Agric Econ 23(2):386–403.

National Environmental Management Authority (NEMA) (2001) State of the environment report. Government of Uganda.

National Environmental Management Authority (NEMA) (2003) State of the environment report. Government of Uganda.

Stoorvogel JJ, Antle JM, Crissman CC, Bowen W (2004) The tradeoff analysis model: Integrated bio-physical and economic modeling of agricultural production systems. Agric Syst 80(1): 43–66.

Sunding D, Zilberman D (2001) The agricultural innovation process: Research and technology adoption in a changing agricultural sector. In: Gardner BL, Rausser GC (eds.), Handbook of agricultural economics, Vol. 1A: Agricultural production. Elsevier, Amsterdam, the Netherlands.

Zhang W, Ricketts TH, Kremen C, Carney K, Swinton SM (2007) Ecosystem services and dis-services to agriculture. Ecol Econ 64(2):253–260.

Chapter 12
Managing Wildlife Damage to Agriculture in Bhutan: Conflicts, Costs and Compromise

Karma Ura, Randy Stringer, and Erwin Bulte

Abstract Conflicts between wildlife and agricultural producers are a dominant problem in Bhutan, with policy debates focusing increasingly on whether most of the conservation costs are borne directly by the small producers and rural poor through crop losses and labor time diverted to guarding crops and livestock. This chapter attempts to quantify the extent of wildlife damage to crops and to livestock in Bhutan. While several important studies document in detail wildlife damage to agriculture in and near protected areas in Bhutan, this chapter aims to provide a more comprehensive assessment of the extent of the problem around the country, presenting the results of a survey of 526 households and outlining the extent of wildlife damage to their crops during a 12-month period.

12.1 Introduction

Managing wildlife is an age-old problem facing agricultural producers around the world. Wild birds consume crops and infect domestic poultry. Predators attack livestock. Herbivores as large as elephants and omnivores as small as mice raid crops, eating planted seeds, budding flowers, ripening fruit, and stored grains. Wildlife populations can pose direct threats to humans too, as snakes, tigers, bears, and wild pigs can attack, harm, and kill. The desire to protect, provide, and prosper means that even those farm families with the strongest philosophical and religious commitment to environmental values resist increasing or even maintaining wildlife populations if the trade-off involves their family's welfare or safety.

Conflicts between wildlife and agricultural producers are a dominant problem in Bhutan; a country considered the conservation centerpiece in a region recognized as one of the planet's ten biodiversity hotspots (World Bank, 2005). Bhutan is recognized internationally for contributing to the global environmental agenda, preserving 72% of its land area as forest and mandating a commitment to maintain at least 60% forest cover in perpetuity. Policymakers are credited for not appropriating short-term economic gains at the expense of the country's pristine environment (International Monetary Fund, IMF, 2004).

L. Lipper et al. (eds.), *Payment for Environmental Services in Agricultural Landscapes*, Natural Resource Management and Policy.
doi: 10.1007/978-0-387-72971-8_12;

In 1993, the country established protected areas on more than a quarter of its total area. The 1995 Forest and Nature Conservation Act provides the legal framework for conservation. These and related conservation-oriented policy initiatives limit grazing, restrict access to forest products, prohibit hunting, and ban many shifting cultivation practices. The policies also raise tensions between those families living in and around the protected areas and park managers, as expanding wildlife populations lead to increased threats to humans, crops, and livestock (Wang & Macdonald, 2006; Wang et al., 2006a).

Many producers argue that too much farm output and too much of their incomes are sacrificed due to the country's commitment to conservation (Wang et al., 2006b). Wildlife cause damage by eating crops and killing livestock, resulting in: (1) lost income and destroyed and damaged assets; (2) large cost in time and money attempting to protect crops and livestock; (3) a disincentive to plant and to invest in rural production; and (4) greater levels of rural-urban migration. The Ministry of Agriculture estimates that on average 21% of rain-fed agricultural land and 8% of irrigated land are left fallow because of either water scarcity or the threat of wildlife damage (Tobgay, 2005). The growing concern is whether and how much conservation benefits are taking place at the expense of basic food security and poverty reduction.

Bhutan's policy debates increasingly focus on whether the cost of conservation is directly borne by the small producers and rural poor through crop loss, labor time diverted to guarding crops and livestock, and indirectly through limited access to forest resources. Integrated conservation development programs in the park areas, including ecotourism and community-based tourism intended to benefit local communities, are still in the theoretical realm (Wang et al., 2006b).

The objectives of this chapter are to quantify the extent of wildlife damage to crops in Bhutan. While several important studies document in detail wildlife damage to agriculture in and near protected areas in Bhutan, this study aims to provide a more comprehensive assessment of the extent of the problem around the country. The following section presents an overview of the conceptual issues with case studies highlighting the potential for payments for environmental services programs to address wildlife–agriculture conflicts. The third section presents the results of a survey of 526 households, outlining the extent of wildlife damage to their crops during a 12-month period.

12.2 Concepts and Issues in Wildlife Management Policies

12.2.1 Wildlife Management Policies and Programs to Protect Agriculture

In the past, biodiversity conservation tended to focus solely on strict protection of crucial biodiversity hotspots. Currently there are more than 100,000 protected areas in the world, covering around 10% of the earth's terrestrial surface (Jenkins et al.,

2004). The critical importance of these areas is well established. In recent years, it has become increasingly clear that, to be successful, conservation must extend beyond protected areas and portray an overall ecosystem approach. However, expanding outside of protected areas to include buffer zones, dispersal areas, and migration corridors has also proven difficult and is constrained frequently by limited public funds and the lack of support from local communities that in many situations receive little direct benefit from the public goods provided by protected areas.

A second wave of policy approaches involved indirect interventions and incentives to protect biodiversity. Indirect interventions aim to: (1) re-direct labor and capital away from uses that are detrimental to habitat and wildlife (e.g., agricultural intensification); and (2) encourage commercial activities that supply ecological services as a by-product (e.g., ecotourism).

While many of these schemes are at the early stages of development, the emerging evidence suggests that in some cases these programs have failed to protect high profile species that remain vulnerable to poaching pressures, or habitat degradation (Madhusan & Karanth, 2000). This perhaps is unsurprising. In the realm of environmental policymaking, because much of the environmental values are not well defined, ascertaining the true demand, or the true opportunity costs, or the true effectiveness of a policy, is difficult.

Ferraro (2001) suggests several disadvantages related to "indirect intervention approaches": (1) they generally generate ambiguous incentives for conservation (the impact of agricultural intensification on the incentive to expand the extensive margin, and the impact of wildlife damage compensation efforts on the incentive to convert more land to production and increase stocking rates) – also see Bulte and Rondeau (2007) and Rondeau and Bulte (2007); (2) indirect intervention programs are often too complex to implement and fail for that reason; and (3) indirect intervention programs do not conform to temporal and spatial dimensions of serious conservation objectives.

The heterogeneity of many environmental resources and households exacerbates these problems. For example, two adjacent farmers may have access to different farming techniques, risk profiles, wealth, or capital, thus rendering their opportunity costs associated with a given incentive scheme very different. It is this informational asymmetry between policymakers and actors that makes many existing policies ineffective. Any scheme that ignores the incentive problems arising from informational imperfections, or the possible heterogeneity of individual responses is likely to be less effective and, at times, even counterproductive.

A more recent market-based conservation program demonstrating promise is payments for ecosystem services (PES). PES attempt to provide incentives to make wildlife conservation or other types of environmental and ecological services provided by agricultural practices or on agricultural landscapes profitable. PES attempt to elicit private information about the opportunity costs of a change in behavior and induce a higher net social benefit from a change in behavior. New economic tools, including bioeconomics, enhance capacity to achieve these goals. Additional conceptual contributions address the asymmetric information and heterogeneity among households that result in wide variations on the impacts of PES on landowning households (Zilberman et al., 2008).

Much of the ecosystem services literature suggests that there is unlikely to be a "one-size-fits-all" remedy to wildlife conservation incentives. Where spatial (or biological) heterogeneity and individual heterogeneity is large, and/or there are severe informational imperfections, effective policymaking requires sharply tailored incentives. The appropriate policy instrument and mechanism therefore depends upon the context of the problem at hand.

12.2.2 Biodiversity PES Projects Focusing on Wildlife Conservation

PES have shown promise in situations where conservation of a certain species or habitat is being sought. According to Wunder (2005), "PES probably has a high potential for achieving real and additional conservation gains in situations where decisions are still on the 'edge,' especially when it is in a use-restricting scheme with service threatened by irreversible loss (e.g., biodiversity)." Additionally, a study focusing on PES schemes attempting to conserve wildlife concluded that PES can be useful when destruction of habitat is a main cause for loss of species and when access to land is critical for harvesting and can be controlled (Pagoila, 2003).

Many biodiversity cases focusing on wildlife protection tend to employ conservation easements or land leases where locals are paid to prevent encroachment or hunting in critical biodiversity areas. In other situations, entrance fees and lodge fees have been used to finance biodiversity conservation areas. Box 1 presents a few examples with a brief description.

12.2.3 Emerging Evidence and Applying Lessons

Perhaps the largest challenge in establishing successful PES projects is convincing those who have received benefits in the past for free to begin paying for ecosystem services. In most cases of successfully established PES schemes, a general deterioration of a previously free environmental service has provoked the establishment of a payment mechanism. Yet, in many situations environmental deterioration does not affect those benefiting as directly as it does when, for example, drinking water is contaminated by poor agriculture practices in an upstream watershed. Often governments or non-governmental organizations (NGOs) are needed to help in the establishment of a PES system. In many cases linking those providing and those benefiting from a given environmental service is costly.

Box 1 PES and Wildlife Conservation Cases

Rio Platano Biosphere Reserve – PES used to provide alternative income source to farmers who were advancing towards Rio Platano Biosphere Reserve. Farmers receive financial assistance from the administration of the protected area for undertaking investments to switch from extensive, wasteful land use to sustainable, more intensive land use. This includes parts of the costs of fencing, new grass seeds, and shade trees to enable them to increase efficiency in cattle production (Hartmann & Peterson, 2004).

Lodge Taxes in Langtang National Park, Nepal – Under the guidance of the Partnership for Quality Tourism Project, the lodge operators in Syabrubensi, one of the main trail heads to the Langtang Valley trek, organized into a Lodge Management Committee and agreed to contribute money to ensure conservation of a critical biodiversity area. A 2-rubies fee per trekker for each night in a lodge or private campground was established. The fees are self-imposed on an honor system, collected by the committee for community development projects, and matched by other Project funds. Projects have included improved water drainage, installation of some litter bins, and latrine construction and maintenance (Preston, 1997).

Ejido Cebadillos Thick-Billed Parrot Conservation – The Wildlands Project, a Tucson-based conservation group, and five Mexican Conservation Groups (Pronatura, Naturalia, Monterrey Tec, Sierra Madre Alliance, and Wildlife Preservation Trust International) formed an agreement with Ejido Cebadillas, a 40,000-acre land cooperative with 74 communal members to protect 6,000 acres of old-growth forest critical for thick-billed parrots. Payments and incentives will cost Wildlands Project about $250,000 over 15 years. Half of the money is paid upfront, and all payments are split evenly among members. Agreement pays Ejido members 50% ($250,000) of the net value of uncut timber within the protected area over 15 years. Wildlands Project will fund a forestry study to create a sustainable logging plan which may allow Ejido to charge a premium for lumber coming from other parts of the Ejido area. Also has included ecotourism initiatives for the construction of three cabins to lodge bird enthusiasts and a cabin for monitoring. Generally has been very successful despite the fact that ecotourism initiatives have not flourished due to difficult access.

Yohultan Land Easement Agreement – Friends of Calakmul, a US/Mexican conservation group, and Yohultan Ejido signed a land easement similar to the Xcupilcab agreement described above. Agreement signed in October of 2003 to protect 34,500 acres of land in the buffer zone of the Calakmul Biosphere Reserve by placing it in a land trust. The conservation group Friends of Calakmul collects donations from individuals and organizations in 15 countries and uses the money to pay the Ejido.

(*continued*)

Box 1 PES and Wildlife Conservation Cases (continued)

Namibia-Lodge Levy - Lianshulu Lodge – In conjunction with the Endangered Wildlife Trust local community has established a Community Development Fund to assist with the education and upliftment of tribal people living outside of the parks. Lodge managers Grant Burton and Marie Holstenson have also initiated the building of the Lizauli traditional village outside Mudumu, which allows visitors an insight into local culture and traditions. This village has been built by, and is managed by, the people themselves with monetary benefits going directly to the community.

In addition to linking those demanding and supplying ecosystem services, functional PES programs may require extensive training, negotiating, monitoring and contracting, organizing payments, and ensuring compliance and other related costs. These setup and operating costs can become very expensive over the long term, increasing substantially when large numbers of ecosystem suppliers are involved. Minimizing these and other transaction costs is essential if PES systems are going to be established. In some situations land acquisition may be a cheaper option although it does not ensure that local's interests will be in line with conservation goals. Land acquisition is normally considered very expensive because one is paying for all of the land-use value; however, if transaction costs (training, negotiation, maintenance) are expensive, that may make acquisition cheaper than providing payments over the long run.

Often transaction costs will have to be financed through grants, subsidies, or donations. These funding mechanisms can be used to initiate important conservation projects that can have positive impacts on attitudes and the socioeconomic well-being of local communities. In particular, they can help to overcome prohibitive transaction costs that prevent markets from establishing on their own. However, it is likely that in many situations without government or NGO intervention, ecosystem service markets will not be established until those benefiting begin to lose the services because of degradation.

12.2.4 Valuing the Benefits of Wildlife Habitat and Agriculture

Like other types of PES programs, valuation of the non-market benefit is a key program component. Much of the literature on wildlife valuation is associated with benefit-cost analysis of conservation policies for specific sites and more commonly focused on consumptive use values such as hunting or fishing or general ecotourism, for example. Surveys extract information from individuals on the willingness to pay (WTP) to provide for the management of species conservation, which as a result does not represent the value of biodiversity itself (Cardoso de Mendoça et al., 2003).

Many studies involve conservation policies as their main motivation for valuation, examining non-consumptive uses, or non-use values of specific animals. These studies are generally framed within a benefit-cost analysis in which the benefits of conservation (mainly based on the WTP of users and non-users) are weighed against the costs (or benefits foregone).

The existing evidence on positive external benefits of wildlife conservation is gathered mainly using contingent valuation (CV) techniques. CV remains controversial among professional economists mainly because it is based on hypothetical questions and not on revealed preferences, which potentially introduces biases (but under some conditions is now also used in U. S. court cases). Since by definition non-use values do not leave clear behavioral traces (and resorting to market data is of little assistance), CV continues to be used extensively, and in the absence of better information its estimates are used as a rough approximation of the true values at stake.

The WTP studies for various species tend to be large, suggesting that the external effects provided by conserving these species is also large. This is confirmed by two recent valuation studies. (1) Kontoleon and Swanson (2003) examine WTP for great panda conservation and conclude that non-use values are large when pandas can live under wild conditions (WTP is small when pandas are conserved in cages or pens).

Certain high-profile species can be used to raise funds to set aside large areas of wild habitat. (2) Swanson and Kontoleon (2003) examine WTP for black rhino conservation, and find that this WTP is affected by the nature of their *in situ* use. When wild rhino populations are also hunted (or sedated to be sustainably culled), WTP for conservation declined sharply. The utilization of wildlife from one constituent affects the production or utility functions of another leading in essence to various forms of production and consumption externalities between these parties.

These types of conflicts between values are at the heart of most disagreements over the direction of conservation witnessed in international wildlife institutions such as the Convention on International Trade in Endangered Species of Wild Fauna and Flora (CITES). Swanson and Kontoleon (2003) found that WTP for strict preservation (zero use but ecotourism) was quite large, and that there is a fundamental conflict between those who enjoy specific uses of a species and those who receive vicarious disutility from this activity by others. This implies that some countries may be able to maximize the total economic value of a particular species by the proscription of specific uses provided that mechanisms are instituted to tap the WTP for such proscriptions.

To what extent can these various estimates be added to arrive at an estimate of the aggregate WTP for wildlife conservation? Arguably, wildlife species are "substitutes" in consumption, and aggregating these estimates causes an overestimate of the true WTP (Knetsch, 1994). Nevertheless, one lesson is that "wildlife conservation" is an important source of positive external benefits, and that certain high-profile species (pandas, rhinos, elephants, whales) must be considered capable of raising large funds for conservation.

Since WTP is effected by (1) the wildness of the species' habitat (panda) and (2) the degree of their use (rhino), it seems that capturing and channeling WTP for certain species to actual conservation efforts may have far-reaching impacts for many other species (and habitat) as well. In many cases, the highly valued charismatic species have the greatest habitat needs and thus shelter a host of other related species in their ecosystems (i.e., umbrella species such as elephants, tigers, lions, and snow leopards). This is fortunate for biodiversity conservation, since it implies that protecting a few key species may have spillover benefits for other less-valued species.

In contrast to the WTP assessments, the Bhutan case presented here examines the costs of conservation imposed on small producers from the country's overall strategy to protect forests and wildlife. While this information provides only part of the picture, it allows for a more adequate assessment of the mix policy options and how to target those policies and programs to compensate effectively small producers.

12.3 The Bhutan Case Study

A predominantly agrarian society, most families in Bhutan live in rural areas, subsisting on a livelihood system integrating crop agriculture and livestock rearing with a wide variety of non-wood forest products. More than 95% of Bhutan's poor live in rural areas, where poverty is 9 times greater than urban areas (World Bank, 2005). In a nationally representative sample, the 2007 Bhutan Living Standard Survey classified 23% of the population as poor. Average household expenditure in urban areas is 1.9 times higher than rural areas. The national poverty line is Nu 1,097 (the Bhutan currency is called ngultrum) per person per month (~US $27.50). Estimated food requirements are Nu 689. About 31% of the rural population is below the poverty line compared to 1.7% in the urban areas (National Bureau of Statistics, NBS, 2007).

Poverty tends to be higher the more remote the rural area. On average, rural households own 3.5 acres of land spread over different agro-ecological zones and altitudes (Tobgay, 2005). The majority of the farmers own little land. Around 33% of farming households own less than 3 acres each. And more than half of the total farming households (55.7%) own less than 5 acres, accounting for one-third of the total agricultural land (Osmani et al., 2007).

Wildlife damage to crops and livestock has gained policy attention since the mid-1990s when the government established more stringent biodiversity conservation and forest protection acts. A 1996 study estimated that farmers lost up to 18% of total household income and, on average, farmers spend about 2 months per year guarding their maize and rice from wildlife such as elephants, porcupines, boars, monkeys, deer, among others (Choden & Namgay, 1996). The estimates suggest that wildlife damage to crops is a significant deterrence to cultivation. A survey on migration, conducted by the Ministry of Agriculture, 2003–2004, estimated that

16% of Bhutanese have migrated, and among the reasons cited include wildlife damage to crops, making farm life risky and unviable.[1]

Wang et al. (2006a) present compelling evidence that Bhutan faces wildlife-human conflicts that were absent two decades ago. Farmers are less tolerate of wildlife damage, demanding action by the government, and that wildlife conflicts impact food security and poverty reduction objectives (Wang & Macdonald, 2006; Wang et al., 2006a). In a series of studies in Bhutan's Jigme Singye Wangchuck National Park, Wang and colleagues examined the nature and scale of human-wildlife conflict, concluding that a significant number of farmers were unhappy with crop and livestock losses and blamed the park for restrictions on resource use.

In interviews with 274 households over a 3-year period from 1999 to 2001, Wang et al. (2006a) found "farmers strongly believed that their livelihoods were in jeopardy because of the Jigme Singye Wangchuck National Park and its policies, and the majority expressed discontent with the restrictions imposed by authorities on access to Park resources." Among the suggestions made by the respondents to address wildlife damage to their crops include improved fencing, technical assistance for husbandry practices, pasture development, controlled hunting of problem animals, more equal distribution of agricultural and grazing land, and providing direct compensation for damage.

Wang et al.'s (2006a) study of livestock damage found that around one fifth of households surveyed reported losses of a total of 2.3% of their domestic animals to wild predators over 12 months. This loss equated to an average annual financial loss equal to 17% (US $44.72) of their total per-capita cash income. Total reported losses during 2000 amounted to US $12,252, of which leopard and tiger kills accounted for 82% (US $10,047). Poor herding practices, inadequate guarding practices, and overgrazing are likely to have contributed to livestock losses. And some 60% of the households were unable to corral livestock due to inadequate stables. They found a significant correlation between the number of livestock lost and the distance between the household and the grazing pasture.

A 2005 study (Sangay, 2005) reviewing Bhutan's wildlife compensation scheme for livestock found compensation for damage in 2003 was paid in 154 cases reported by 115 farmers; 764 compensation cases were paid to 654 farmers in 2004; and 624 cases were paid to 594 farmers in 2005. The total compensation paid to the 1,361 farmers for 1,542 livestock amounted to Nu 5,454,950 for the 3-year period (US $126,859 with exchange rate of US $1.00 = Nu 43). Leopards accounted for 64% of the claims and tigers for 24%. Table 12.1 presents data from Sangay's report, showing the compensation rates for domestic animals when killed by wildlife. Among the study's recommendations include improved prevention through building corrals and guard dogs to switching cattle breeds and more productive pasture grasses to reduce reliance on forest grazing.

[1] But the sample was limited: 990 urban households and 2,300 rural households. In descending order the reasons were: Lack of education facilities (46%), lack of employment or job search (17%), inadequate service facilities (15%), small holdings (7%), drudgery of farm work (5%), labor contributions (3%), and crop damage (3%).

Table 12.1 Market value of different domestic livestock breeds in US dollars

	Young (<12 months)	Young (1–3 years)		Adult (>3 years)		Prevailing compensation rate in US $
		Male	Female	Male	Female	
Pure mithun	116	279	279	465	349	174
Mithun cross (50%)	93	140	163	233	279	105
Exotic cross-breed (Jersey/ Brown Swiss)	116	140	163	279	349	105
Yak	70	116	116	163	186	174
Local cattle (Thrabam, Drangla, Yangku, Yangkum, Deeob, Deeodam, buffaloes)	70	116	116	186	233	70
Horse	70	116	116	279	186	81
Mule	93	233	233	581	581	163
Donkey				465		163
Sheep	12	26	26	46	46	16

Source: Table 7 in Sangay (2005)

12.4. Survey of Wildlife Damage to Crops and Livestock

12.4.1 Wildlife Damage Survey

This study adds to the existing evidence on wildlife damage to crops and livestock in Bhutan with a survey on a representative sample of 526 rural households from October 18, 2006, to January 26, 2007. Other studies focused on specific areas, including parks and parts of the country more prone to predator strikes on livestock. The survey presented here covers a random sample from 13 *dzongkhags* across the country (Bumthang, Lhuentse, Mongar, Paro, Pemagatshel, Punakha, Samdrup-jongkhar, Samtse, Sarpang, Trashiyangtse, Trashigang, Trongsa, and Zhemgang).

The survey also gathered information on farm income from a subset of 166 households from seven *dzongkhags* (Bumthang, Lhuentse, Mongar, Paro, Punakha, Sarpang, and Trongsa). The average household size is 6 (standard deviation = 2.3), with a range from 1 to 15. Farm income averaged Nu 43,755 (standard deviation = 92,511) with a range from Nu 500 to Nu 1,040,000. The income data are skewed with the single highest income-earning household accounting for 14% of the total income. The top 10% income-earning households account for 48% of the total income. About 70% of the households had farm incomes below the mean.

12.4.2 Perceptions and Estimates of Wildlife Damage: The Sample Survey

Table 12.2 presents the survey data, profiling the number of producers, the crops cultivated, and the total and average area planted. The survey aimed to determine

Table 12.2 Summary of crop production

Crop	No. of farmers	Average size (acres)	Total area (acres)	Percentage of farm households	Percent of total cultivated area
Maize	368	1.3	475	71	24
Paddy (red + white)	301	1.8	545	58	28
Buckwheat (sweet + bitter)	128	1.9	240	25	12
Wheat	147	1.2	177	28	9
Total cereals	505	2.9	1,491	97	76
Potato	233	0.9	220	45	11
Chili	153	0.5	72	29	4
Total main crops	515	3	1,553	99	79
Total minor crops	339	1.2	411	65	21
Total cropped area	516	3.8	1,964	99	100
Total uncultivated area	149	3.4	507	29	21
TOTAL	526	4.7	2,471	100	100

Source: Authors' survey

how much land was cultivated and harvested during the previous 12-month period. This includes multiple crops on the same fields during the year. The survey also asked respondents how much land they left fallow during the year specifically due to the risk of wildlife damage. (This means in some agricultural ecosystems one field could potentially be left fallow for up to 3 times during a 12-month period.) The total farmland managed by the 526 respondents is 2,471 acres, or 4.7 acres per farmer.

This total area includes 1,964 cultivated acres and 507 acres left fallow. The average area cultivated is 3.7 acres and the average area left fallow due to wildlife damage risk is about 96 acres per farmer. Examining these data in more detail reveals 149 producers (29% of the total sample) account for the 507 fallow acres. The average land left fallow for those 149 farmers is 3.4 acres per producer (standard deviation = 6.9). One relatively large landowner accounted for more than 10% of the total uncultivated area.

Cereals dominate crop production in the sample, with the two major subsistence crops, maize and rice, dominating the cereal crops. Overall, cereals account for 1,491 acres, 76% of the sample's total cropped area of 1,964 acres. More farmers produced maize than any other crop (n = 368), and rice was grown on more area (545 acres) than any other crop. Rice was produced on 58% of the farms, accounting for 28% of the total cropped area with an average area of 1.8 acres per farm. Maize was grown by 71% of the farms planting on average 1.3 acres and totaling 24% of the cropped area. Producers also planted millet, buckwheat, wheat, and barley.

By far, the sample's major cash crop is potatoes with 45% of the respondents (n = 233) planting on average nearly 1 acre each. Among the important minor cash crops are apples, chilies, ginger, mustard, and oranges, each representing roughly 10% of the number of producers.

Table 12.3 Respondents' perception of crop vulnerability

	Crop vulnerability to wildlife damage				Total
	Highly vulnerable	Slightly vulnerable	Not vulnerable	Do not know	
Maize	298 (84.2%)	32 (9.0%)	4 (1.1%)	20 (5.6%)	354 (100%)
Potato	161 (53.1%)	39 (12.9%)	10 (3.3%)	93 (30.7%)	303 (100%)
Paddy (white)	134 (43.1%)	46 (14.8%)	3 (1.0%)	128 (41.2%)	311 (100%)
Paddy (red)	87 (29.4%)	54 (18.2%)	4 (1.4%)	151 (51.0%)	296 (100%)
Wheat	46 (15.7%)	71 (24.2%)	5 (1.7%)	171 (58.4%)	293 (100%)
Sweet buckwheat	33 (11.5%)	69 (24.0%)	5 (1.7%)	180 (62.7%)	287 (100%)
Bitter buckwheat	24 (8.7%)	71 (25.8%)	4 (1.5%)	176 (64.0%)	275 (100%)
Chili	21 (9.0%)	42 (18.0%)	20 (8.6%)	150 (64.4%)	233 (100%)
Barley	19 (7.2%)	56 (21.1%)	5 (1.9%)	185 (69.8%)	265 (100%)

Source: Authors' survey

One aim of the questionnaire is to solicit farmers' perceptions of how vulnerable different crops are to wildlife damage. Table 12.3 highlights how vulnerable respondents consider eight key crops: maize, potatoes, white and red paddy, wheat, sweet buckwheat, bitter buckwheat, chili, and barley. Of those respondents growing maize, 84% ranked it as a crop highly vulnerable crop to wildlife damage, and 9% as slightly vulnerable. Potatoes are rated highly vulnerable by 53% and slightly vulnerable by 13%. White and red rice are rated highly vulnerable by 43% and 29%, respectively. Less than 15% of the respondents rated wheat, buckwheat, chili, and barley as highly vulnerable, with more than half responding that they "do not know" whether these crops are vulnerable or not.

Most farmers reported that crops are raided during their mature stage, just before and during harvest. Potatoes and chilies are also raided at the "seedling stage." Wild pigs, monkeys, porcupines, and pygmy hogs are considered the animals causing the greatest damage to maize, rice, wheat, and potatoes. The sambar, barking deer, and sloth bears are all considered as animals that cause slight damage.

Maize and potato producers indicated that the frequency of wildlife raids on their crops is "daily" once the crop begins to mature and near harvest, with most damaging raids occur primarily at "late night." Farmers in the sample guarded crops both day and night, however. Potatoes, the main cash crop, received more guarding attention on average than did other crops, with 100 days and 96 nights. Farmers averaged 48 days and 52 nights guarding their maize fields, and 40 days and 46 nights guarding their white paddy rice fields.

The survey results suggest livestock losses across the range of farming contexts throughout the country: 89% ($n = 469$ out of 526) of the respondents report owning livestock. Of that percentage, 20% reported livestock lost to wildlife predation, a somewhat skewed distribution. Before loss, the average livestock owner had ~8 head of livestock. This was reduced to an average 7.5 head due to wildlife depredation. The average livestock farmer loses 3.4% of the herd to predation each year, a significant pressure on agricultural households throughout the country.

Table 12.4 Livestock damage due to wildlife attacks

Livestock species	Average no. of livestock lost	Average depredation rate per farmer (%)
Per poultry farmer ($n = 286$)	0.9	10.5
Per horse farmer ($n = 82$)	0.2	5.1
Per Jatsham farmer ($n = 133$)	0.2	4.8
Per Swiss cow farmer ($n = 23$)	0.04	4.3
Per local cattle farmer ($n = 340$)	0.3	3.3
Per Jersey farmer ($n = 83$)	0.1	3.0
Per other cattle farmer ($n = 42$)	0.1	2.0
Per oxen farmer ($n = 334$)	0.7	1.9
Per pig farmer ($n = 70$)	0.03	0.7
Total for all livestock species (excluding poultry) ($n = 469$)	0.4	3.4

Source: Authors' survey

Table 12.4 presents depredation rates per livestock species. In general, respondents reported fewer pigs and oxen losses. Based on this sample, local cattle breeds do not demonstrate a significantly higher capacity to ward off predators as is commonly speculated. The Swiss and Jersey breeds suffered depredation rates at 4.4% and 3%, respectively. In addition to Swiss breeds, horse and Jatsham cattle are among the most vulnerable livestock species with 5.1% and 4.8% respective depredation rates. Local cattle breeds suffer depredation rates of 3.3%.

While it is difficult for farmers to ascertain the wildlife species responsible for farm animal deaths, the questionnaire prompted them to identify which species they held responsible for killing their livestock. Most animals were identified. However, of the 197 reported losses of livestock, leopard and wild dog are the the most commonly suspected of attacking wildlife species, with 26 cases attributed to each of them. Tigers are held responsible for 13 cases, sloth bears 5 cases, and wolves 3 cases.

The highest farm animal depredation rate is for poultry. Owned by 53% of the respondents, poultry registers a depredation rate of 10.5%. In addition, the majority of farm animals killed are poultry comprising 57% of all reported losses. The civet was listed as the animal most responsible for the loss of poultry with 29 out of 266 cases attributed to it.

12.4.3 Estimating the Costs of Wildlife Damage to Crops

In addition to soliciting information on area planted by crop, the animal species responsible for damage, and the frequency of wildlife raids, each farmer was asked to provide information to help estimate the costs wildlife impose on specific crops in at least three ways: lost revenue due to crop damage, lost revenue due to leaving land fallow specifically because of wildlife risks, and the opportunity costs of the producers' time spent guarding against wildlife. The survey data capture information that provides a rough indication of these costs; however, the data do not allow calculating net income losses to producers from wildlife damage.

For example, respondents were asked to estimate their actual crop production in kilograms (kg) during the past year and then to estimate the perceived crop damage in kg, i.e., what the crop production would have been without wildlife damage. More than four-fifths of the farmers reported some wildlife damage to crops during the previous year. Losses by crop and by farm vary greatly, for instance, 21% of maize producers had no damage, while 8% had lost their entire crop.

Table 12.5 presents the farmers' response to actual output and the perceived output per farm for five key crops, with calculations estimating the average kg loss per farm and average kg loss per acre. The average estimated production loss for the 233 farmers producing potatoes is 776 kg per farm. Maize losses are 460 kg per farm, wheat is 263 kg, red rice is 493 kg, and white rice is 428 kg.

As a baseline reference to compare the respondents' perceived output estimates, Table 12.6 presents national averages from the Ministry of Agriculture for annual yield data for 2 years, 2000 and 2003, by crop. In other words, the national average-yield data in Table 12.6 are an average of the 2000 and 2003 crop years.

The data in Table 12.6 demonstrate that for all crops except for maize, the perceived yield estimates are below the national averages. And even in the maize case, the respondents' perceived output is well within the norm for maize yields in

Table 12.5 Production losses due to wildlife damage

Crop and number of farms	Acres	Production (000 kg)			Average loss (kg)	
		Actual output	Perceived output	Wildlife damage	Per farm	Per acre
Maize (n = 368)	475	296.5	465.8	169.3	460	356
Wheat (n = 147)	177	61.4	100.0	38.6	263	218
Red rice (n = 156)	245	234.8	311.7	76.9	493	314
White rice (n = 207)	300	269.2	396.4	127.2	428	424
Potatoes (n = 233)	220	337.2	518.1	180.9	776	822

Source: Authors' survey

Table 12.6 Estimated yields and national averages

Crop	Yields (kg per acre)			
	Output	Expected	National average (2000–2003)[a]	Percentage loss relative to expected output
Maize	624	981	870	36
Wheat	347	565	668	39
Red rice	958	1,272	1,194	25
White rice	897	1,321	1,194	32
Potatoes	1,533	2,355	3,765	35

Source: Authors' survey

[a]National averages cited in Osmani et al. (2007) from Bhutan Ministry of Agriculture and Forestry

Table 12.7 Revenue losses from wildlife damage per farm and per acre

Crop and number of farms	Mid-range crop price 2006–2007 (Nu)			Estimated revenue loss as percent of actual crop production	
	Nu per kg	Revenue loss per farm	Revenue loss per acre	Per farm (%)	Per acre (%)
Maize (*n* = 368)	8	3,680	2,851	57	57
Wheat (*n* = 147)	15	3,938	3,271	63	63
Red rice (*n* = 156)	20	9,858	6,277	33	33
White rice (*n* = 207)	16	9,833	6,785	47	47
Potato (*n* = 233)	9	6,988	7,401	54	54

Source: Authors' survey

Bhutan; during the 2000 crop year, the national average yield for maize was 1,012 kg per acre compared with the respondents' estimate of 981 kg per acre (the 2003 national average maize yield was 728 kg per acre). In the survey, maize output was on average 36% lower due to wildlife damage, potato yields were 35% lower, red paddy rice 25% lower, white paddy rice 32% lower, and wheat 39% lower.

Finally, Table 12.7 presents estimates of the revenue lost due to wildlife damage caused to the five main crops. The prices used in the calculations represent a mid-range price for 2006 and 2007. The estimates for the average revenue lost per farm for maize is Nu 3,680, for potatoes the loss is Nu 6,988, and for red and white rice the loss is more than Nu 9,800. The crop damage represents more than a 50% reduction in revenue for the maize, wheat, and potato crops. As a percentage of farm income, these losses are non-trivial. The average farm income from the survey subsample is Nu 43,755. Thus, for a wheat producer, the lost revenue is equivalent to 23% of farm income; for a potato producer, the lost revenue is equivalent to 16% of farm income; and for a maize producer, the lost revenue is equivalent to 8% of farm income.

Calculating the monetary costs from time spent guarding against wildlife damage and from leaving land fallow is even more difficult. For example, potatoes, the main cash crop, received more guarding attention on average than did other crops, with 100 days and 96 nights. Farmers averaged 48 days and 52 nights guarding their maize fields, and 40 days and 46 nights guarding their white paddy rice fields. Rural wages may provide a rough guide of how to value guarding time (daily wages range from Nu 75 in the southern part of the country to Nu 125 in the North). However, using wages assumes that the farmers would either hire someone or, if they did not have to guard their crops, they would work as rural wage laborers. Guarding during the day is often done simultaneously with other on-farm activities. The most direct cost from guarding at night is sleep deprivation and the inevitable lower productivity resulting from multiple nights with little sleep.

Similarly, an accurate value for fallow land is difficult to measure. On average, each farm household in the sample left just under 1 acre (.97) fallow

Table 12.8 Costs of livestock depredation

Livestock species	Unit value (US $)	Number of kills	Average household loss for those who own livestock species (US $)	Total costs (US $)
Mithun cross (Jatsham cattle) (50%)	279	23	306	6,417
Exotic cross-breed (Jersey/ Brown Swiss)	349	9	449	3,141
Yak	186	8	744	1,488
Other local cattle	233	113	387	26,329
Horse	279	18	359	5,022
Sheep	$46	17	156	782
Total for those affected ($n = 94$)	–	188	459	43,179
Total for those who own livestock ($n = 469$)	–	188	92	43,179
Total for all respondents ($n = 526$)	–	188	82	43,179

Source: Authors' survey with livestock values based on Table 12.1

during the 12-month period. These numbers suggest that the average Bhutan farmer may leave fallow more than 20% of their farmland each year because of wildlife raids. However, the sample has 149 producers, or 29% of the total farmers, who left land fallow because of the threat of wildlife damage, each leaving 3.4 acres fallow on average. These results suggest that the costs associated with leaving land fallow are very high for a specific subgroup of producers. To assess the net loss for these producers requires more detailed information on production costs, revenue, and profit than is available.

Table 12.8 summarizes costs of livestock depredation using 2005 data (Sangay, 2005); as values differ by the animal's sex, the cost of replacement is based on the most economically valuable sex of each livestock species. Of reported livestock kills, the economic cost to a livestock-owning household (n = 469 out of 526) averages at US $92.10. The total cost borne by those farming households actually affected by livestock depredation ($n = 94$) averages US $459.40. Local cattle kills (113 cases) were the most cited, averaging a cost to the average local cattle-owning household of US $387.20. Eighteen kills of horses were reported with the average horse-owning household bearing a cost of US $358.70. When adjusted as an average financial loss across all farmers in the survey, the average farming household in Bhutan loses US $82.10 in livestock assets to wildlife predation. Sufficient income data were not available to determine the percent of income lost to livestock depredation.

12.5 Discussion and Summary

The conflict between nature conservation and wildlife destruction of crops and livestock is a major policy debate in Bhutan. A large part of this debate involves forests as a habitat, providing wildlife benefits for many and costs borne by a small group of farmers. The survey asked farmers about their observations and uses of local forests to determine perceived benefits and losses. The forest coverage pattern over the last 10 years varies depending on region. Certain *dzongkhags* (Bumthang, Trongsa, Zhemgang, Mongar, and Trashiyangtse) registered significant increases in forest cover. At the national level, 47% of the respondents reported that the forest within 30 min walking distance from their residence has increased over the last 10 years, and some 32% noted a decrease in forest size.

A PES view would suggest that a number of ecological services, primarily the use of forest products, could be framed as benefits that mitigate the costs imposed by increasing wildlife threats. In the sample survey, 95% of the respondents use timber resources, nearly all of those citing firewood as a product harvested (99%). Along with bamboo, fodder, and cane shoot, firewood is increasingly difficult to access, as evidence by 85% reporting that they face an increased distance in accessing timber resources. Most respondents are required to walk either within "less than 1 hour" (45%) or "1 to 3 hours" (51%) to collect firewood. Cited as causes for the increase in collecting firewood are "increasing population" (65%) and "strict forest rules" (18%). Due to government regulations and increasing needs of the population, the rate of increase in nearby forests does not directly translate to increases in timber product accessibility.

Non-timber forest resources were harvested by 59% of the respondents. While this would seem to argue for a heavy usage of forest resources, the number's significance should be adjusted to account for the frequency of use: 51% of those who use non-timber forest resources typically harvest only once per year. Another 30% harvest 2–3 times a year. Those who harvest such resources "3 to 4 times a year" or "more than 4 times a year" are much lower, respectively, 9.4% and 8.4% of those utilizing non-timber forest resources.

Respondents were asked about their use of 25 different non-timber forest products. As products range significantly across Bhutan's diverse bioregions, the "don't know" responses were typically high. Excluding that category, very few products registered much of an increase. Bamboo, at the highest, had 27% mention an increase and fodder a 25% increase, with the rest of those who knew the status of those products stating that the availability of the crop had either decreased or stayed the same. In other words, no products registered a significant increase. Isolating the *dzongkhags* with high responses for forest coverage increase, the results remain low. It appears that even for the sub-group reporting increases in forest cover nearby their landholdings, there has not been a corresponding increase in non-timber forest products. Likewise, respondents suggest that these products do not significantly benefit rural income. More than 97% of respondents for both timber and non-timber forest products use products for self-consumption only.

The survey results presented here suggest that producers do face high costs in terms of: (1) damaged and eaten crops; (2) time spent guarding crops during the day and the night; and (3) forgone production with many producers unwilling even to plant due to the threat of wildlife damage. The data from this study's countrywide survey estimate that wildlife damage to crops may cause production to be lower by

25–39%. Additional estimates suggest that, without wildlife damage, average crop income could be increased by 33% to more than 60%.

Other studies focusing on wildlife damage to livestock in and near protected areas estimate damage equal to an average annual financial loss of 17% of the household's total per-capita cash income (Wang et al., 2006a). While the data did not provide for a calculation of percent of income lost per household, the estimation of the cost to each agricultural household as projected by this sample is US $82.

The survey attempts to provide information to better inform the debate about the extent of crop damage and livestock depredation caused by wildlife and the impact of such costs on producer revenue. Much of the debate argues that the cost of conservation should not fall adversely on rural producers who already represent the lower income group in Bhutan's economy. Imposing a conservation burden on this group is regressive and compounds their poverty.

Not only do crop damage and livestock depredation limit severely the agricultural livelihoods of many small producers, but the increased presence of forest cover near agricultural land due to conservation policies has not necessarily provided significant increased benefits to local residents, as judged by their responses. While other benefits, such as improved watershed quality, could still prove to be a significant benefit, the increase in services due to growth in forest coverage is unlikely to outweigh the increased costs observed for Bhutan's small-scale producers.

In addition to seeking a balance between benefits and costs borne by communities, the type of cost borne by the community may have significant impacts on both conservation and the viability of small-scale agricultural production. In this case, the cost is a loss of production efficiency and livelihood potential. A different type of cost, for example, requiring communities to pay in part for fencing or corralling solutions, might directly benefit both rural livelihoods and conservation efforts while still operating within the framework of PES programs.

The data presented here suggest that small producers in Bhutan are paying a high price for national and global conservation benefits. How can these costs, so far borne by the small producers, be distributed more equitably across society, indeed the global community who are benefiting from nature conservation? Among the proposed solutions include direct compensation, culling wildlife, and subsidizing fences. A proper study assessing the trade-offs between compensation, culling, and prevention may help better direct which areas, which crops, and which producers to target.

The survey results suggest that the problem is so widespread that compensation is unlikely to provide appropriate incentives and prohibitively expensive to monitor. At the same time, compensation for catastrophic losses can be appropriate. The survey results indicate that up to 10% of producers lose their entire crop.

Culling is also an unlikely option given the spiritual and religious nature of Bhutan's rural producers. Research to develop low-cost fencing and subsidizing fences (wooden and non-lethal electric) to protect crops appears as the most appropriate solution. The initial fencing costs may be high relative to the benefits in the first few years, but over a half decade or more the crop revenue benefits are likely to mean a positive benefit cost ratio. The subsidy may be justified on the basis of the social value of conservation and the potential benefits of lower food prices – as crop damage is reduced, greater food supplies are delivered to local markets.

Acknowledgments The authors thank the United Nations Food and Agriculture Organization and the Roles of Agriculture Project for funding support. The authors are very grateful to Karma Kesang, Raghavan Sampath, and Ryan Slunaker for research assistance, survey support, and data analysis. We wish to dedicate this chapter to the late Karma Kesang who died in a tragic river accident while completing the survey. He and his wife collected the data throughout Bhutan.

References

Bulte EH, Rondeau R (2007) Wildlife damage and agriculture: A dynamic analysis of compensation schemes. Am J Agric Econ 89:490–507.

Cardoso de Mendoça MJ, Sachisida A, Loureiro PRA (2003) A study on the valuing of biodiversity: The case of three endangered species in Brazil. Ecol Econ 46(1):9–18.

Choden D, Namgay K (1996) Report on the findings and recommendations of the wild boar survey. Unpublished report. National Plant Protection Center, Ministry of Agriculture, Royal Government of Bhutan, Thimphu.

Ferraro PJ (2001) Global habitat protection: Limitations of development interventions and a role for conservation performance payments. Conserv Biol 15:990–1000.

Hartmann J, Peterson L (2004) Marketing environmental services: Lessons learned in German development cooperation. Tenth Biennial Conference of the International Association for the Study of Common Property.

IMF (International Monetary Fund) (2004) Bhutan poverty reduction strategy paper. IMF Country Report No. 04/246. Washington, DC.

Jenkins M, Scherr S, Inbar M (2004) Markets for biodiversity services. Environment 46(6):32–42.

Knetsch JL (1994) Environmental valuation: Some problems of wrong questions and misleading answers. Environ Values 3(4):351–368.

Kontoleon A, Swanson T (2003) Property rights for the Great Panda. Land Econ 79:483–499.

Madhusan MD, Karanth KU (2000) Hunting for an answer: Is local hunting compatible with wildlife conservation? In Robinson JG, Bennett, EL (eds.) Hunting for sustainability in tropical forests. Columbia University Press, New York.

NBS (National Bureau of Statistics) (2007) Bhutan living standard survey (BLSS). National Statistic Bureau, Thimphu.

Osmani SR, Bajracharya BB, Tenzing S, Wangyal T (2007) Macroeconomics of poverty reduction: The case study of Bhutan. UN Development Programme, New York, Thimphu.

Pagoila, S (2003) Can programs of payments for environmental services help preserve wildlife? Convention on International Trade in Endangered Species of Wild Fauna and Flora. Workshop on Economic Incentives and Trade Policies.

Preston L (1997) Investing in mountains: A participatory electronic conference of the mountain forum. Mt Res Dev 17(2):182–185.

Rondeau D, Bulte EH (2007) Compensation for wildlife damage: Habitat conversion, species preservation and local welfare. J Environ Econ Manag 54(3):311–322.

Sangay T (2005) Tiger conservation program, species conservation research and monitoring. Section Nature Conservation Division, Department of Forests, Thimphu, Bhutan.

Swanson T, Kontoleon A (2003) Conflicts in wildlife conservation: Aggregating total economic values. Department of Economics, UCL. Paper presented at 4th BIOECON workshop on the Economics and Biodiversity Conservation, Venice, Italy.
Tobgay S (2005) Small farmers and the food system in Bhutan. Agricultural Marketing Services Ministry of Agriculture, Royal Government of Bhutan.
Wang SW, Mcdonald DW (2006) Livestock predation by carnivores in Jigme Singye Wangchuck National Park, Bhutan. Biol Conserv 129(4):558–565.
Wang SW, Curtis PD, Lassoie JP (2006a) Farmer perceptions of crop damage by wildlife in the Jigme Singye Wangchuck National Park, Bhutan. Wildl Soc Bull 34(2):359–365.
Wang SW, Lassoie J, Curtis PD (2006b) Farmer attitudes towards conservation in Jigme Singye Wangchuck National Park, Bhutan. Environ Conserv 3(2):148–156.
World Bank (2005) Bhutan land management project. World Bank, Washington, DC.
Wunder S (2005) Payments for environmental services: Some nuts and bolts. CIFOR Occasional Paper No. 42.
Zilberman D, Lipper L, McCarthy N (2008) When could payments for environmental services benefit the poor? Environ Dev Econ 13:255–278.

Chapter 13
Conclusion

David Zilberman

This book identifies numerous ways that ecological stewardship can be a source of income for producers. Payments for ecosystem services (PES) do not only consist of hydrological or soil carbon sequestration projects but include payments to enhance wildlife that provide recreational benefits, payments for cultural practices that will reduce likelihood of disease, and payments for protection against deforestation. PES programs can benefit both crop and livestock farming, forestry and range management, and in cases not presented here (Waibel & Zilberman, 2007) aquaculture. The benefits of PES can include pollution reduction, preservation of natural resources, and generation of recreational and ecological amenities. The diversity of possibilities emphasizes the importance of ecological and environmental entrepreneurship, namely, the capacity to identify opportunities for increasing the resource base of farmers while improving environmental qualities.

The ecological entrepreneurship required to establish PES programs cannot be obtained without a multidisciplinary scientific base, combining understanding of human and animal behavior, knowledge of biological and ecological processes, adherence to financial principles and constraints, and ingenuity to design technologies for monitoring and measurement. Thus, further expansion and ingenuity in designing PES programs will benefit both from growing experience and research in their use and from appropriate education in environmental management. While, in some cases, environmental entrepreneurs may be born, in many other cases they will be a product of experience and education, so one of the challenges in expanding PES programs is to provide the environmental management training combining economic and natural resource systems to managers and leaders in the field.

The book demonstrates that it is insufficient to identify opportunities for establishing PES programs, without identifying sources of funding for these programs. Thus far, the majority of PES programs have been supported by the public sector, and many of these programs will continue to be supported by governments and international organizations because they provide public goods. Yet, at the same time, various private agents may benefit from environmental services provided by the rural sector, including water utilities, providers and users of recreational services, environmentalists and philanthropists, pharmaceutical companies, and natural resource-based industries. Leaders of PES program initiatives should be

L.M. Lipper et al. (eds.), *Payment for Environmental Services in Agricultural Landscapes*, Natural Resource Management and Policy.
doi: 10.1007/978-0-387-72971-8_13;

able to identify possible sources of funding and negotiate support. In many cases, the provision of environmental services is accomplished by a large number of uncoordinated, independent economic agents, and again, leaders of PES initiatives should also be able to establish cooperation and collective actions by individuals who may benefit from the programs.

PES programs vary in their context and dimension. Some are fully local, where producers within a region are providing benefits primarily to other local stakeholders, and thus the arrangement is, in essence, a sort of local subsidy. In other cases, a local activity has national or global implications, for example, when farmers in Ghana sequester carbon, or producers in Brazil do not clear-cut rainforest. In these cases, the PES program may require coordination between a local provider and global beneficiaries. Sometimes, as occurs with greenhouse gases, the coordination occurs through market-like mechanisms, or else it may occur through international agreements. Developing such coordination mechanisms is one of the major challenges in the design of new PES programs.

The book emphasizes the importance of program design in a manner to induce farmer provision of desired environmental amenities. The design of incentives has to be based on understanding farmers' behavior, as well as the relationship between agricultural practices and environmental and ecological objectives. Since desirable outcomes are not always observable and are sometimes affected by random factors, as is the case with soil carbon sequestration, a key challenge is to identify observable measures that have a stable and reliable relationship to environmental outcomes. Therefore, PES programs may require reassessment in terms of actual objectives, and flexibility of design to update and incorporate new information and new knowledge as they are accumulated. Furthermore, the feasibility and design of PES programs depend on the performance and availability of measurement technologies, and improvement in remote sensing, computerization, and other monitoring technologies may lead to redesign of PES programs already in place. The spread of information technology to the developing world, and the fast diffusion of wireless technology in particular, suggests new opportunities for better measurement and, thus, better design of PES programs.

A major objective of the book is to better understand the relationship between PES and poverty reduction. Many of the chapters identify situations where PES schemes could actually harm the poor and result in undesirable distributional outcomes. Certain land diversion PES programs may reduce the resources available for small farmers, increase the price of food, thus negatively affecting the urban poor, and mostly benefiting landowners and larger farmers. Similarly, payments to reduce dangers to wildlife may also be mostly beneficial to larger farmers. Lack of property rights and ability to control the environment may also lead to design of programs harming the poor. Modification of PES programs in an attempt to meet both environmental and distributional objectives may result in inefficient and awkward designs. There are cases where pursuit of PES may lead to improved distributional objectives, but they are dependent on positive correlations between poverty and ability to provide environmental services and a legal system that allows the poor to take advantage of this capacity. However, these correlations do not hold in general, so the pursuit of ecological objectives and distributional objectives may require introducing multiple policies, including compensatory policies to protect against losses for the poor and disadvantaged.

PES programs are an institutional innovation. Like any other innovation, their diffusion is a matter of time, and depends on awareness and early successes. This book, as part of the ongoing research on PES, aims to provide lessons to improve program design and to provide examples in the case studies that will inspire imitation and improvement. Further research on PES should continue to explore new opportunities and arrangements, and critically evaluate their various impacts on environmental quality, market conditions, and social outcomes. Future research may also pursue more experimental routes, where the performance of alternative designs is compared, or where the best such design is compared to other environmental policies like cap and trade, taxation, or direct control.

Reference

Waibel H, Zilberman D (eds.) (2007) International research on natural resource management: Advances in impact assessment, FAO, Rome. CAB International, Oxfordshire, UK; Cambridge, MA.

Index

Printed in the United States of America